Lecture Notes on Coastal and Estuarine Studies

Managing Editors:
Richard T. Barber Christopher N.K. Mooers
Malcolm J. Bowman Bernt Zeitzschel

3

Synthesis and Modelling of Intermittent Estuaries

A Case Study from Planning to Evaluation

Edited by W.R. Cuff and M. Tomczak jr.

Springer-Verlag Berlin Heidelberg GmbH

Managing Editors

Richard T. Barber
Coastal Upwelling Ecosystems Analysis
Duke University, Marine Laboratory
Beaufort, N.C. 28516, USA

Malcolm J. Bowman
Marine Sciences Research Center, State University of New York
Stony Brook, N.Y. 11794, USA

Christopher N. K. Mooers
Dept. of Oceanography, Naval Postgraduate School
Monterey, CA 93940, USA

Bernt Zeitzschel
Institut für Meereskunde der Universität Kiel
Düsternbrooker Weg 20, D-2300 Kiel, FRG

Contributing Editors

Ain Aitsam (Tallinn, USSR) · Larry Atkinson (Savannah, USA)
Robert C. Beardsley (Woods Hole, USA) · Tseng Cheng-Ken (Qingdao, PRC)
Keith R. Dyer (Taunton, GB) · Jon B. Hinwood (Melbourne, AUS)
Jörg Imberger (Western Australia, AUS) · Akira Okubo (Stony Brook, USA)
William S. Reeburgh (Alaska, USA) · David A. Ross (Woods Hole, USA)
S. Sethuraman (Upton, USA) · John H. Simpson (Gwynedd, UK)
Absornsuda Siripong (Bangkok, Thailand) · Robert L. Smith (Corvallis, USA)
Matthias Tomczak (Cronulla, AUS) · Paul Tyler (Swansea, UK)
Michitaka Uda (Tokyo, Japan)

Editors

Dr. Wilf Cuff
Research Scientist
Maritimes Forest Research Centre, Environment Canada
Canadian Forestry Service
P.O. Box 4000, Fredericton NB, E3B 5P7, Canada

Dr. Matthias Tomczak jr.
CSIRO Marine Laboratories, Division of Oceanography
P.O. Box 21, Cronulla N.S.W. 2230, Australia

ISBN 978-3-540-12681-2   ISBN 978-3-642-49991-3 (eBook)
DOI 10.1007/978-3-642-49991-3

2131/3140-543210

PREFACE

This book reports on the findings of, and summarizes the conclusions from, the Port Hacking Estuary Project, a model–guided, multidisciplinary study of an estuarine ecosystem. The Project began in 1973, at a time when it was thought that environmental problems could be solved readily by assembling a multidisciplinary team of research scientists and having them co–ordinate their research around the construction of an ecosystem model. But a decade has passed and time has not been easy on this approach. The anticipated predictive dynamic models have not been produced and bitter argument has often marred the course of such studies.

Yet the need to anticipate the flow of various chemical species (carbon, oxygen, nitrogen, phosphorus, toxicants) through the environment remains: the evidence is everywhere, from fertilization of urban lakes to acid rain. The magnitude of the problem ensures that funds will continue to be made available – although with short–term variations as perceptions swing. It is thus clear that although the difficulties are great, so is the need. It is from this background that we present this book.

The Port Hacking Estuary Project involved some 15 – 20 research scientists over a period of 5 years. The goal was to research the flow of carbon into, within, and out of a small unpolluted estuary chosen for convenience rather than for its social significance. The idea was to use the information obtained from these studies to build a predictive dynamic model.

Emphasis was placed on the South West Arm of Port Hacking: this Arm is broadly characterized by exhibiting two states. In one state the salinities are typical of an arm of the sea. In the other, which occurs for short periods following rainstorms, a stratified water column is set up when fresh water sits on the surface and saline tidal water enters over a sill and falls down into the basin of South West Arm to reach water of a similar density. In fluid dynamics terms, South West Arm can be described as a fjord; but the intermittency of freshwater inflow has such a marked effect on the biota that a different classification is warranted on ecological considerations. Our choice is reflected in the title of this book. "Intermittent estuaries" do exist in various places around the world but as yet do not seem to have been recognized as a class.

Another feature of this book, as reflected in the subtitle, is that both scientific and organizational aspects are discussed. In model–guided multidisciplinary studies they are not independent, and in this case study we use a description of both to draw lessons about this genre of scientific endeavour.

The Port Hacking Estuary Project was an umbrella for such a diverse range of investigations that it is not possible to ascribe overall success or failure to the Project. Suffice it to say that the Project did not reach its goal of making a predictive dynamic ecosystem model of South West Arm. The (by now expected) personal animosities associated with this genre of scientific endeavour were present. The unique aspect of the project is that participants agreed to continue to work together in this analysis of their efforts. Their willingness to submit their activities to close scrutiny speaks highly of them, and the resulting book is a fitting close to the Port Hacking Estuary Project. Our hope is that it can also help to clarify some of the problems that plague other multidisciplinary studies, and thereby contribute to a solution.

The book begins with a chapter by Radway Allen, former Chief of the Division of Fisheries and Oceanography, and under whose initiative this Project was begun. He describes the motivation and hopes for the Project, as seen in 1973 and 1981. Then Parker and Tranter, who were largely responsible for the organizational aspects, combine with Rochford, Chief of the Division during the latter part of the Project, to give a history of research in Port Hacking before 1973, and to describe the initial organization and direction. The body of the text includes papers covering a wide variety of observational, experimental, and modelling aspects relating to South West Arm. These papers are of interest in their own right but have been selected to provide the reader with an understanding of South West Arm. (They do not, of course, represent the total published output of the Project: a list of other publications is appended to the paper by Parker *et al.*) We also include a paper by Vaudrey *et al.* on the data set obtained by the Project and on the data base management system used to store and retrieve the data. An attempt to interconnect the facts presented so far is made in the chapter where many of the Project participants (Cuff *et al.*) join together to synthesize the static information for South West Arm. Then one of us (W.R.C.) synthesizes the dynamic information that has been obtained by the Project (not resorting to the general literature for missing information). These two papers represent the culmination of our synthesis efforts but a number of other models were constructed throughout the Project and a gradual evolution of ideas about ecosystem modelling occurred; these experiences are described by Sinclair *et al.* Their chapter builds up to the last part of the book where we evaluate the Project (as a representative of the genre), in two separate contributions, each of us along his own line of thought. While the two papers may not be totally complementary, we feel that the full story emerges only when both contributions are taken together.

In 1981 the CSIRO Division of Fisheries and Oceanography was reorganized into the Division of Fisheries Research and the Division of Oceanography, both Divisions now being known as the CSIRO Marine Laboratories. This change is reflected in the affiliation of authors given in their contributions.

Over the past decade of work so many people have played their role and we begin by thanking all for their part. But specifically we thank the Chiefs of the Division of Fisheries and Oceanography (David J. Rochford) and of Computing Research (Peter J. Claringbold) for supporting this evaluation of the Project and Bob Kelly and Stephen Kessell for reviewing the book. We also thank the following individuals for contributions of various sorts: Peter Sands, Peter Benyon, Rob Hurle, Gay Watt and Joan Brown.

This book was typeset on the COM (Computer Output on Microfilm) of the CSIRO's CSIRONET computer network, using COMTEXT, a computer typesetting program developed by the CSIRO Division of Computing Research. We thank Bob McKay for advice with COMTEXT when we had difficulties, and the following members of the CSIRO Division of Computing Research and of the CSIRO Marine Laboratories for assistance:
computer file production: Sue Aynsley, Narelle Hall, Maryanne Nicholas,
drafting: Neridah Charlesworth, Ian Hamilton, Bea Lindsay,
photography: Charles Purday,
technical editing: Phil Hindley, Jan Somers.

Wilfred R. Cuff, Canberra
Matthias Tomczak, Cronulla

CONTENTS

VIII

Synthesis and Modelling of Intermittent Estuaries
(W.R. Cuff and M. Tomczak jr. eds) Berlin, Heidelberg,
New York: Springer (1983), pp. 1–6.

Introduction to the Port Hacking Estuary Project

K. Radway Allen

Division of Fisheries Research

CSIRO Marine Laboratories

P.O. Box 21, Cronulla, N.S.W. 2230, Australia

Summary. Estuaries are exposed to a variety of human activities
and are susceptible to stress resulting from these activities. In
subtropical areas, many estuaries are affected by freshwater inflow
only occasionally and often quite irregularly. The Port Hacking
Estuary in south–east Australia falls into this category. It was
studied by the CSIRO Division of Fisheries and Oceanography
during 1973–78.

Key words: estuaries, Port Hacking Estuary Project, South West
Arm

It is probably a truism that in the biological world the interfaces between widely differing environments are commonly the site of complex communities whose study can be particularly rewarding to the understanding of the structure and dynamics of biological systems. The marine environment has many such interfaces. These include interfaces between different water masses, between the water and the atmosphere, and between the water and the substrate. The coastal interface is, in a sense, a special case of the substrate interface, but it has many characteristics which distinguish it from the interface with the seabed in deeper water. The coastal interface also presents a great variety of different conditions: open and exposed rocky coasts, open but more sheltered coasts with sandy or muddy substrates, and the still more sheltered conditions of embayments and estuaries. Considered as an interface the estuary has a dual character; on the one hand it involves an interface between the sea and the land, and on the other an interface between sea water and fresh water. The complexity of estuarine systems is increased by their three–dimensional structure, and by their variability in the time dimension. The important changes with time include not only the regular tidal cycles, with daily and longer components, but also the less frequent and less regular changes associated with the amount of fresh water entering the system. These also may have seasonal components. For all these reasons the study of estuarine systems at the holistic level presents problems which are not only of great intrinsic interest but also call for co–ordination between a number of scientific disciplines.

In addition to their scientific interest, estuaries often have such great social and economic values that there are good practical reasons for their study. These reasons are of particular importance in Australia because of the concentration of the population along the coastal fringe, especially in the south–eastern area. This concentration leads, on the one hand, to increased significance of the contribution of estuaries to the desirable way of life, and on the other hand, to increased risks that human activities will bring about changes in the estuaries which will reduce the value of that contribution. As well as their simple aesthetic values, estuaries have a great social value because of the opportunities they provide for aquatic sports, *e.g.* sailing, fishing, water–skiing etc, which now attract a large proportion of the population. Estuaries are also important as breeding grounds for many species of fish which are valuable to commercial or recreational fisheries.

It is unfortunate that the very attractiveness of estuaries as places near which to live makes them also attractive for other human activities which have adverse effects on them. They are convenient places in which to dump wastes, they may be naturally good sites for ports and docks, while their shores present the temptation to reclaim land for commercial or residential development. In addition to such activities on the shores of the estuaries themselves, their location at river mouths renders them susceptible to the effects both of wastes dumped upstream and of interference with the flow of fresh water into them.

Estuaries are not only peculiarly attractive to deleterious human activities, but in important ways they are also more susceptible to the results of these activities than are the waters of the open coasts. This applies particularly to the discharge of harmful materials, since in estuaries these are not dispersed nearly as rapidly as in the open ocean. Estuaries are also unique in their susceptibility to interference with the normal discharge of fresh water into them.

There is thus an expanding need for the scientific study of estuarine systems to provide a basis for the advice required by the administrative authorities which have the responsibility of determining whether proposals which could affect estuaries should be rejected or approved, and in the latter case what modifications may be necessary to minimize any possible adverse effect.

For most such proposals, scientific investigations of particular sites are needed and generally these have to be concentrated on the special features that are most likely to be affected. Such studies are very commonly constrained within narrow time and budgetary limits, and typically have to tailor their programs to the specific questions for which short–term answers are required. In general, the responsibility for the conduct of such studies lies with the agencies which, directly or indirectly, have responsibility for carrying out the proposed work. In Australia these responsibilities usually lie within the ambit of the State governments. Many such studies have been carried out at various levels by these agencies. Two of the more comprehensive of these have been the study of Westernport Bay in Victoria, to provide a background for future decisions on the industrial development of the area (Ministry for Conservation, 1975), and the work on the Blackwood River in Western Australia, to determine the probable effects of dredging for mineral sands (Hodgkin, 1978).

These *ad hoc* studies need, however, to be able to draw on a general pool of knowledge of the structure and behaviour of estuarine systems, so that work in the particular project can be concentrated on a study of those particular aspects which general knowledge indicates as being of most immediate importance. While the contributions of the *ad hoc* studies to the general knowledge of estuarine systems cannot be ignored, there is also a need for more studies having as their direct aim increased understanding of the principles underlying the structure and dynamics of estuarine systems, without having to divert effort to the answering of immediate practical problems.

The Commonwealth Scientific and Industrial Research Organization (CSIRO) of Australia had, in the 1950s, through the work of D.J. Rochford, already made important contributions to the general classification of the particular types of estuaries most characteristic of Australia, and had accumulated much information on their physical and chemical structure (Parker *et al.,* 1983). In about 1973, however, it was felt that it would be valuable to build on this basis by undertaking wider studies which would be concerned not only with the physics and chemistry of estuaries, but also with their biological communities, and especially with the relation of these communities to the physico–chemical environment.

It was realized early that, at least in the first stages, it would be necessary to concentrate such a project on a single estuary. If a detailed continuing study of an estuarine system were to be undertaken, there were obvious advantages in selecting a site as close as possible to the Division's laboratory, to minimize time wasted in travelling and to have the full laboratory facilities as available as possible to the field studies. This pointed clearly to the selection of a site in the Port Hacking area since the laboratory is situated on the shores of this estuary. However, the concept of studying the selected system in as much detail as possible indicated that the actual study area should be small, so as to avoid increasing the quantity of observations required beyond practicable limits, and to simplify the analysis. It was therefore decided to concentrate on only a part of Port Hacking, and South West Arm was finally selected as the site of the study. This arm is simple in structure, consisting of a single deep basin with one inflowing stream. Its boundaries could be quite well defined since it is separated from the main estuary by a shallow bar.

During much of the year very little fresh water flows into South West Arm (SWA) but, as a result of heavy rains, there are occasional and irregular major inflows of fresh water, which generally last for only a few days. Estuaries with this type of freshwater input are relatively common in much of Australia, particularly in the south–east. SWA also resembles many Australian estuaries with small or intermittent freshwater inflows in having a comparatively deep basin cut off from the ocean by a shallow bar, although the basin is unusually deep in comparison to its extent. SWA is therefore a fairly typical, although small–scale, representative of an important type of Australian estuary.

A further advantage presented by SWA was that both the Arm itself and the whole of its catchment area lay within the boundaries of the Royal National Park (Fig. 1). It appeared, therefore, to be in a virtually pristine condition, without significant modification due to human activities and it was hoped that it could later be used as a basis for comparison with conditions in other estuaries which had been modified by man.

While it was decided to concentrate the study on SWA it was clearly not possible to ignore entirely the rest of Port Hacking, and some aspects of the work had to extend into it. In particular, any estuarine study needs information on the water movements into and out of the lower end of the system. For an estuary as a whole these movements are driven by the ocean tides. For SWA, however, these tides have been modified by passage up Port Hacking and the water may also be modified chemically by freshwater discharges from other tributaries of Port Hacking.

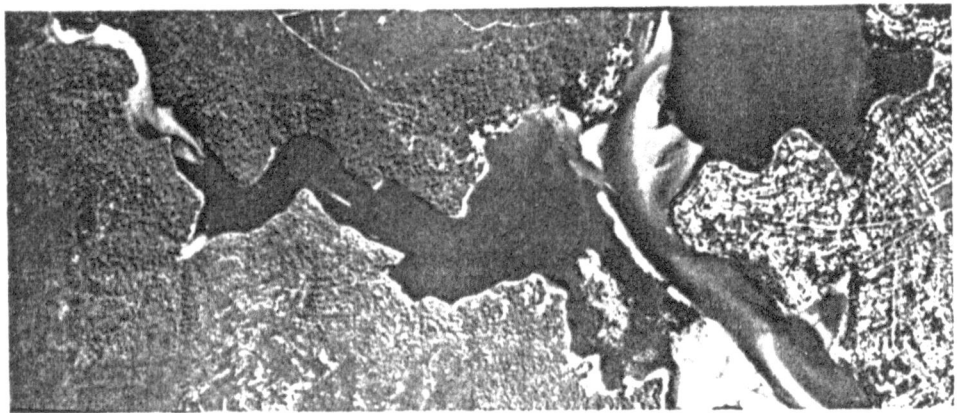

Fig. 1: An aerial view of South West Arm looking inland, showing South West Arm to the left, Port
Hacking in the centre and Lilli Pilli to the right.

It was obvious that a comprehensive study along the lines proposed would need substantial contributions from scientists expert in appropriate fields of the biological, physical and chemical sciences. It would also require support in such other disciplines as mathematics and geology. From the earliest stages of the planning of the Project it was therefore necessary to bring in a team of scientists capable of covering at least the essential fields involved. The CSIRO Division of Fisheries and Oceanography was fortunate in having among its staff biologists, physicists and chemists who could take part in such a project, but, as the present volume shows, assistance was also sought from scientists in other institutions, who were expert in disciplines not represented in the Division.

Early in the development of the Project it was decided that the study would be greatly helped by centering it on an appropriate model. Use of a model would help to ensure that all the important components of the system were identified and the relations between them defined. To be useful such a model would have to be quantitative. It would, therefore, be necessary to define the quantities which should, if possible, be measured. These would include both the amounts of material in the various compartments of the model at different times and the rates (and their dependencies) at which material was flowing between the compartments. These quantities must all be measurable in the same units.

In other ecosystem models, carbon, phosphorus, nitrogen, oxygen and energy have all been used on occasion as the basis for measurement. In the SWA study it was decided early to use carbon for this purpose, largely because it is common to all the sub–systems involved, and several techniques for tracing its flow through components of the system are well established. In spite of this, however, it was found necessary during the study to develop further methods for measuring carbon in particular situations.

During the progress of the study it has become apparent that the use of the carbon model has produced some successes and some disappointments. Perhaps the major success was in helping to ensure the comprehensive nature of the study. Perhaps the major disappointment has been that the use of the model to predict changes in state resulting from environmental changes is much more difficult than was hoped initially. Essentially this is because of the large amount of effort required to obtain a complete set of parameter values for a dynamic ecosystem model. In practice, the most quantified models are those with simple (usually linear) functions relating the rate of transfer of material only to the amounts in the donor and recipient compartments; such models are best suited to simple man–made ecosystems and a good example is that of Martin *et al.* (1976). Yet some of the transfer functions are known to be complex non–linear. The amount of information which is required to identify with any reliability such a function is enormously greater than that required for a linear function, and therefore a study capable of producing a credible predictive model would require much greater resources over a much longer period than were available for the SWA study. Nevertheless, it is believed that the work which was done on SWA has been well worth while and has added substantially to our understanding of the structure and relationships of estuaries.

REFERENCES

Hodgkin, E.P.: *An Environmental Study of the Blackwood River Estuary, Western Australia 1974–75*. Perth: Department of Conservation and Environment (1978)

Martin, G.D., Mulholland, R.J., Thornton, K.W.: Ecosystem approach to the simulation and control of an oil refinery waste treatment facility. *Journal of Dynamic Systems, Measurement and Control* **March 1976**, 20–29 (1976)

Ministry for Conservation: *A Preliminary Report on the Westernport Bay Environmental Study. Report for the Period 1973–4*. Melbourne: Ministry for Conservation (1975)

Parker, R.R., Rochford, D.J., Tranter, D.J.: History and organization of the Port Hacking Estuary Project. In: W.R. Cuff and M. Tomczak jr, eds *Synthesis and Modelling of Intermittent Estuaries*. Berlin, Heidelberg, New York: Springer (1983)

Synthesis and Modelling of Intermittent Estuaries
(W.R. Cuff and M. Tomczak jr. eds) Berlin, Heidelberg,
New York: Springer (1983), pp. 7–16.

History and Organization of the Port Hacking Estuary Project

Robert R. Parker[†], David J. Rochford[‡], David J. Tranter[†]

[†] Division of Fisheries Research
CSIRO Marine Laboratories
P.O. Box 21, Cronulla, N.S.W. 2230, Australia

[‡] Division of Oceanography
CSIRO Marine Laboratories
P.O. Box 21, Cronulla, N.S.W. 2230, Australia

Summary. The history of research into Port Hacking before the
Port Hacking Estuary Project of 1973–1978 is summarized. The
different steps of the organization of the Project are then described:
project initiation, problem definition and refinement, staffing,
facilities, field work. A list of publications resulting from the
Project is appended.

Key words: estuaries, project organization, Port Hacking Estuary
Project, South West Arm

In Australia, research in estuaries began in 1941. The initial emphasis was on hydrology and on the chemistry of sediments and water. Special attention was given to their phosphorus and organic content. These studies were considered at the time to be relevant to problems of the oyster growing industry, particularly to the spatfall characteristics and the fattening potential of East Australian estuaries . One of the study areas was Port Hacking which was known at the time to have good spatfall but poor fattening characteristics (Rochford, 1952). Among the more important controlling factors that were identified were tidal scouring and the nature of the intertidal sediments. In this respect, Port Hacking stood out as a marine dominated estuary with limited freshwater discharge, factors which appeared to explain the poor fattening characteristics of the system.

These studies led to a classification of Australian estuarine systems by Rochford (1951). The application of this classification to estuaries in other countries was explored at an international symposium (Rochford, 1959) and the conclusion was reached that the Rochford classification was peculiar to the Australian situation.

During these early studies, anoxic conditions were observed periodically in the South West Arm basin of Port Hacking, which led to large increases in the nutrient load in the water column. The same thing happened occasionally in other Australian estuaries. Rochford & Newell (1974) explored the significance of this phenomenon in relation to studies of estuarine pollution where oxygen is commonly used to diagnose the "health" of systems subject to organic pollution. Unless natural oxygen cycles are known, the oxygen parameter is of limited usefulness.

Rochford (1974) attributed the de–oxygenation cycle in South West Arm to periodic topping of the water column by fresh water and subsequent mortality and decay of the organisms there. At a critical level of oxygen, nutrients were thought to be released by the sediments into the overlying water column.

Baas Becking and Wood made extensive associated observations on the microbiology of Australian estuaries, including Port Hacking. Baas Becking & Wood (1955) and Baas Becking et al. (1959) endeavoured to characterize the biological processes in the estuarine environment and the chemical reactions governing these processes in terms of pH and Eh. These authors believed that the water environment is controlled by photosynthesis at the surface and the sulphur cycle below.

Wood (1964a–g) distinguished three communities: plankton, epontic, and sediment and coined the term "protoplankton" to include all unicellular organisms. He found this useful in that some of the diatoms and most of the dinoflagellates are myxotrophic, while some of the dinoflagellates do not consistently contain chlorophyll. The soft–bodied flagellates are notoriously inconsistent in this regard, defying classification into autotrophic or heterotrophic types.

Phytoplankton records of Port Hacking were kept for a number of years (1939–1950) and can, to some extent, be correlated with hydrological observations. This data series and studies of periphyton settling on glass slides impressed Wood with the unpredictable dynamic status of the protoplankton community. The annual cycles differed so markedly from year to year that the term "cycle" was declared a misnomer. There were trends of a sort in that a dinoflagellate maximum could be expected from September through February, but it could not be predicted with certainty. Sometimes such a bloom would come in August, sometimes not until November.

It was clear that each estuary has its characteristic protoplankton, to the extent that the origin of a sample could be recognized from its taxonomic mix. (The taxonomy of the diatoms and dinoflagellates was fairly well known (Wood, 1951; 1954; Wood *et al.*, 1959; Crosby & Wood, 1958, 1959).) The distributions of species of diatoms and dinoflagellates in both time and space led Wood to the concept of indicator species, *i.e.* species which indicated the origin of the water mass in which they were found. In food chain studies, he concluded that the important estuarine food resources are concentrated in such areas as seagrass flats where epontic forms are readily available and where the sediments are relatively stable and have a high organic content.

This was the state of knowledge which existed when the present Port Hacking Estuary Project began, and we now turn to a description of this Project.

In June 1973 the Chief (K. Radway Allen) of the Division of Fisheries and Oceanography announced a re–organization of research programs into five disciplinary Groups (water movement and properties, ecosystems, crustacean biology, fish biology, and population dynamics), each with its Group Head. He also foreshadowed the setting up of interdisciplinary projects involving more than one Group, with each project having a Co–ordinator to ensure that project participants co–operate effectively, to bring them together fairly frequently to talk about the progress of the work, and to keep in touch with the day to day operations of all the components of the project, so as to ensure that they fit in well together and make the best use of facilities and opportunities. Two existing projects were formally identified as interdisciplinary projects and the Port Hacking Estuary Project followed (10 October 1973) as the first new interdisciplinary project.

As explained in Allen (1983) the broad aim of the Project was to obtain increased understanding of the principles underlying the structure and dynamics of Australian estuaries. It was decided early in the study to centre it on an appropriate model, in this case a model of the flow of carbon into, within, and out of South West Arm.

Interested staff collectively began to plan the details of the project. The first step was to identify the parts of the system and how they fitted together, as then understood (Fig. 1). Boundaries between the estuary and each of substrate, sea, land and air were recognized. Living carbon C_L (*i.e.* carbon contained in living organisms)

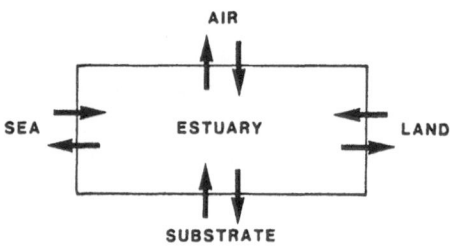

Fig. 1: Basic scheme of estuary interactions.

was visualized as measurable in terms of ATP (adenosine triphosphate) with autotrophic carbon C_A (*e.g.* phytoplankton) measurable in terms of chlorophyll *a* (CHL *a*). Heterotrophic carbon C_H (*e.g.* zooplankton) is then simply the difference between C_L and C_A. Detritus also used C_L in its estimation but only as subtracted from total particulate organic carbon, which was to be measured directly. A matrix was constructed to describe the system in a qualitative way (Fig. 2). The illustration shows generalized compartments, expected flows of carbon between them, and the expected state dependencies of these flows. Flows *within* the heterotroph compartment were also recognized as being likely, arising from its internal heterogeneity.

The physical boundaries of South West Arm were then defined and gains and losses across these boundaries were considered. Then, making a distinction between water column and sediment sub–systems, and between aerobic and anaerobic

	Recipient					
	DIC	AUT	HET	DOC	DET	
DIC		R				Dissolved inorganic carbon
AUT	D		B	D	D	Autotrophs
HET	D		●	D	D	Heterotrophs
DOC		R	R		D	Dissolved organic carbon
DET			R	D		Detritus

Fig. 2: A generalized interaction matrix of the estuarine ecosystem. Material fluxes are shown according to whether they were determined by the donor compartment (D), the recipient compartment (R), or by both compartments (B). ● – see text for details.

heterotrophs, a diagram of boxes and arrows was produced to show the flow of carbon (Fig. 3). This conceptual model, together with controlling factors such as light, temperature, nutrients, oxygen, precipitation, mixing, and transport became a statement of the second level of objectives.

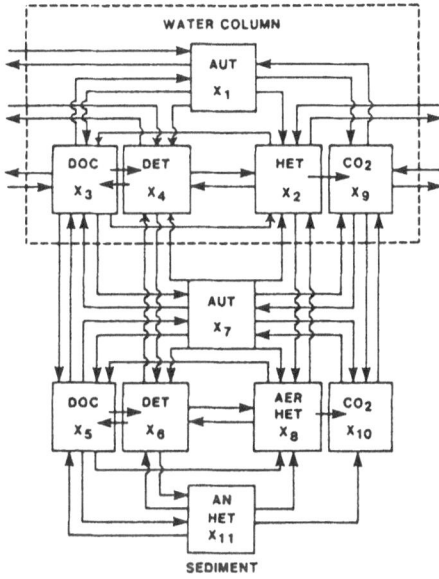

Fig. 3: Carbon flow diagram based on the interaction matrix of Fig. 2 but including inflows and outflows across the boundaries, sediment and water column subsystems (including flows between them) and aerobic (AER) and anaerobic (AN) sediment heterotrophs.

The next step was to make a reconnaissance of the study area so that the conceptual components could be given a concrete interpretation: AUT became phytoplankton, benthic micro–algae, macro–algae and seagrasses, and mangroves; HET became bacteria, zooplankton, nekton, and epifauna; and so on.

Staffing was largely by redeployment. Some scientists from other laboratories were attracted by the project, enlarging the scope of expertise available.

Individual scientists selected projects from a list of goal–oriented problems identified from the flow diagram of Fig. 3. The studies which proceeded were a compromise between the interests of the participants, the support available for the project, and priorities dictated by the need to run segments of the project in parallel or in sequence. For example, water movements and mixing had to be understood before a plankton sampling program could be designed; phytoplankton and zooplankton had to be measured at the same time because of their interactive relationship. However, studies on benthic algae and seagrasses could proceed quite independently. A central monitoring program was then begun with special emphasis on supposed forcing functions and on compartment size and taxon composition.

The Project had to establish support facilities *de novo*. Laboratories were created out of existing warehouse space or brought prefabricated to the site. A vessel was designed and built. The "state of the art" of measurement was at times inadequate and new techniques had to be worked out before a study could proceed. Nevertheless, field observations were underway on a "gumboot and rowboat" basis by late 1974. This phase of work is described in greater detail in CSIRO (1976).

Simulation modelling was an ongoing process. It was assumed at first that a predictive model would emerge within a reasonable time but as work proceeded it became apparent that this goal was not to be reached in the foreseeable future. With the passage of time, it eventually happened that the overriding need was for a model to collate and synthesize the knowledge that had been gained and to ascertain the completeness of this knowledge (see Sinclair *et al.*, 1983).

A re–evaluation of the project took place in 1977. Some of the scientists felt constrained by the model and by the way that the support available was dictated by the model (more detail in Cuff, 1983). This led to less emphasis on the modelling of carbon flow and some broadening of scope in the associated research. For example, work was begun on adenylate energy charge as an index of organism 'vitality' (Rainer *et al.*, 1979).

Before the project reached fruition, the overall priorities of the Division changed. New projects began elsewhere in Australia to which several of the participants in the Port Hacking Estuary Project were seconded. Much of the research then underway was never completed or fully utilized.

A list of papers published as a result of the Project is appended.

REFERENCES

Allen, K.R.: Introduction to the Port Hacking Estuary Project. In: W.R. Cuff and M. Tomczak jr, eds *Synthesis and Modelling of Intermittent Estuaries* Berlin, Heidelberg, New York: Springer (1983)

Baas Becking, L.G.M., Wood, E.J.F.: Biological processes in the estuarine environment. *Proceedings of the Koninklijke Nederlandse Akademie van Wetenschappen* **B58**, 160–181 (1955)

Baas Becking, L.G.M., Thompson, J.M., Wood, E.J.F.: Some aspects of the ecology of Lake Macquarie N.S.W. with regard to an alleged depletion of fish. *Australian Journal of Marine and Freshwater Research* **10**, 269–278 (1959)

Crosby, L.H., Wood, E.J.F.: Studies on Australian and New Zealand diatoms. I. Planktonic and allied species. *Transactions of the Royal Society of New Zealand Biological Sciences* **85**, 483–536 (1958)

Crosby, L.H., Wood, E.J.F.: Studies on Australian and New Zealand diatoms. II. Normally epontic and benthic genera. *Transactions of the Royal Society of New Zealand Biological Sciences* **86**, 1–58 (1959)

CSIRO: *Estuarine Project Progress Report 1974–1976*. Sydney: CSIRO Division of Fisheries and Oceanography (1976)

Cuff, W.R.: An evaluation of the Port Hacking Estuary Project from the viewpoint of applied science. In: W.R. Cuff and M. Tomczak jr, eds *Synthesis and Modelling of Intermittent Estuaries*. Berlin, Heidelberg, New York: Springer (1983)

Rainer, S.F., Ivanovici, A.M., Wadley, V.A.: Effect of reduced salinity on adenylate energy charge in three estuarine molluscs. *Marine Biology (Berlin)* **54**, 91–99 (1979)

Rochford, D.J.: Studies in Australian estuarine hydrology. I. Introduction and comparative features. *Australian Journal of Marine and Freshwater Research* **2**, 51–59 (1951)

Rochford, D.J.: The application of studies on estuarine hydrology to certain problems in Australian oyster biology. *Rapports et Procès-Verbaux des Réunions Conseil International pour l'Exploration de la Mer* **131**, 35–37 (1952)

Rochford, D.J.: Classification of Australian estuarine systems. *Archivio di Oceanografia e Limnologia (Supplemento)* **11**, 171–177 (1959)

Rochford, D.J.: Sediment trapping of nutrients in Australian estuaries. *CSIRO Division of Fisheries and Oceanography Report* **61** (1974)

Rochford, D.J., Newell, B.S.: Water quality attributes of N.S.W. In: Australian UNESCO Committee for Man and the Biosphere, *Report of Symposium on the Impact of Human Activities on Coastal Zones* Canberra: Australian Government Publishing Service (1974)

Sinclair, R.E., Cuff, W.R., Parker, R.R.: Ecosystem modelling of South West Arm, Port Hacking. In: W.R. Cuff and M. Tomczak jr, eds *Synthesis and Modelling of Intermittent Estuaries*. Berlin, Heidelberg, New York: Springer (1983)

Wood, E.J.F.: Phytoplankton studies in eastern Australia. *Proceedings of the Indo–Pacific Fisheries Council 1950*, 69–72 (1951)

Wood, E.J.F.: Dinoflagellates in the Australian region. *Australian Journal of Marine and Freshwater Research* **5**, 171–351 (1954)

Wood, E.J.F.: Studies in microbial ecology of the Australasian region. I. Relation of oceanic species of diatoms and dinoflagellates to hydrology. *Nova Hedwigia* **VIII**, 5–20 (1964a)

Wood, E.J.F.: Studies in microbial ecology of the Australasian region. II. Ecological relations of oceanic and neritic diatom species. *Nova Hedwigia* **VIII**, 20–35 (1964b)

Wood, E.J.F.: Studies in microbial ecology of the Australasian region. III. Ecological relations of some oceanic dinoflagellates. *Nova Hedwigia* **VIII**, 35–54 (1964c)

Wood, E.J.F.: Studies in microbial ecology of the Australasian region. IV. Some quantitative aspects. *Nova Hedwigia* **VIII**, 453–461 (1964d)

Wood, E.J.F.: Studies in microbial ecology of the Australasian region. V. Microbiology of some Australian estuaries. *Nova Hedwigia* **VIII**, 461–527 (1964e)

Wood, E.J.F.: Studies in microbial ecology of the Australasian region. VI. Ecological relations of Australian estuarine diatoms. *Nova Hedwigia* **VIII**, 527–548 (1964f)

Wood, E.J.F.: Studies in microbial ecology of the Australasian region. VII. Ecological relations of Australian estuarine dinoflagellates. *Nova Hedwigia* **VIII**, 548–568 (1964g)

Wood, E.J.F., Crosby, L.H., Cassie, V.: Studies on Australian and New Zealand diatoms. III. Descriptions of further discoid species. *Transactions of the Royal Society of New Zealand Biological Sciences* **87**, 211–219 (1959)

APPENDIX: Publications arising from the Port Hacking Estuary Project

Batley, G.E., Gardner, D.: A study of copper, lead and cadmium speciation in some estuarine and coastal marine waters. *Estuarine and Coastal Marine Science* **7**, 59–70 (1978)

Batley, G.E., Gardner, D.: Sampling and storage of natural waters for trace metal analysis. *Water Research* **11**, 745–756 (1978)

Batley, G.E., Giles, M.S.: Solvent displacement of sediment interstitial waters before trace metal analysis. *Water Research* **13**, 879–886 (1979)

Batley, G.E., Giles, M.S.: A solvent displacement technique for the separation of sediment interstitial waters. In: R.A. Baker, ed. *Contaminants and Sediments, Vol. 2.* Michigan: Ann Arbor (1980)

Bull, J.D., Burchmore, J.J., Pollard, D.A.: Feeding ecology of the sympatric species of leatherjackets (Pisces: Monacanthidae) from a *Posidonia* seagrass habitat in New South Wales. *Australian Journal of Marine and Freshwater Research* **29**, 631–643 (1978)

Bulleid, N.C.: Adenosine triphosphate analysis in marine ecology: a review and manual. *CSIRO Division of Fisheries and Oceanography Report* **75** (1977)

Bulleid, N.C.: An improved method for the extraction of adenosine triphosphate from marine sediment and seawater. *Limnology and Oceanography* **23**, 174–178 (1978)

Caperon, J., Smith, D.F.: Photosynthetic rates of marine algae as a function of inorganic carbon concentration. *Limnology and Oceanography* **23**, 704–708 (1978)

Colquhoun-Kerr, J.S.: Carbon flux through the South West Arm populations of *Crassostrea commercialis* and *Trichomya hirsuta. CSIRO Division of Fisheries and Oceanography Report* **79** (1977)

CSIRO: *Estuarine Project Progress Report 1974–1976.* Sydney: CSIRO Division of Fisheries and Oceanography (1976)

Cuff, W., Sinclair, R., Parker, R.R.: The development of an ecosystem model of South West Arm (Port Hacking, N.S.W.). *Simulation Modelling Techniques and Applications, Proceedings of SIMSIG–78 Simulation Conference, Canberra,* 33–38 (1978)

Cuff, W., Sinclair, R., Parker, R.R.: Carbon flow within South West Arm of Port Hacking (N.S.W., Australia). In: P.A. Trudinger, M.R. Walter and B.J. Ralph, editorial committee. *Biogeochemistry of Ancient and Modern Environments.* Canberra: Australian Academy of Science (1980)

Godfrey, J.S., Parslow, J.: Description and preliminary theory of circulation in Port Hacking estuary. *CSIRO Division of Fisheries and Oceanography Report* **67** (1976)

Griffiths, F.B., Caperon, J.: Phytoplankton loss by estuarine zooplankton grazing. *Marine Biology (Berlin)* **54**, 301–309 (1979)

Griffiths, F.B., Rimmer, D.W.: A description of a paired sampler suitable for quantitative plankton studies. *CSIRO Division of Fisheries and Oceanography Report* **94** (1977)

Hutchings, P.A., Rainer, S.F.: A key to the estuarine polychaetes of New South Wales. *Journal of the Linnean Society of N.S.W.* **104**, 35–48 (1979)

Hutchings, P., Rainer, S.: Designation of a neotype of *Capitella filiformis* Claparède, 1864, type species of the genus *Heteromastus* (Polychaeta: Capitellidae). *Records of the Australian Museum* **34**, 373–380 (1981)

Ivanovici, A.M.: A method for extraction and assay of adenosine triphosphate nucleotides from molluscan tissue. *CSIRO Division of Fisheries and Oceanography Report* **118** (1981)

Ivanovici, A.M., Rainer, S.F., Wadley, V.A.: Free amino acids in three species of estuarine mollusc: responses to factors associated with reduced salinity. *Comparative Biochemistry and Physiology* **70A**, 17–22 (1981)

Kirkman, H.: Growth of *Zostera capricorni* Aschers. in tanks. *Aquatic Botany* **4**, 367–372 (1978)

Kirkman, H., Griffiths, F.B., Parker, R.R.: The release of reactive phosphate by a seagrass community. *Aquatic Botany* **6**, 329–337 (1979)

Kirkman, H., Reid, D.D.: A study of the role of the seagrass *Posidonia australis* in the carbon budget of an estuary. *Aquatic Botany* **7**, 173–183 (1979)

Kirkman, H., Reid, D.D., Cook, I.H.: Biomass and growth of *Zostera capricorni* Aschers. in Port Hacking, N.S.W. Australia, *Aquatic Botany* **12**, 57–67 (1982)

Parker, R.R., Sibert, J.: Studies on a production system using a large volume floating pond. *10th European Symposium on Marine Biology, Ostend Belgium*, **2** 457–466 (1975)

Parker, R.R.: The CSIRO Fisheries and Oceanography Estuarine Project *Australian Marine Sciences Bulletin* **58**, 11–14 (1977)

Parker, R.R.: Guidelines for ecosystem research in coastal lagoons. In: Coastal Lagoon Research, Present and Future *UNESCO Technical Papers in Marine Science* **32**, 305–314 (1981)

Rainer, S.F.: The benthic macrofauna of Gunnamatta Bay, Port Hacking, N.S.W.: Biological and physiological data, 8 January 1975 and 18 February 1975. *CSIRO Division of Fisheries and Oceanography Microfiche Data Series* **3** (1979)

Rainer, S.F.: The benthic biotopes of South West Arm, Port Hacking, N.S.W., 1975. *CSIRO Division of Fisheries and Oceanography Report* **109** (1980)

Rainer, S.F.: Temporal patterns in the structure of macrobenthic communities of an Australian estuary. *Estuarine Coastal and Shelf Science* **13**, 597–620 (1981)

Rainer, S.F.: Trophic structure and production in the macrobenthos of a temperate Australian estuary. *Estuarine Coastal and Shelf Science* **15** (1982, in press)

Rainer, S.F., Fitzhardinge, R.: Benthic communities in an estuary with periodic de oxygenation *Australian Journal of Marine and Freshwater Research* **32**, 227–243 (1981)

Rainer, S.F., Griffiths, F.B.: Hydrology of an estuary with periodic de-oxygenation. *CSIRO Division of Fisheries and Oceanography Report* **117** (1980)

Rainer, S.F., Hutchings, P.A.: Nephytidae (Polychaeta: Errantia) from Australia. *Records of the Australian Museum* **31**, 301–347 (1977)

Rainer, S.F., Ivanovici, A.M., Wadley, V.A.: Effect of reduced salinity and adenylate energy charge in three estuarine molluscs. *Marine Biology (Berlin)* **54**, 91–99 (1979)

Sandland, R.L., Young, P.C.: Probabilistic tests and stopping rules associated with hierarchical classification techniques. *Australian Journal of Ecology* **4**, 399–406 (1979)

Scott, B.D.: Phytoplankton distribution and light attenuation in Port Hacking estuary. *Australian Journal of Marine and Freshwater Research* **29**, 31–44 (1978)

Scott, B.D.: Nutrient cycling and primary production in Port Hacking, New South Wales. *Australian Journal of Marine and Freshwater Research* **29**, 803–815 (1978)

Scott, B.D.: Seasonal variations of phytoplankton production in an estuary in relation to coastal water movements. *Australian Journal of Marine and Freshwater Research* **30**, 449–461 (1979)

Smith, D.F.: Quantitative analysis of the functional relationships existing between ecosystem components. I. Analysis of the linear intercomponent mass transfers. *Oecologia* **16**, 97–106 (1974)

Smith, D.F.: Quantitative analysis of the functional relationships existing between ecosystem components. II. Analysis of non–linear relationships. *Oecologia* **16**, 107–117 (1974)

Smith, D.F.: Quantitative analysis of the functional relationships existing between ecosystem components. III. Analysis of the ecosystem stability. *Oecologia* **21**, 17–29 (1975)

Smith, D.F.: Feeding and food–webs. In: T. Platt, K.H. Mann, and R.E. Ulanowics, eds. *Mathematical Models in Biological Oceanography.* Monographs on Oceanographic Methodology 7, Paris: UNESCO Press (1981)

Smith, D.F.: Measuring rates of cycling of elements. In: T. Platt, K.H. Mann, and R.E. Ulanowics, eds. *Mathematical Models in Biological Oceanography.* Monographs on Oceanographic Methodology 7, Paris: UNESCO Press (1981)

Smith, D.F., Bulleid, N., Campbell, R., Higgins, H., Rowe, F., Tranter, D., Tranter, H.A.: Marine food web analysis: An experimental study of demersal zooplankton using isotopically labelled prey species. *Marine Biology (Berlin)* **54**, 49–59 (1979)

Smith, D.F., Higgins, H.: An interspecies regulatory control of dissolved organic carbon production by phytoplankton and incorporation by microheterotrophs. In: M. Loutit and J.A.R. Miles, eds. *Microbial Ecology.* Berlin: Springer–Verlag, 34–39 (1978)

Smith, D.F., Wiebe, W.J.: Constant release of photosynthate from marine phytoplankton. *Applied and Environmental Microbiology* **32**, 75–79 (1976)

Tranter, D.J., Bulleid, N.C., Campbell, R., Higgins, H., Rowe, F., Tranter, H.A., Smith, D.F.: Nocturnal movements of phototactic plankton in shallow waters. *Marine Biology (Berlin)* **61**, 317–326 (1981)

Wadley, V.A.: Spatial and temporal heterogeneity in the epibenthic fauna of estuarine sand and seagrass beds. (Thesis abstract). *Australian Journal of Ecology* **6**, 217 (1981)

Wadley, V.A., Ivanovici, A.M., Rainer, S.F.: A comparison of techniques for the measurement of adenine nucleotides in three species of estuarine molluscs. *CSIRO Division of Fisheries and Oceanography Report* **129** (1980)

Weiner, P., Kirkman, H.: Continuous recording technique to measure oxygen release from a seagrass community within an acrylic insulation chamber. *CSIRO Division of Fisheries and Oceanography Report* **96** (1979)

Wiebe, W.J., Smith, D.F.: Direct measurement of dissolved organic carbon release by phytoplankton and incorporation by microheterotrophs. *Marine Biology (Berlin)* **42** , 213–223 (1977)

Wiebe, W.J., Smith, D.F.: ^{14}C–labeling of the compounds excreted by phytoplankton for employment as a realistic tracer in secondary productivity measurements. *Microbial Ecology* **4**, 1–8 (1977)

Young, P.C.: Temporal changes in the vagile epibenthic fauna of two seagrass meadows (*Zostera capricorni* and *Posidonia australis*). *Marine Ecology Progress Series* **5**, 91–102 (1981)

Synthesis and Modelling of Intermittent Estuaries
(W.R. Cuff and M. Tomczak jr. eds) Berlin, Heidelberg,
New York: Springer (1983), pp. 17–26.

Geological Aspects of the Port Hacking Estuary

Alberto D. Albani[†], Peter C. Rickwood[†]
James W. Tayton[‡], B. David Johnson[‡]

[†] School of Applied Geology
University of New South Wales, Kensington, N.S.W. 2033, Australia

[‡] School of Earth Sciences
Macquarie University, North Ryde, N.S.W. 2113, Australia

Summary. The geology of Port Hacking, a small estuary on
Australia's east coast, is reviewed and results of a survey based on
continuous seismic profiling for the determination of the depth of
bedrock are reported. This depth is between 40 and 60 m below
the present bottom of the highly silted estuary. An estimate is
derived for the amount of construction sand that could be dredged
from Port Hacking.

Key words: estuaries, geology, bedrock topography, sand deposit,
Port Hacking, South West Arm

1. INTRODUCTION

Port Hacking is a highly silted tidal estuary of dendritic shape, which marks the present south–eastern limit of development of the Sydney metropolitan area (Fig. 1). Its northern shores are suburbs within the Sutherland Shire whereas on the south it is largely bounded by the Royal National Park. It is occupied by the Hacking River which has a length of 42 km from its most distant source, 2 km west of Stanwell Tops, to Port Hacking Point. The navigable portion, the last 12.6 km up to Audley Causeway, constitutes Port Hacking which is fed by many small tributary streams usually occupying narrow valleys and draining a dissected plateau. Our study encompassed the whole of the estuary that was navigable by our survey vessel and a preliminary account of our work was reported by Albani (1976).

The geology of Port Hacking is reviewed in Section 2 and a survey, based on continuous seismic profiling for the determination of the depth of bedrock, is described in subsequent sections.

2. GEOLOGY

Within the catchment area of the Hacking River are rocks of Triassic age (about 230 million years), of which the oldest are siltstones and sandstones of the Narrabeen Group; these are overlain by the Hawkesbury Sandstone Group that forms many of the cliffs in the Sydney region. The sequence is terminated by the Wianamatta Group composed predominantly of shales which have yielded much of the clay for the local brickpits, one of which is within the catchment area.

All of these units are of Triassic age and, with one exception (see below), there are no known representatives of younger units except for the Quaternary (less than 2 million years) unconsolidated sediments filling the estuary.

South of Audley, the Hacking River has cut a valley down into Lower Triassic sediments of the Narrabeen Group, but throughout the final 18 km of its course the banks are composed of Middle Triassic sediments, chiefly of the Hawkesbury Sandstone Group. Unconsolidated sand and mud bound the river and estuary at many places, but otherwise the only variation in rock type is the occurrence of a deeply–weathered, undated, igneous dyke very close to the wharf at Bundeena

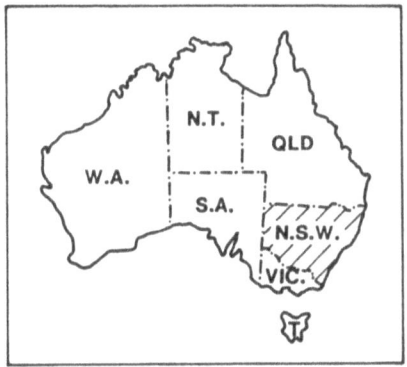

Fig. 1: Locality map.

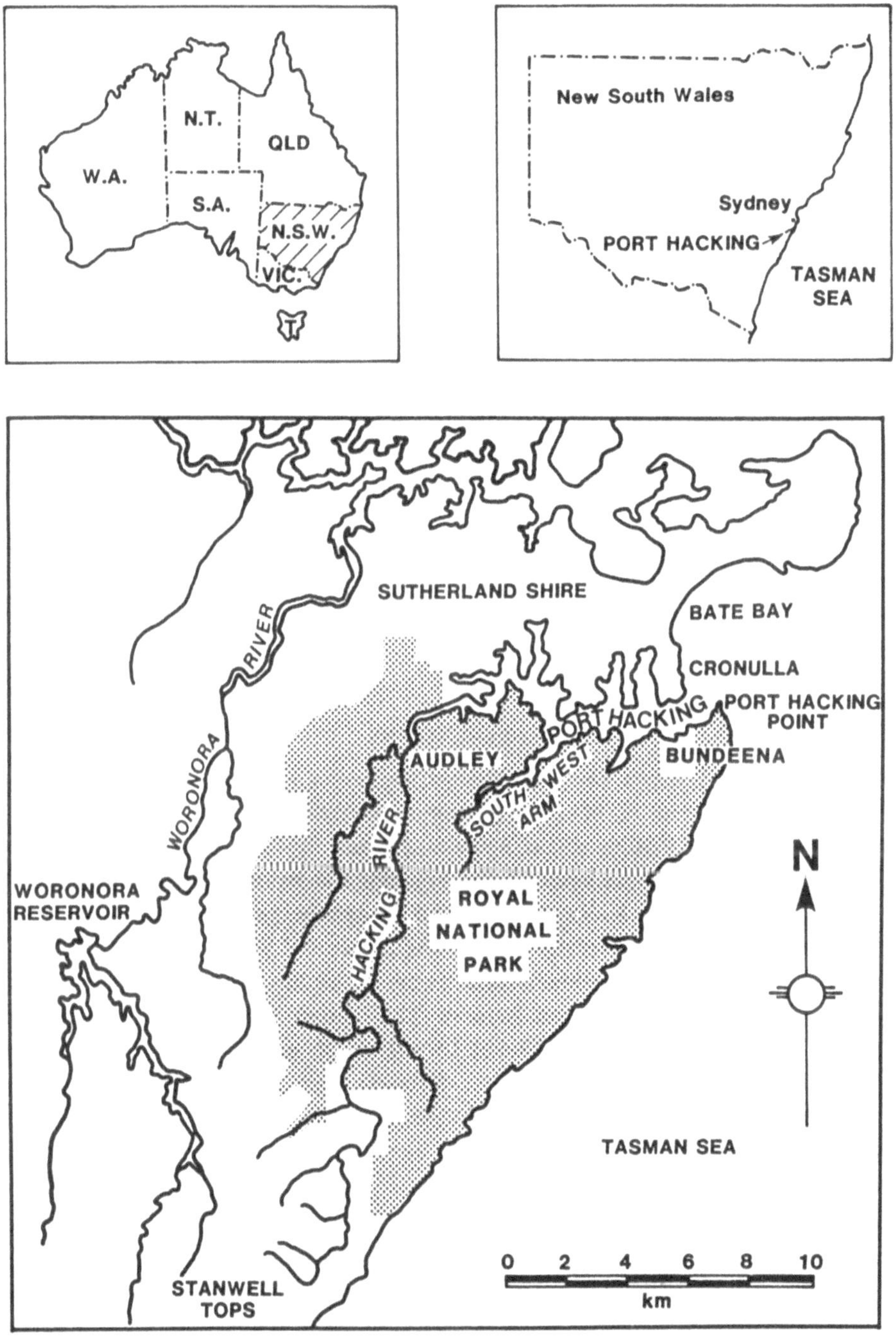

(Fig. 2). This baked the surrounding rock so that for a short distance the sandstone has polygonal fractures and is said to contain tridymite (Osborne, 1948), the high temperature polymorph of quartz.

Upstream from North West Arm the river is silted and very shallow, but downstream to Lilli Pilli the water depth is considerable, being in excess of 20 m over much of this distance (Fig. 2). Between Lilli Pilli and Bundeena, as well as in Gunnamatta Bay, the estuary is almost completely filled by unstable sand banks, which periodically shift and present a navigational problem to the community using this waterway. For this reason, the water depths shown on Fig. 2 must be considered approximate although they have been taken from the most recent bathymetric charts (Department of Public Works, N.S.W., 1968). Seaward of Bundeena the ocean swell keeps the estuary open and at times the water becomes extremely rough. The landward tributaries, *e.g.* North West Arm and South West Arm, are silted mainly in their upper reaches but the sediment is predominantly mud rather than sand.

3. METHOD OF STUDY

The depth to bedrock has been determined using a continuous seismic profiling system, with a sparker of 80–200 J (Tayton, unpublished data). The sound waves developed by the exploding steam bubbles, as well as their subsequent reflections, were detected by a set of six hydrophones contained in an oil–filled, neutrally–buoyant polythene tube. Both the electrode and the hydrophone eel were towed about 10 m behind a survey vessel. A specially designed amplifier and filter system enabled the six electronic signals to be combined selectively to obtain maximum suppression of multiple reflections from the sea bottom. The output signals were supplied to a modified Furuno depth recorder (Model 85014) which provided a continuous record, in graphic form, of the travel times of reflections from the sea surface, sea bottom, bedrock, and sometimes intermediate layers. Not all multiple reflections could be suppressed but the records were generally clear. Travel times could be determined with an accuracy of ± 0.5 ms, which is equivalent to a linear accuracy of approximately 1 m.

Because of the narrowness of the estuary, position fixing presented few problems. Surveys were always performed close to slack high tide conditions, so that lateral drift between stations was minimal. Straight line courses (Fig. 2) were steered between prominent features and intermediate points were determined by the photographic technique described by Albani (1980). The total length of survey lines was 65 km.

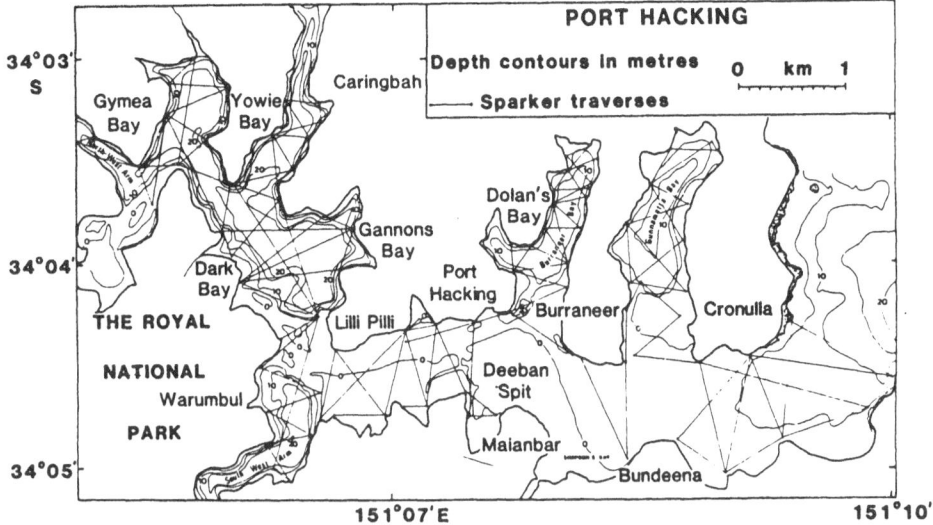

Fig. 2: Bathymetry and traverse locations.

The vessel used for sparker studies has to be one whose engine noise is essentially outside of the frequency band used for measurement. It is also important that the vessel should produce as little water disturbance (and hence noise) as possible, so those of very shallow draft are precluded. The vessel used for most of this work, SS Maluka II, drew 1.5 m and for safety needed 2 m of water depth. Accordingly, some places within the survey area could not be examined because of the degree of siltation. Hence data are lacking for :

(a) a large, almost terrestrial area, south and south–west of Burraneer Bay encompassing Deeban Spit and Simpson's Bay.

(b) Dolan's Bay, the northern half of Yowie Bay, North West Arm and South West Arm upstream of the termination of traverses (Fig. 2).

Area (a) is bounded by traverses along which satisfactory sparker records were obtained and interpolations were made. In locations of category (b), the bedrock depth can generally be estimated with confidence, for the valleys are narrow and frequently bounded by rock outcrops. Nevertheless, we have been cautious in not extrapolating over the entire areas of Yowie Bay and South West Arm.

To interpret the sparker records we assumed sound velocities of 1700 m s^{-1} for the uppermost 30 m of sediment, and 2150 m s^{-1} for sediment beneath this level: justification for these values has been given by Johnson *et al.* (1977). On all records, the uppermost reflector has been interpreted as the sea bottom and the lowermost reflector has been assumed to be bedrock of Hawkesbury Sandstone. Occasionally the bedrock constitutes the sea floor but generally it is capped by unconsolidated sediment within which other reflecting surfaces can usually be detected. In some regions, *e.g.* west of Lilli Pilli, the unconsolidated sediments are almost acoustically opaque and in these areas it proved necessary to re–run traverses after equipment modifications had been made to yield greater than normal sensitivity.

4. BEDROCK TOPOGRAPHY

Two unusual topographic features have been found. On the eastern side of Burraneer Bay the bedrock drops very steeply at locations only a few metres from the shore. The appearance of the contours resembles those of a fault scarp but further substantive evidence is lacking. On the northern side of Yowie Bay a very small, sharp rock pinnacle rises to within 5 m of sea level. It is covered by less than a metre of sediment, and is too small in area to plot on Fig. 3.

Due to additional data, the interpretation of the bedrock drainage pattern at the entrance of Port Hacking is now (Fig. 3) slightly different to that shown by Johnson *et al.* (1977, Fig. 3) and Albani *et al.* (1978). Opposite Port Hacking Point the depth to bedrock is about 95 m and it progressively diminishes as the channel is traced upstream. The ancient river channel was deflected from a straight course by ridges
(a) extending in a north–easterly direction from Maianbar;
(b) extending north–east from Warumbul;
(c) extending eastwards from the southern headland of Dark Bay; and
(d) extending southwards from Yowie Bay.

Of all these features the most significant is probably the ridge near Dark Bay; its upper surface is less than 20 m below sea level and in places it is covered by less than 5 m of sediment. To the north and east of it the water is up to 20 m deep, but to the south there is considerable silting so that in places nil depth of water is charted. The smaller ridge extending north–east from Warumbul, near South West Arm, is a similar divide; deep water lies to its south but little to its north and east. The unconsolidated sediment lying between these ridges and extending to Bundeena is predominantly sand of marine origin. However, to the south–west of the Warumbul ridge and to the north–west of the Yowie Bay/Dark Bay divide, the sediment is predominantly mud of terrestrial origin. In the progressive silting of Port Hacking these bedrock ridges would have been significant sediment barriers until they became submerged when the sea

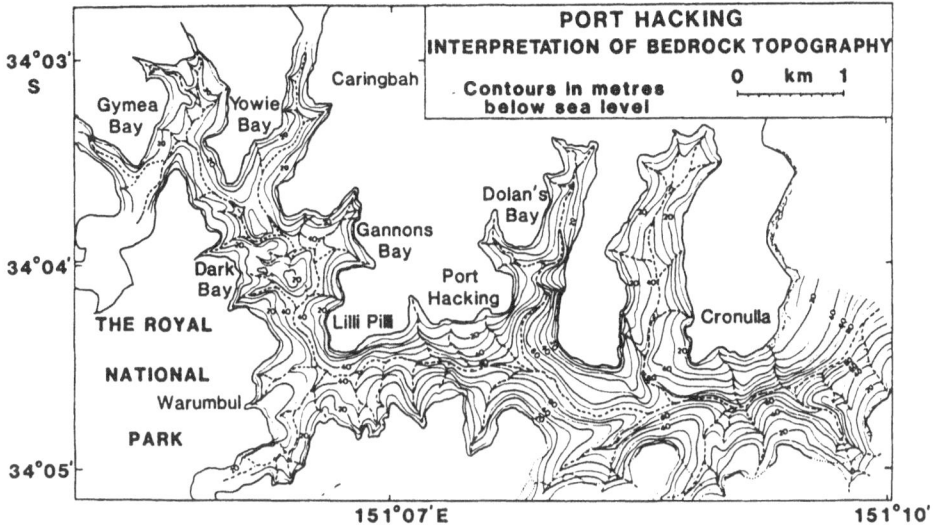

Fig. 3: Bedrock topography.

level was formerly at 20 m, approximately 9000 years before present (B.P.). according to the data of Smart (1977).

At its seaward extremity, the ancient Hacking River drainage channel joined that of the combined Cooks – Georges Rivers midway between Port Hacking Point and Osborn Shoal (Johnson *et al.,* 1977). Their combined channel can be traced for several kilometres into the present continental shelf.

5. UNCONSOLIDATED SEDIMENTS

The unconsolidated sands, silts, and muds that occur within the Hacking estuary are believed to have been derived from rocks within the catchment area.

Most of the erosion would have occurred during periods when the sea level was much lower than at present and the estuary was occupied by a river system. The depositional area of this fluvial activity would have been eastward of the present entrance to Port Hacking, where silts and sands would have been accumulated.

As the sea level rose the sediments were transported upstream by waves and currents and thus deposited by a marine mechanism so justifying our subsequent usage

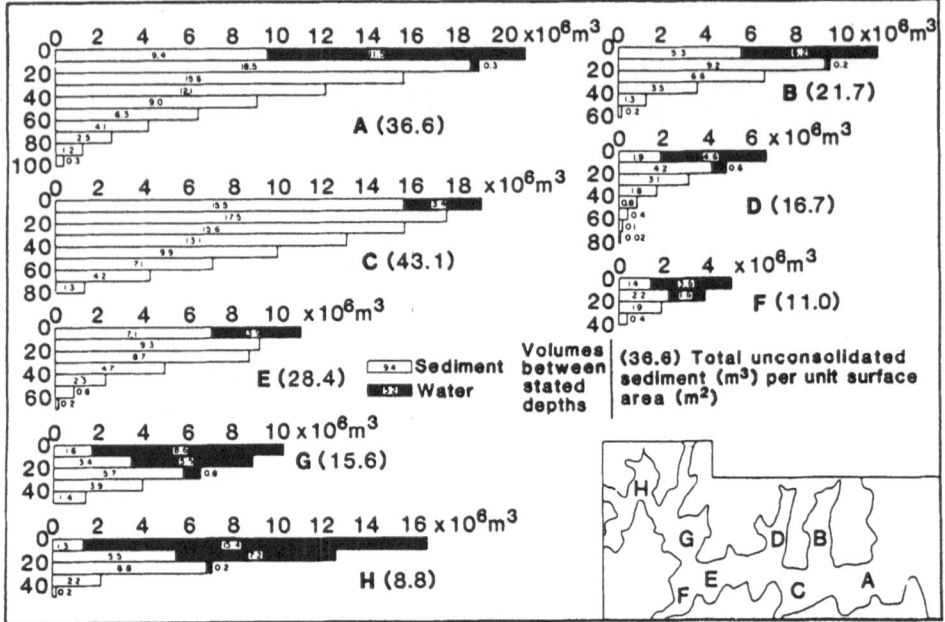

Fig. 4: Sediment volumes.

been proven by Coleman (1979) who observed silt and fine sand migrating upstream from areas of dredging.

The characteristics of the unconsolidated sediment vary both horizontally and vertically within the survey area. However, within the uppermost 20 – 25 m no drastic lithological changes are indicated by the seismic records.

The Port Hacking survey area contains approximately 2.7 x 10^8 m^3 of sediment (Table 1) and its distribution with depth is shown in Fig. 4.

Table 1.
Some physical parameters of Port Hacking.

Physical Parameter	Value
Total surface area	10.7 x 10^6 m^2
Total water surface area	8.9 x 10^6 m^2
Total water volume	72.8 x 10^6 m^3
Total sediment infill	273.7 x 10^6 m^3

For convenience of discussion, the area has been arbitrarily divided into 8 sections (A–H) based on bedrock morphology and sediment characteristics. Silt and clay predominate as the topmost sediments in sections B, D, F, G, and H whereas in sections A, C, and E it is sand with grain size ranging from 0.93ϕ (0.5mm) to 2.20ϕ (0.2mm) together with irregularly distributed shell fragments which locally may constitute 45% (Coleman, 1979).

Sand in sections C and E may be regarded as being commercially attractive to the building industries. Some 32 Mt of sand exists above 10 m below sea level (b.s.l.), but if dredging were carried out to 20 m b.s.l. a total of 69 Mt could be procured. Sand in section A was excluded from these calculations assuming the desirability of leaving a barrier to the open sea. However, if section A is included the respective total resources are 45 Mt (to 10 m b.s.l.) and 108 Mt (to 20 m b.s.l.). The largest of these figures is roughly three times that of the estimated sand resources of the Kurnell Peninsula (35 Mt) from which 25% of the construction sand used in Sydney is currently procured (N.S.W. Planning and Environment Commission, 1979). The estimated rate of depletion of the Kurnell resources is 1.2 Mt yr^{-1} (N.S.W. Planning and Environment Commission, 1979, p. 34), of which 92.5% is used for construction purposes. Accordingly, Port Hacking could yield enough construction sand to supply the population of Sydney at this rate for 93 years.

ACKNOWLEDGEMENTS

We thank Mr F. Potts for supplying one of the vessels used in this investigation, and also Mr and Mrs C. Dransfield for permitting use of their jetty.

This project was made possible through the support of the Sutherland Shire Council and we are most grateful for the encouragement given by the successive Shire Presidents, Councillors K.M. Skinner, P.C. Lewis, and M.T. Tynan and particularly Mr A.G. Hill, the Shire Clerk.

REFERENCES

Albani, A.: Preliminary report on the bedrock topography in South West Arm. In: *Estuarine Project Progress Report 1974–1976.* Sydney: CSIRO Division of Fisheries and Oceanography (1976)

Albani, A.D.: A vessel positioning method for surveys in coastal waters. *Journal of the Royal Society of New South Wales* **113**, 31–33 (1980)

Albani, A.D., Johnson, B.D., Rickwood, P.C., Tayton, J.W.: The bedrock morphology of Botany Bay, Bate Bay and Port Hacking. In: *A Geological Investigation of the Seaboard Area of the Sutherland Shire.* Sydney: Unisearch Ltd (1978)

Coleman, H.: *A Study of the Calibre and Movement of Sediments in the Vicinity of Maianbar, Port Hacking, N.S.W.* B.Sc. (Hon.) thesis, University of New South Wales (1979)

Department of Public Works, N.S.W.: Port Hacking Sounding: 5 sheets (1968)

Johnson, B.D., Albani, A.D., Rickwood, P.C., Tayton, J.W.: The bedrock topography of the Botany Basin, New South Wales. *Journal of the Geological Society of Australia* **24**, 403–408 (1977)

N.S.W. Planning and Environment Commission: *Kurnell Planning Study.* Sydney: New South Wales Planning and Environmental Commission (1979)

Osborne, G.D.: Note on the occurrence of tridymite in metamorphosed Hawkesbury Sandstone at Bundeena and West Pymble, Sydney District, New South Wales. *Journal and Proceedings of the Royal Society of New South Wales* **82**, 309–311 (1948)

Smart, J.: Late Quaternary sea–level changes, Gulf of Carpentaria, Australia. *Geology* **5**, 755–759 (1977)

Synthesis and Modelling of Intermittent Estuaries
(W.R. Cuff and M. Tomczak jr. eds) Berlin, Heidelberg,
New York: Springer (1983), pp. 27–54.

Tidal Flushing and Vertical Diffusion in South West Arm, Port Hacking

J. Stuart Godfrey

Division of Oceanography

CSIRO Marine Laboratories

P.O. Box 21, Cronulla, N.S.W. 2230, Australia

Summary. South West Arm (SWA), a small Australian estuary, is hydrodynamically a small fjord with highly intermittent river discharge; tidal inflow sinks into it in a thin turbulent sheet. An existing water quality model is adapted to the situation in SWA. It assumes horizontal homogeneity and allows for entrainment and interleaving of the tidal inflow, passive convective cooling, and vertical eddy diffusion, and it predicts running–mean values over a tidal cycle. Both the kinetic energy of the tidal inflow and the potential energy released by the turbulent sheet of sinking water are considered as possible energy sources for the diffusing eddies.

Application to the response of SWA to a rainstorm results in energy conversion efficiencies of 0.025 – 0.05, comparable to those found in a reservoir and in a Norwegian fjord. However, reasonable simulations of flood response can be obtained for a rather wide range of parameter values. Application to spring warm–up in SWA needed slightly lower conversion efficiencies – around 0.025 – to get satisfactory results; but these efficiencies are in any case uncertain to within a factor of 3, due to lack of knowledge of the kinetic energy of the inflow. The fact that a marked spring–neap cycle is observed in the rate of temperature increase at 16 m during spring warm–up in SWA suggests that kinetic energy influx is the major contributor to eddy diffusion there.

Order–of–magnitude estimates for dissolved oxygen show that (a) during spring warm–up, dissolved oxygen concentration at the bottom of SWA is principally a balance between eddy diffusion and biological consumption; and (b) estimates of the rate of diffusion through 13 m depth, using diffusivities calculated from observed temperature structure, agree well with measured consumption rates. Oxygen response to a rainstorm is modelled reasonably well.

Key words: estuaries, fjords, nutrient regeneration, oxygen, diffusion, Port Hacking, South West Arm

1. INTRODUCTION

Port Hacking estuary consists of four deep (15–25 m) basins, connected to the ocean by a narrow, 2 km long tidal channel that is only 2 m deep (Fig. 1 of Vaudrey *et al.*, 1983). Small creeks flow in at the landward ends of the two innermost basins; these creeks generally discharge less than 1 m^3 s^{-1} to the estuary, comparable to the evaporative loss from the basins. Consequently, the Port Hacking basins are for much of the time simply enclosed arms of the sea, with near–marine salinity throughout. Following heavy rainstorms the discharge can rise to 100 m^3 s^{-1}, but it will return to 1 m^3 s^{-1} or less within a few days of the storm; surface salinity then decays towards marine conditions over a few weeks.

South West Arm (SWA) is one of the Port Hacking basins. Its hydrology has been extensively studied by the CSIRO Marine Laboratories (Parker *et al.*, 1983). In geomorphological terms, SWA is a deep "coastal lagoon" (Bird, 1968), and estuaries of this geographic type are common. For the purposes of this chapter, however, a hydrodynamic classification of the estuary is more relevant; in Hansen & Rattray's (1966) classification scheme, SWA would be described as a small, and highly intermittent, fjord. In the few days of strong river discharge following a rainstorm it behaves as a typical "active", or Type 1, fjord (Pickard, 1961), with a sharply defined surface layer 1–2 m deep, the salinity of which increases markedly towards the mouth; this layer overlays nearly 20 m of marine water. As the flow decreases, the surface layer and halocline deepen and become more saline (Godfrey & Parslow, 1976); during this phase SWA can be described as a Type 2 fjord. It finally approaches a purely marine condition after a few weeks. It is then a Type 3 fjord in Pickard's notation; Pickard refers to such marine fjords as "passive", but as seen below, this is a misnomer for SWA. Large temperature differences can occur between water inside and outside SWA, and this leads to density flow phenomena entirely analogous to those seen under Type 1 or Type 2 conditions. Furthermore, considerable bottom de–oxygenation can occur due to the influence of thermal stratification alone.

Except in the active surface layer under the few days per year of strong river flow, most of the basin is surprisingly close to horizontal homogeneity, at least in water density (Godfrey & Parslow, 1976). Only on a rising tide within a few tens of metres of the steep sandbanks at the seaward end of SWA are there marked departures from horizontal homogeneity (*e.g.* Fig. 1). Whenever the well mixed water entering from the tidal channel is denser than ambient surface water in SWA, it sinks in a turbulent sheet down the sandbanks into SWA, entraining SWA water as it falls. A very sharp front, typically a metre or less across, separates the inflowing water from ambient SWA water. The inflowing sheet eventually moves horizontally out into

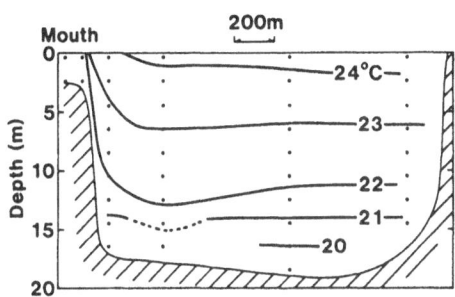

Fig. 1: Temperature structure along SWA just before high tide on 11 January 1982.

SWA, at or near the depth where its density (after entrainment) equals that of its surroundings. This behaviour is found throughout the spring and summer, regardless of salinity within SWA: even under purely marine conditions, the tidal inflow is colder by as much as 3^0 C than SWA surface water in spring and summer. The resulting density difference between tidal inflow and SWA water is sufficient for front formation and sinking to occur. In autumn and winter, front formation is confined to the recovery periods after rainstorms.

Fig. 1 illustrates the temperature structure found in SWA just before high tide on a day of near–marine conditions. A sharp temperature front can be seen near the mouth; the water is sinking to between about 8 and 12 m and flowing up the estuary at this depth. Figures similar to Fig. 1 can be obtained for salinity following a rainstorm. These rising tide fronts at the entrance to SWA are crucial to the estuary's dynamics throughout the spring and summer and also in autumn and winter whenever there has been significant river flow. If one is to understand the biological and chemical variability in SWA, it is important to start with an understanding of how these fronts and the associated sinking plumes affect vertical mixing and tidal flushing within SWA. This chapter discusses the dynamics of the estuary from this point of view.

2. TIDAL FLUSHING IN PORT HACKING ESTUARY

If a parameter q (which may be heat, salt, dissolved oxygen, etc.) is being exchanged with the ocean, the amount of q lost from the inner two basins (Port Hacking Basin and SWA) per tidal cycle can be written

$$(Q_{in} - Q_o)V_{ex} \tag{1}$$

where Q_{in} is a representative value of the concentration of q in the water that (permanently) leaves the two–basin system on the ebb tide, and Q_0 the concentration of q in the ocean water that replaces it on flood; V_{ex} is the volume of water that leaves on the ebb and is replaced by ocean water on the flood. Experience suggests that if the water inside the estuary is significantly density stratified, only the top 2 m flow out on the ebb; if it is homogeneous in density (unstratified), then it is also close to homogeneity in the other variables considered here. In either case we may take Q_{in} as the horizontal average concentration in Port Hacking Basin and SWA in the top 2 m. Furthermore, because of the high degree of homogeneity horizontally it should suffice to use an area–weighted mean of one observation near the centre of each basin; this or minor variations of it is the procedure adopted here. Q_o is taken as the concentration at 1 m depth between the CSIRO laboratory and Bundeena (see Fig. 1 of Vaudrey et al., 1983).

In Eq. 1, V_{ex} will be less than the total tidal prism, because much of the water leaving the basin on the ebb returns to it on the flood. Godfrey & Parslow (1976) estimated V_{ex} by examining the response of salinity inside the estuary for three weeks after a strong, short rainstorm. They found that the two–basin horizontal average salinity $S(z,t)$ at depth z and time t after the storm was given approximately by

$$S(z,t) = S_0 - S_i \, f(z) \, exp(-t/\lambda) \tag{2}$$

where S_0 is ocean salinity, S_i the initial salinity disturbance, λ a decay time and $f(z)$ a time–independent profile function, normalized to unity at 1 m depth. (The normalized salinity distribution actually deepened somewhat over the three–week observation period, but this was a minor effect compared to the overall exponential decay.) They showed that the exponential decay in Eq. 2 could be obtained from a simple salt conservation model with the tidal flux of salt given by Eq. 1, and with V_{ex} a constant. The model yields the condition

$$V_{ex} = \int f(z) \, \hat{A}(z) \, dz \, T_t/\lambda \tag{3}$$

where $\hat{A}(z)$ is the combined area of Port Hacking Basin and SWA at depth z, and T_t the tidal period. After considering several storms, λ was found to be 8 ± 2 days, and $\int f(z) \, \hat{A}(z) \, dz = \hat{A}(0) \times (4 \pm 1)m$, whence $V_{ex} = \hat{A}(0) \, h_{ex}$, with $h_{ex} = 0.25 \pm 0.1$ m.

In other words, tidal flushing following a flood is roughly equivalent to the removal of 0.25 ± 0.1 m from the top 2 m of the two–basin system on the ebb tide, and its replacement – with vertical mixing to regenerate the profile function f(z) – with the same depth of ocean water on the flood.

Furthermore, SWA and Port Hacking Basin generally have closely similar temperature and salinity, so we will use a rule of the form of Eq. 1 to describe tidal flushing of SWA. For SWA alone V_{ex} will be assumed to be A(0) h_{ex}, where A(z) is the area of SWA at depth z. However, it should be noted that a rule such as Eq. 1 should be used with caution in estimating the flushing of quantities other than heat and salt (*e.g.* Kjerfve & Proehl, 1979); for other quantities, horizontal variations in concentration within the two–basin system may be comparable to the concentration differences between the estuary and the adjacent ocean. Even for salinity and temperature, there is much room for improvement beyond the simple rule of Eq. 1 for accurate studies of SWA. For example, although Godfrey & Parslow (1976) found V_{ex} to be constant, it is in fact likely to vary with the spring–neap cycle, being greater (by some unknown factor) at spring than at neap. This has not been allowed for in the numerical model developed below. However, most of the results presented below are believed to be fairly insensitive to details of the description of tidal flushing.

3. VERTICAL DIFFUSION IN SOUTH WEST ARM

SWA is an unusually quiet body of water for an estuary; dye streaks in it 0.05 m across generally retain their integrity for 15 minutes or more, while the bottom sediments are extremely soft and flocculent below about 10 m depth. Available descriptions of vertical diffusion in estuaries (*e.g.* the review in Bowden & Hamilton, 1975) are generally motivated by the idea that turbulence is generated by bottom friction; it is shown below that bottom friction is a relatively small energy source for turbulence in SWA, so these "conventional" descriptions seem inappropriate.

Fischer *et al.* (1979) give a detailed description of physical processes contributing to mixing in reservoirs, with application to the Wellington Reservoir, Western Australia. There are several qualitative similarities between the hydrodynamics of SWA and Wellington Reservoir (WR). It has already been noted that SWA is surprisingly homogeneous horizontally, at least in temperature and salinity; this is also true of WR. Furthermore, river inflow to WR often sinks at sharply defined fronts and continues into WR at some intermediate depth, just as tidal inflow does in SWA. The two water bodies have similar maximum depths (20 m for SWA, 25 m for WR), though the area of SWA is only about 5% of that of WR. Both salinity and temperature contribute to the density stratification of both water bodies, as can be seen by comparing Figs 2, 3, which show annual cycles of temperature, salinity and

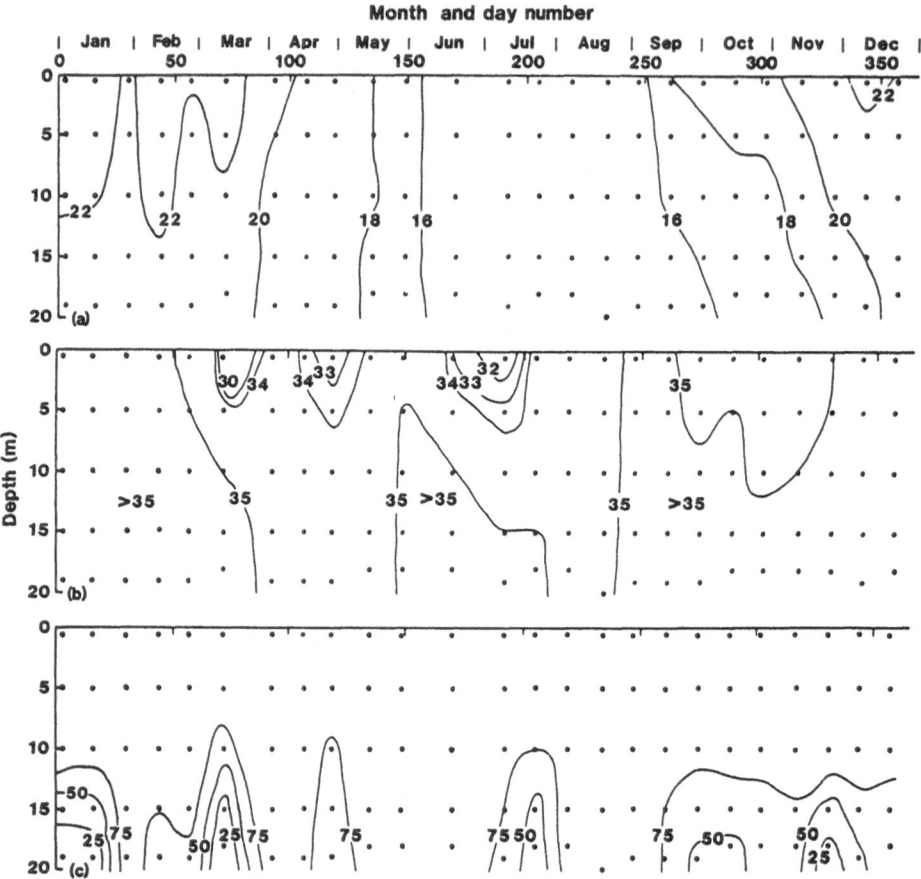

Fig. 2: a, Water temperature (°C); b, salinity (10^{-3}); c, dissolved oxygen (% saturation) in SWA during 1975. From Scott (1978b).

oxygen in SWA and WR, respectively. However, at this stage some important differences between the two water bodies can be seen.

WR has a pycnocline throughout the spring, summer and autumn, primarily due to thermal stratification; a hypolimnion occurs below it, nearly homogenous in temperature, salinity and oxygen. This kind of behaviour is seen only intermittently in SWA after heavy rain; during the dry spring and early summer of 1975, thermal density stratification developed in SWA, but it was relatively weak and no hypolimnion developed. This suggests more vigorous mixing, particularly near the bottom, in SWA than in WR. De–oxygenation proceeds much more rapidly following

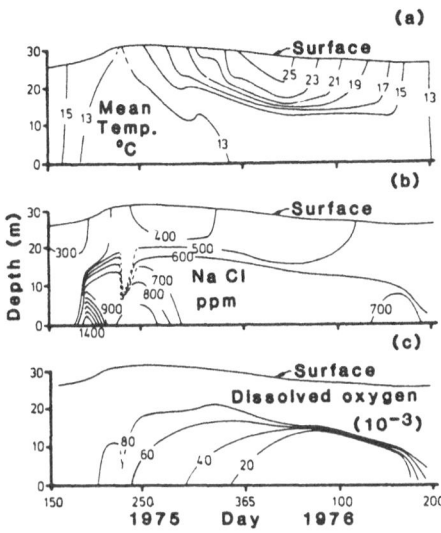

Fig. 3: a, Water temperature (°C); b, salinity (10⁻³); c, dissolved oxygen (% saturation) in WR during 1975–76. From Fischer *et al.* (1979).

stratification in SWA than in WR, presumably because of the strongly reducing bottom sediments at the bottom of SWA.

Wind–mixing is insufficient to bring heat to the bottom of WR in spring and summer, and, since SWA is of nearly the same depth and probably less windy than WR, some mechanism other than wind–stirring is presumed to be responsible for the mixing of heat in SWA. It seems plausible that tidal rise and fall create mixing and interleaving in SWA in much the same way that river inflow and withdrawal create them in the reservoir, but that the much greater vigour of tidal processes compared to river inflow and withdrawal contribute to the greater degree of mixing in SWA. Furthermore, the spring warm–up of bottom water in SWA is rather smooth and steady, which suggests a constant source of mixing such as tidal motions, rather than spasmodic events such as wind–mixing. The present chapter discusses mixing processes in SWA, relying fairly heavily on the analogy with reservoir mixing dynamics and with Fischer *et al.*'s (1979) discussion of them. It is useful to subdivide the various mixing processes in SWA into "shallow" and "deep" processes; the former correspond fairly closely to those discussed as "vertical mixing in the epilimnion" and "mixing of inflows" in Fischer *et al.* (1979), while the latter correspond to "vertical mixing in the hypolimnion". After qualitative discussion of mixing processes under these two headings, a numerical model will be introduced and used to simulate variations of salinity and temperature in SWA and to draw some conclusions regarding the behaviour of dissolved oxygen.

3.1 Shallow mixing processes in SWA

3.1.1 Interleaving

Fischer *et al.* (1979) found that in WR the front between inflowing river water and surface reservoir water occurred at a particular water depth, determined by a condition on the internal Froude number. Huzzey (1981) found that a somewhat similar condition held in the inflow to Port Hacking Basin – though the front was typically 5 – 10 km from the river mouth in WR, but only a few tens of metres from the sandbank at the entrance to Port Hacking Basin. The difference occurs because the entrances to Port Hacking Basin (and to SWA) have slopes of order 10^{-1} or greater, compared to 10^{-3} in WR.

At the start of the rising tide, the inflow to SWA has only just left SWA or Port Hacking Basin, so it is generally of nearly the same density as the surface water inside. However, as the tide rises, inflow density gradually changes until it is close to oceanic values at high tide. Consequently, when ocean water is denser than surface water in SWA, the inflow sinks steadily deeper into SWA as the tide rises, until it reaches a maximum depth Z_m at high tide. On the falling tide, outflow is drawn from the top 1 – 2 m whenever there is significant stratification.

Only a few direct observations of this process have been made, so we cannot give an authoritative parameterization of the net effect of these processes over a tidal cycle. In the model discussed later, the depth Z_m to which the tidal inflow sinks (with entrainment) just before high tide is first estimated through the equation

$$\sigma_t(Z_m) = (\sigma_0 + E/h \int_0^z m\sigma_t(z) \, dz) \, / \, (1 + E \, Z_m/h) \qquad (4)$$

where E is an entrainment ratio (*e.g.* Turner, 1973) and h the thickness of the sinking plume; σ_0 is the density anomaly of the ocean water, and $\sigma_t(z)$ the density anomaly of ambient SWA water at depth z. Physically, the right hand side of Eq. 4 estimates the density anomaly of the incoming water after it has sunk through a depth Z_m, and at each depth entrained a fraction ($E\Delta z/h$) of its own volume of water from depth interval Δz, at density anomaly $\sigma_t(z)$. Eq. 4 states that the plume stops following the bottom and interleaves with the water inside SWA at the depth where its density (after entrainment) equals that of the ambient fluid. Fig. 1 suggests that this last assumption may not be entirely adequate – some overshoot probably takes place but we ignore this.

It is then assumed that the net effect of tidal flushing and interleaving over a tidal cycle is the same as if each layer of thickness Δz above Z_m were mixed with a constant volume $V\Delta z/Z_m$ of density $\sigma_t(Z_m)$, the volume being chosen so that the net buoyancy flux entering SWA is given by Eq. 1. The combination E/h is not well known from observations, but values of order 0.1 m^{-1} are in accord with the observations of Godfrey & Parslow (1976) and Huzzey (1981). The range $0.025 < E/h < 0.4$ m^{-1} covers the range compatible with these observations.

3.1.2 Convection

Whenever cooling causes surface water to become denser than the water below it, convective instability occurs, and a surface mixed layer develops to some depth H_0. In "passive convection", the density is undisturbed below H_0, is continuous at H_0, and is constant above it; the surface heat loss is all absorbed in the depth H_0. However, the convective motion may be vigorous enough to carry the mixing below H_0 ("penetrative convection"). A density discontinuity then develops at the bottom of the surface mixed layer, and the density of the mixed layer increases somewhat over the "passive" value. Order of magnitude estimates based on Fischer *et al.* (1979) suggest that penetrative convection is not likely to be very important in SWA; but ordinary, passive convection is, at least in the autumn cooling period, when it leads to complete vertical homogeneity under marine conditions.

3.1.3 Wind mixing

Winds introduce mechanical energy to the surface mixed layer, which can cause the layer to deepen. This process is important in reservoirs, because the deepening effect can accumulate over some months in dry weather; but in SWA, interleaving of tidal inflows tends to destroy wind–mixed layers within days of their formation. We will therefore ignore the formation of surface mixed layers by the wind.

3.2 Deep mixing processes in SWA – eddy diffusion

When stratification is present, the water below the depth Z_m to which the tidal inflow penetrates is essentially isolated. Under these conditions, "eddy diffusion" becomes a major contributor to the slow variations with time that are observed in the deeper parts of SWA. It is usual to parameterize the eddy diffusive vertical transport F of a property q across a surface at depth z by

$$F = -\kappa(z,t) \, A(z) \, \partial Q(z,t)/\partial z \qquad\qquad (5)$$

where $A(z)$ is the area of the basin at depth z, κ is the "eddy diffusivity", and Q the concentration of q. Eq. 5 is based on the idea that eddies of small vertical extent occasionally develop within the water body concerned, and create local mixing; the cumulative effect is to transport q down the gradient of Q. However, $\kappa(z,t)$ is a purely empirical quantity, which differs by many orders of magnitude from situation to situation. It is a function of the vertical size, velocity and intermittency of the eddies. Bowden & Hamilton (1975) review several formulae for κ that have been tried in density–stratified fluids; however, all these formulae are based on a bulk Richardson number, which is not a physically illuminating parameter in SWA. Fischer et al. (1979) adapted a formula derived by Ozmidov (1965), and suggested that if eddies dissipate energy at a rate ϵ per unit mass, in a stratified fluid of density $\rho(z)$, the eddy diffusivity should be of the form $C\epsilon/(g/\rho_0 \, \partial\rho/\partial z)$, where C is a dimensionless constant primarily dependent on reservoir geometry. McEwan (in press) carried out laboratory experiments which showed that when friction against side walls is allowed for, C has a value of 0.24 ± 0.1 in a uniformly stratified fluid; C can be interpreted as the efficiency of conversion of turbulent kinetic energy to potential energy of the water column, or as a flux Richardson number. Thompson (1980) suggested that McEwan's value may have a fundamental physical meaning, being related to the critical Richardson number 0.25 for Kelvin–Helmholtz instability. Fischer et al. (1979) used for ϵ the volume average of the total energy dissipation rate within the water body, since this is a quantity that can be estimated from external variables; with this assumption, they found a value of 0.048 for C in WR. Stigebrandt (1976) found a flux Richardson number in a Norwegian fjord of 0.05; his result can also be interpreted as saying that the constant C was 0.05. The difference between McEwan's (in press), Fischer et al.'s (1979) and Stigebrandt's (1976) values for C is probably due mainly to the fact that the energy consumption rates ϵ used by Fischer et al. (1979) and Stigebrandt (1976) are not corrected for frictional dissipation against the bottom and sides. We shall use a minor adaptation of Fischer et al.'s (1979) formula:

$$\kappa(z,t) = \Sigma_i C_i \epsilon_i/(g/\rho_0 \, \partial\rho/\partial z) \qquad\qquad (6)$$

where a separate constant C_i has been introduced for the i'th physically distinct energy source, ϵ_i, each assumed to lie between (say) 0.025 and 0.25. Four possible energy sources will be considered namely:

(i) Influx of potential energy by tidal motions;
(ii) Influx of kinetic energy by tidal motions;
(iii) Work done by bottom stresses, associated with tidal motions;
(iv) Work done by wind stresses.

Corresponding energy dissipation rates are ϵ_1, ϵ_2, ϵ_3 and ϵ_4, respectively. It will be shown that (iii) and (iv) are probably relatively unimportant, so only the first two terms contribute to the sum in Eq. 6.

3.2.1 Influx of potential energy into SWA

The rate of input of potential energy to SWA during the rising tide, averaged over a full tidal period, can be written

$$dP/dt = A(0) \int \partial H/\partial t \ (g \ \delta\rho(t) \ Z(t)) \ dt \ / \ T_t \tag{7}$$

Here H is the tide height and T_t the tidal period; $A(0) \ \partial H/\partial t$ is the rate of influx (volume) on the rising tide; $g \ \delta\rho(t) \ Z(t)$ is the amount of potential energy released per unit volume; g is gravity, Z the depth through which the inflow sinks to reach its equilibrium level, and $\delta\rho$ the difference between the inflow density and the average density of the water above depth Z inside SWA. The integration is over the rising tide only. If it is assumed that both Z and $\delta\rho$ increase linearly with H during the rising tide from zero at low tide to maximum values Z_m, $\delta\rho_m$ at high tide, then

$$\epsilon_1 = dP/dt \ (\rho_0 V)^{-1} = \tfrac{1}{3} \ \Delta H \ (g \ \Delta\rho \ Z_m) \ / \ (T_t \ D \ \rho_0) \tag{8}$$

where D is the maximum depth of SWA (the volume V of SWA is close to $\tfrac{1}{2} A(0)D$), and ΔH is the tidal range. We have also used the assumption (confirmed by model calculations below) that $\delta\rho_m \approx \Delta\rho/2$, where $\Delta\rho$ is the density difference from top to bottom of SWA. Under summer thermal stratification conditions, reasonable values of $\Delta\rho/\rho_0$ and Z_m are 10^{-3} and 10 m respectively, and the tidal range $\Delta H \approx 1$ m, so $\epsilon_1 \approx 4 \times 10^{-8}$ watts kg^{-1}. Shortly after a rainstorm, $\Delta\rho/\rho_0$ may be as high as 10^{-2}, though Z_m will be rather less – perhaps 5 m – so $\epsilon_1 \approx 2 \times 10^{-7}$ watts kg^{-1} at this time.

After it has sunk to its equilibrium level, the inflow also creates internal waves in the process of spreading horizontally within the basin; if these break and are dissipated at the sides of the basin, as assumed by Stigebrandt (1976), a further contribution to potential energy influx to SWA should be included. However, order–of–magnitude estimates based on Stigebrandt's Eq. 16 suggest that this term makes a contribution to ϵ_1 comparable to Eq. 8 only when stratification is quite small. Since it is relatively difficult to estimate, this term is not included in the calculations given below. In any case, according to Eqs 6 and 8, potential energy input to SWA makes a contribution to eddy diffusivity κ given by

$$\kappa_1 = C_1 Z_m \Delta H / (3 T_t S(z)) \tag{9a}$$

where we have written $d\rho/dz = S(z) \Delta\rho/D$; $S(z)$ is a non–dimensional stratification of order unity. In fact the energy dissipation is likely to be concentrated in regions of high stratification rather than uniformly throughout the volume, as assumed here. Furthermore in actual trials Eq. 9a sometimes leads to unrealistic results, because the depth–dependence of $S(z)$ can cancel the depth–dependence of $\partial Q/\partial z$ in Eq. 5. The factor $S(z)$ in Eq. 9a has been omitted in the results reported below, i.e. eddy diffusivity is calculated from

$$\kappa_1 = C_1 Z_m \Delta H / (3T_t) \tag{9b}$$

where Z_m is obtained from Eq. 4.

Eq. 9b shows no explicit dependence on stratification $\Delta\rho$, so that κ_1 might be thought to be roughly constant apart from a minor spring–neap variation through ΔH. In practice Z_m is found to increase from 2 m to the full depth, 20 m, of the basin following a rainstorm, as the stratification weakens; so κ_1 increases by a factor of about 10 during the flushing of fresh water from Port Hacking – i.e. κ_1 in fact increases during decreasing stratification, as one expects intuitively of the eddy diffusivity. In the actual runs of the model, the tidal range ΔH was estimated as daily average tide range at Fort Denison tide gauge, 30 km north of Port Hacking (from the Australian National Tide Tables, 1975).

3.2.2 Influx of kinetic energy into SWA

The water entering SWA via the entrance channel on the flood tide comes in with speeds of up to 0.5 m s^{-1} on a tide of average range (1 m). However, most of the water comes in over sandflats, with maximum speeds of about 0.15 m s^{-1}. This results in the import of kinetic energy into SWA at a rate dK/dt, averaged over a full tidal period, given by

$$dK/dt = \rho_0 A(0) \int \partial H/\partial t \, (\tfrac{1}{2} u^2(t))dt/T_t \tag{10}$$

where $(\tfrac{1}{2} u^2)$ is the kinetic energy density averaged over the cross–sectional area of the entrance to SWA. The integral in Eq. 10 is taken over the rising tide only. Available observations are inadequate to estimate $(\tfrac{1}{2} u^2)$ over a tidal cycle, but from the above rough figures for maximum inflow speeds, it can be estimated that

$$\epsilon_2 = (dK/dt) / \rho_0 V \approx 2 \times 10^8 (\Delta H/T_t)^3 / D \tag{11}$$

where the coefficient of 2×10^8 may be in error by as much as a factor of three.

The kinetic energy release depends on the cube of the tidal amplitude ΔH, so it is strongly dependent on the spring–neap cycle – it will be about eight times larger at spring tide than at neap. Fischer *et al.* (1979) did not include a kinetic energy term in their expression for eddy diffusivity; this is reasonable for river flow into the gently sloping entrance of WR, since the kinetic energy with which the river water enters will almost certainly be dissipated by bottom friction before it has a chance to affect the reservoir's stratification. However, this may not be true at the steeply sloping entrance to SWA. Eq. 11 yields $\epsilon_2 \sim 8 \times 10^{-8}$ watts kg^{-1} for $\Delta H = 1$ m. This is rather larger than the potential energy release rates during the summer thermally–stratified conditions, and spring tide values of ϵ_2 are still greater.

Eqs 6 and 11 imply a contribution κ_2 to eddy diffusivity of

$$\kappa_2 = 2 \times 10^8 \, C_2 \, (\Delta H/T_\iota)^3 \, \rho_0 \, / \, (g \, \Delta \rho S) \tag{12a}$$

where, as in Eq. 9, the factor of S is uncertain, because it is not known if energy dissipation is uniformly distributed throughout SWA. It is omitted in the results presented below, *i.e.* eddy diffusivity is calculated from

$$\kappa_2 = 2 \times 10^8 \, C_2 \, (\Delta H/T_\iota)^3 \, \rho_0 \, / \, (g \, \Delta \rho) \tag{12b}$$

Eq. 12b becomes infinite in the limit of zero stratification; this is evidently unrealistic. If the estuary were nearly unstratified, Bowden & Hamilton's (1975) estimate of eddy diffusivity should be of about the right order of magnitude. One of the estimates they give, from Pritchard's (1960) studies of the James River, is

$$K = 0.537 \times 10^{-3} \, U \, D \tag{13}$$

where U is a typical tidal flow speed; they considered this estimate uncertain by a factor of 5, and this seems reasonable also in SWA. With $U > 0.02$ m s^{-1} inside SWA and $D = 20$ m, this yields $K = 2 \times 10^{-4}$ m^2 s^{-1}. Whenever κ_2 as estimated from Eq. 12b exceeded K, it was reset to K. In a few model runs, a different value of the cutoff K was used.

3.2.3 Work done by bottom stresses

In the absence of stratification, tide–induced average bottom velocities u_b inside SWA are readily estimated from mass continuity to be of order 0.02 m s^{-1}. The work done by bottom stresses is about $C_B \rho_0 u_b^3 \sim 10^{-5}$ watts m^{-2}, for a drag coefficient C_B of 10^{-3}; the corresponding energy dissipation term ϵ_3 is roughly $2 C_B \rho_0 u_b^3 / D \rho_0 \sim 10^{-9}$ watts kg^{-1} – small compared to typical values of ϵ_2 obtained from Eq. 11.

If there is marked stratification, velocities will decrease in the lower half of SWA and increase in the top half, perhaps by as much as a factor of 2; in this case $\epsilon_3 <$ 4×10^{-9} watts kg^{-1}, which is still small compared to the uncertainties in ϵ_1 and ϵ_2. For stratifications such that the internal seiche period of SWA is of the same order as the tidal period T_t there may be resonance, and the work done by bottom stresses may then be appreciable; but for this to occur, a crude estimate of the seiche period as $2L/((g/\rho_0) \Delta\rho D)^{1/2}$ with L the length of SWA implies $\Delta\rho/\rho_0 \sim 3 \times 10^{-5}$ for resonance. This is only slightly different from zero, to the limit of the measurement techniques used in SWA, so this effect is unlikely to be important. Consequently we ignore the direct contribution of bottom stresses to eddy diffusion.

3.2.4 Work done by wind stresses

The wind does work $W = \tau \, u_w \, A(0)$ on SWA, where the wind stress $\tau = C_D \, \rho_A$ U^2; C_D is the drag coefficient, ρ_A is air density and U is wind speed. u_w is the resulting downwind surface drift, of order $(\tau/\rho_0)^{1/2}$. Hence $\epsilon_4 = W/V \approx$ $2 \, (C_D^3 \rho_A^3/\rho_0)^{1/2} \, U^3/D$. Daily average values of U^3 were obtained for a wind recorder at the CSIRO laboratory (Fig. 1 of Vaudrey et al., 1983); it was found, for a 9–month record with a number of gaps in it, that ϵ_4 exceeded 3×10^{-8} watts kg^{-1} on only 20 out of 153 days, and it was less than 0.3×10^{-8} watts kg^{-1} on 60 out of 153 days. SWA is more sheltered than the laboratory site, so that these estimates of the work done by wind stresses are probably too high. Since they are generally smaller than the kinetic energy input due to the tides, we shall ignore ϵ_4 in modelling eddy diffusivity in SWA; only on a few days of strong winds is this likely to introduce serious error.

4. A NUMERICAL MODEL OF SOUTH WEST ARM

Since observations were made without regard to the phase of the tide, a simple model simulating the running–mean properties of SWA over a tidal cycle was developed. Horizontal homogeneity was assumed, and tidal flushing was assumed to be given by Eq. 1, with $V_{ex} = A(0) \, h_{ex}$: h_{ex} was taken as 0.25 m. Vertical mixing processes are as discussed in the preceding section; summarizing that section, we assume that three processes are important:
a. *Eddy diffusion.* This creates vertical transports of a parameter q according to Eq. 5, where $\kappa = \kappa_1 + \kappa_2$ is obtained from Eqs 9b and 12b, with the cut–off of Eq. 13. This diffusivity is supposed to act throughout the entirety of SWA. The constants C_1 and C_2 are to be determined by fitting to the data, but both should be less than 0.25 and

probably about 0.05 as found by Fischer *et al.* (1979) and by Stigebrandt (1976).

b. *Flushing and interleaving of tidal inflows.* The procedure to be used in describing the net effect of this process over a tidal cycle is described after Eq. 4. The constant E/h is to be determined by fitting to the data; direct observations limit it to a range of about $0.025 < E/h < 0.4$ m^{-1}.

c. *Convection.* Whenever a vertical region of the model becomes convectively unstable due to surface cooling it is mixed, conserving all water properties, according to the "passive convection" scheme described in Section 3.1.2.

The 3 rules a, b and c were used to build a tide–averaged numerical model of SWA, with layers $\Delta z = 2$ m deep from surface to bottom at 20 m depth. At the n'th time step, the effects of diffusion and of any internal sources and sinks were estimated through the equation

$$Q_k^{n+1} - Q_k^n = ((F_k^{n-1} - F_{k+1}^{n-1}) / \Delta z + G_k^n (A_k + A_{k+1})/2) \Delta t \qquad (14)$$

where F_k^n is the diffusive transport of q, from Eq. 5, through a surface at depth $k\Delta z$ at time $n\Delta t$, and G_k^n is the source term for q between depths $k\Delta z$ and $(k+1)\Delta z$. (Actual model runs shown here are only for salinity and temperature; in the former case G_k^n is river inflow (usually zero), while in the latter, it is the net heat input into the k'th layer.) Next the effects of tidal flushing and interleaving were allowed for at each time step, using the procedure described and finally the resulting profiles were tested for convective instability at the end of each time step, and mixed convectively where necessary.

A fixed time step of 1 h was used in all but one test run for which a time step of 0.5h was used; the difference in model results from this reduction of time step was about 1% – small compared to the other substantial error sources in this model. A time step of 2h led to Courant–Friedrichs–Loewy instability. With such long time steps, a year's run of the model cost about $Aust.1 on a CDC 7600 computer, taking 2 seconds CPU time.

5. MODEL RESULTS

The first aim of the numerical work must be to calibrate the various empirical constants used – namely C_1, C_2, E/h and K. It is unfortunate that there are so many empirical constants, each a function of the geometric details of SWA, but they all seem to be necessary for a comprehensive description of the system. Luckily, there are large periods of time when the model results are sensitive to only one or two of these constants, and furthermore some independent constraints on the values of each parameter have already been mentioned. Hence the results are thought to be physically meaningful.

The model is first used to simulate the response to a single, heavy rainstorm; it is found that quite good simulations can be obtained for a range of the parameters listed above, within the allowed limits, *i.e.* the rainstorm response by itself cannot give a unique calibration of the model. Simulations of the spring warm–up of SWA provide slightly different constraints on the model parameters. The results reported below provide some "feel" for the physical processes which are important in SWA, and what parameter values should be used in describing those processes.

5.1 Model simulation of the response to a rainstorm

During 10–13 March 1975 an extremely heavy rainstorm fell on the catchment of Port Hacking Estuary; river inflow reached about 60 m^3 s^{-1} before falling to values of about 1 m^3 s^{-1} within 3 days. Fig. 4a shows the resulting salinity at various depths in SWA.

The model was run with various choices of parameters, to simulate Fig. 4a; initial salinity was taken as $5.5 \cdot 10^{-3}$ in the top 2 m, and as $35.5 \cdot 10^{-3}$ below. River inflow was ignored after 10 March. The model was first run with no eddy diffusivity ($C_1 = C_2 = 0$), and various choices of E/h in $0.025 < E/h < 0.4$ m^{-1}; none of these runs gave good simulations of Fig. 4a.

In view of the fact that Fischer *et al.* (1979) and Stigebrandt (1976) both found a constant C of 0.05, the first simulation shown is for $C_1 = C_2 = 0.05$ (Fig. 4b); E/h was 0.1 m^{-1}. Evidently, quite good agreement with observation is obtained. It is not physically significant that the rate at which the fresh water is flushed from SWA is correctly modelled in Fig. 4b – the data of Fig. 4a were in fact used to estimate the flushing parameter V_{ex} of Eq. 1 (Godfrey & Parslow, 1976) and hence in the generation of Fig. 4b. However, the rather sudden $1.5 \cdot 10^{-3}$ decrease in bottom salinity observed about 30 March, and the accompanying collapse of stratification, are moderately well reproduced by the model. They occur, in the model, because the eddy diffusivity contribution κ_2 increases strongly a few days earlier (light line, inset to Fig. 4b), until it reaches the limiting value K. This increase is in turn due partly to the large reduction in vertical salinity difference between 10 and 27 March, and partly to the spring tide about 27 March (heavy line, inset to Fig. 4b). According to Eq. 12b, these will both tend to create a large increase in κ_2. Godfrey & Parslow (1976) suggested that strong winds about 30 March may have caused the stratification collapse; Fig. 4b suggests that the spring tide may have been the main cause.

The other term, κ_1, in the eddy diffusivity was less than 0.13×10^{-4} m^2 s^{-1} throughout the run shown in Fig. 4b, so one would expect that the results would not be sensitive to moderate changes in C_1. When C_1 is varied from 0 to 0.1 with C_2 and E/h held at the values of Fig. 4b, the resulting salinity pattern is very similar to Fig. 4b.

The dashed line in Fig. 4b shows the salinity of the tidal inflow, after it has sunk with entrainment into SWA; the depth to which it sinks can be inferred from the points at which it crosses the lines for salinity at constant depth. It is seen that the tidal inflow sinks only a small depth into SWA immediately following a rainstorm, but the penetration depth increases steadily thereafter. The details of this behaviour might be expected to be sensitive to the entrainment parameter E/h, but this was not found to be the case. The model results – both the salinities at constant depth and the final inflow salinities – were surprisingly insensitive to variations in E/h in the range $0.025 < E/h < 0.4$ m^{-1}, at least for $C_1 = C_2 = 0.05$.

Runs with the parameters of Fig. 4b, but with the cutoff K of Eq. 13 being varied from 1 to 4×10^{-4} m^2 s^{-1} also made little change to the results.

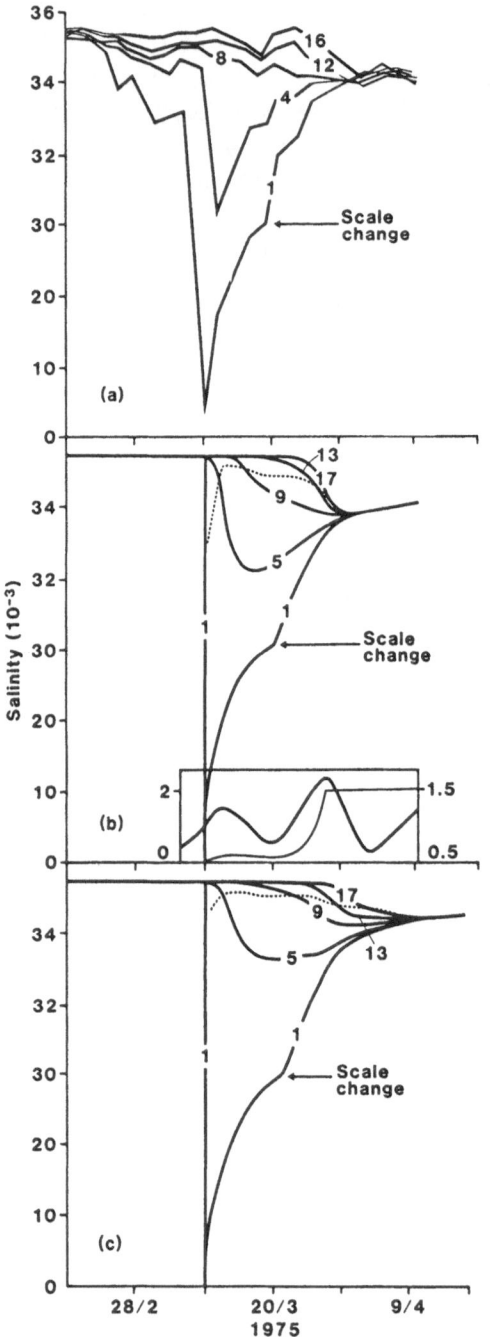

Fig. 4a: Observed salinity near the centre of SWA before and after the rainstorm of 10–13 March 1975 at 1, 4, 8, 12 and 16 m depths (note the scale change at $30 \cdot 10^{-3}$) After Godfrey & Parlow, 1976.

Fig. 4b: Calculated salinity in SWA for $C_1=C_2=0.05$, $E/h=0.1$ m^{-1}, at 1, 5, 9, 13 and 17 m depths. The dotted curve gives the salinity of the inflow after it has sunk into SWA. The right hand scale and heavy line in the inset, show the daily mean tidal range (m) while the left hand scale and light line in the inset show the eddy diffusity (10^{-4}m^{2}s^{-1}).

Fig. 4c: Calculated salinity in SWA for $C_1=C_2=0.025$, $E/h=0.1$m^{-1} at 1, 5, 9, 13 and 17 m depths. The dotted curve gives the salinity of the inflow after it has sunk into SWA.

Thus for parameter values near those of Fig. 4b, the results prove to be insensitive to changes in C_1, E/h and K. However, the results *are* sensitive to changes in C_2; Fig. 4c shows the calculated salinity pattern for $C_1 = C_2 = 0.025$, with E/h and K as for Fig. 4b. Evidently, the freshwater influence does not penetrate as deeply as in Fig. 4b, and the collapse of stratification occurs later, due to the fact that the dominant term κ_2 in the diffusivity has been halved in magnitude.

Runs with $C_1 = C_2 = 0.0125$, and with $C_1 = C_2 = 0.1$, were definitely not in good agreement with Fig. 4a so adequate simulations are confined to the ranges $0.025 < C_2 < 0.05$ and $0 < C_1 < 0.1$. However, if the factor of 3 uncertainty in the coefficient 2×10^{-8} in Eq. 12a is accounted for, the range of possible values of C_2 is correspondingly increased.

5.2 Model simulation of thermal stratification during spring warm–up

As remarked earlier, it seems likely that the smooth warm–up of the bottom layers of SWA during a dry spring is due to tide-induced eddy diffusion of heat downwards, and the main purpose of this section is to test this hypothesis. However this cannot be done without some attention to the heat budget of SWA, so some crude information on this will be introduced as model input.

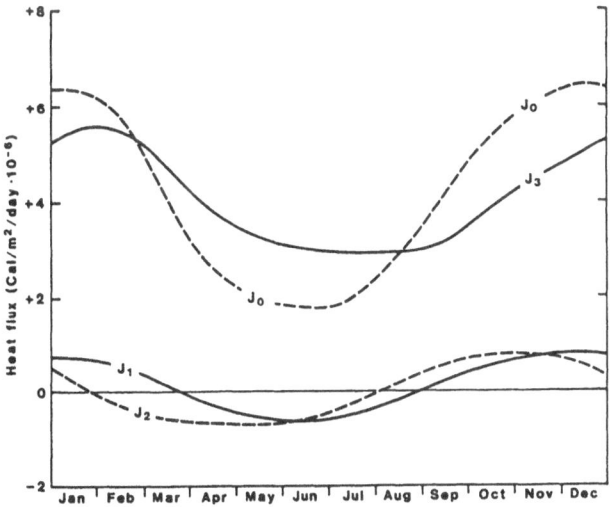

Fig. 5: Estimated seasonal cycle of heat fluxes in SWA. J_0 is visible light energy, J_1 is tidal flushing of heat out of SWA, J_2 is the rate of change of heat content in SWA, and $J_3 = J_0-J_1-J_2$ is the estimated heat loss due to sensible and evaporative heat transfer and infra-red radiation across the surface.

Fig. 5 shows a sinusoidal seasonal curve drawn by eye through available data on downward short wave radiation, J_0, from a radiometer at the Cronulla laboratory (Scott, 1979). A similar smooth curve through data on the temperature difference between SWA and adjacent ocean regions is used, with Eq. 1, to estimate tidal flushing of heat J_1 from SWA; and the area average rate of change of heat content of SWA, J_2, is also estimated from the data of Fig. 2a. Fig. 5 shows J_1 and J_2 as well as J_0; J_1 and J_2 are of similar magnitude and phase, and both are small compared to J_0. The difference $(J_0 - J_1 - J_2) = J_3$ is also shown in Fig. 5. It presumably represents surface heat loss from SWA due to evaporative and sensible heat exchange, and to infra–red radiation. While the estimates of Fig. 5 are crude, they should permit calculation of the spring warm–up of the deeper parts of SWA, which are expected to be more sensitive to the magnitude of eddy diffusion than to details of the heat budget. J_3 is assumed to be lost from the top 2 m of SWA. Half of J_0 is supposed to be absorbed in the top 2 m, and the remainder attenuates with depth as $\exp(-z/D_0)$, with $D_0 = 2.6$ m: this value of D_0 is quite well–defined in dry weather, from Scott's (1978a) observations.

The model was run with J_3 and J_0 as source terms in Eq. 14, distributed in depth as mentioned; ocean temperature T_0 was assumed to vary sinusoidally over a year between 22^0C in February and 15^0C in August, as observed.

For this calculation, salinity was assumed to have a marine value $(35.5 \cdot 10^{-3})$ throughout the year, so the result can be compared with observations only in dry weather conditions.

Fig. 6a shows observed temperatures at various depths in SWA, during a long dry period in 1975 (Vaudrey *et al.*, 1983); surface salinities remained above $34 \cdot 10^{-3}$ on all but two observation days, so density stratification was due primarily to thermal effects. Days on which temperature differences accounted for less than 70% of the density differences from top to bottom are marked by arrows in Fig. 6a.

Fig. 6b shows corresponding temperatures at similar depths from the model, with the same parameters $C_1 = C_2 = 0.025$, $E/h = 0.1$ m^{-1}, $K = 2\times10^{-4}$ m^2 s^{-1} as in Fig. 4c; while Fig. 6c has the same parameters as in Fig. 4b. The order of displaying the two simulations is reversed relative to Fig. 4 so that Fig. 6b, which is obviously a better simulation than Fig. 6c, lies next to the observations in Fig. 6a. In Fig. 6b,c, and in all other spring warm–up runs, the runs were started on 22 May to permit the model to settle down before the spring warm–up commenced. In all cases, the water column was homogeneous before 8 August, due to convective cooling.

Daily average tidal range at Fort Denison is shown in Fig. 6d. A strong spring–neap cycle can be seen in the top–to–bottom temperature difference, throughout the warming period in Fig. 6c and in the first one–third of it in Fig. 6b. The reason for this is that according to Eq. 12b, spring tides result in markedly increased eddy diffusivities; these reduce top–to–bottom temperature differences, which in turn tend to increase the eddy diffusivity still further. This is an instability mechanism which has already been encountered in the collapse of stratification following a flood (Fig. 4). However, in Fig. 6b the spring–neap cycle becomes much less obvious after 9 October, presumably because the stratification becomes too strong for this kind of instability to be important.

There is a clear suggestion of similar behaviour in the observations: from 5 September to 5 November, stratification seems to be weakest near spring tide. To test this hypothesis statistically, the rate of increase of temperature at 16 m was estimated from Fig. 6a, and plotted against the simultaneous value of the daily average tidal range, for data between 7 September and 2 November. The result, Fig. 7, shows a positive correlation which is significant by t–test at the 99.999% level. The correlation is still significant from 3 November to 28 December, but considerably less so.

The dotted lines in Fig. 6b,c show the temperature of the tidal inflow, after it has sunk (with entrainment) into SWA. In Fig. 6b, the inflow penetrates to 13 m by 3 October; it then remains at almost constant depth until 16 January, when it rapidly deepens to reach the bottom of SWA. In Fig. 6c, the inflow reaches the bottom by 1 October, and stays close to the bottom from then on. Unfortunately we have almost no direct data on this inflow depth: it would provide a useful test of the model.

The arrows between vertical lines in Fig. 6b,c show times when the eddy diffusivity κ_2 is at the limiting value (from Eq.13) of 2×10^{-4} m^2 s^{-1}. The results in Fig. 6c are evidently much influenced by this cutoff, but those in Fig. 6b are not.

Several runs of the model were made, analogous to those discussed previously for the rainstorm response, to test the sensitivity of the results of Fig. 6b to moderate changes in model parameters. Two runs with parameters as in Fig. 6b, except that $C_1 = 0$ in one and $C_1 = 0.05$ in the other, showed rather little change from Fig. 6b. Two further runs with parameters as in Fig. 6b except that E/h was 0.025 m^{-1} in one and 0.4 m^{-1} in the other also produced little change. Finally two runs with parameters as in Fig. 6b but with K = 10^{-4} m^2 s^{-1} in one and 4×10^{-4} m^2 s^{-1} in the other caused little change. To express these results quantitatively, all 6 runs just referred to had their largest temperature stratification in a two–week interval starting 25 November, and the values of this maximum stratification all lay between 2.27 and 3.14°C. However, comparison of Figs 6b and 6c shows that the results are strongly sensitive to the choice of C_2; presumably the instability referred to earlier acts to keep stratification small throughout the spring, for the parameters of Fig. 6c.

Three other runs are of interest. First, as a test that the eddy diffusion is primarily due to the kinetic energy term κ_2, as suggested by the earlier results, this term was omitted ($C_2 = 0$) and runs with various values of C_1 were tried. Reasonable simulations of average levels of stratification were obtained for $C_1 = 0.1$ and 0.2; in view of the discussion of Section 3.2, such high conversion efficiencies are unlikely, but not impossible. However, there was almost no spring–neap cycle in the predicted

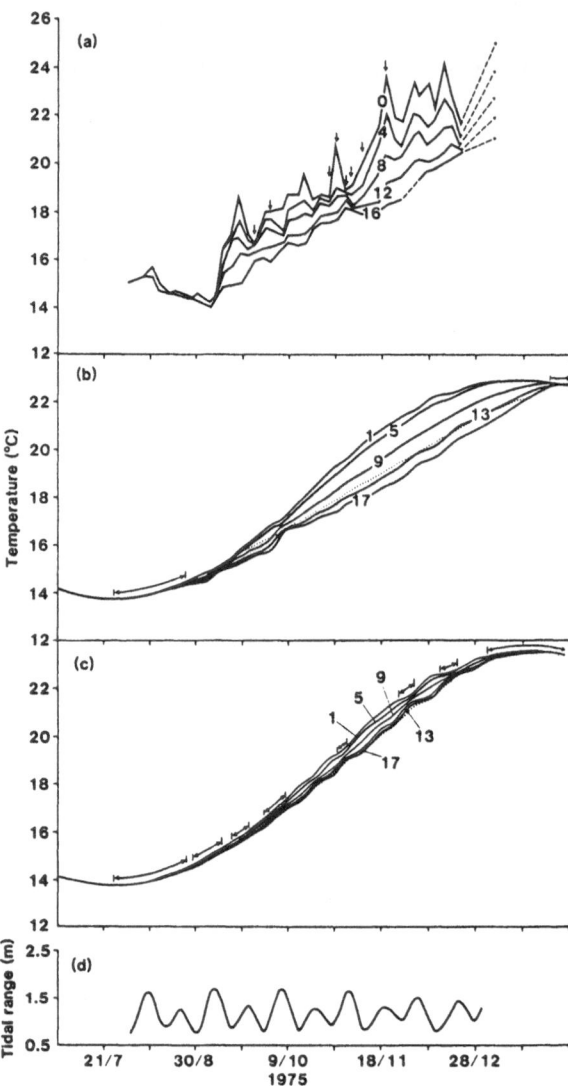

Fig. 6a: Observed temperature at different depths (m) near the centre of SWA, during the spring warm–up of 1975. Arrows indicate days on which temperature differences accounted for less than 70% of the density difference from top to bottom of SWA.

Fig. 6b: Simulated temperature in SWA for the heat fluxes of Fig. 5, and $C_1=C_2=0.025$, $E/h = 0.1$ m^{-1}. Arrows between vertical bars indicate times for which κ_2 has the limiting value K of Eq.13. The dotted line indicates the temperature of the tidal inflow, after sinking and entrainment.

Fig. 6c: As for Fig. 6b, but with $C_1=C_2=0.05$.

Fig. 6d: Daily average tidal range at Fort Denison, 30 km north of Port Hacking, August–December 1975.

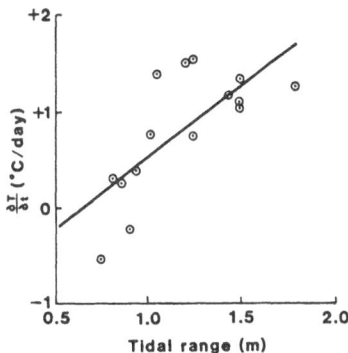

Fig. 7: Rate of change of temperature at 16 m in SWA plotted against the simultaneous value of daily average tidal range at Fort Denison, from the data of Figs 6a,d, between 7 September 1975 and 2 November 1975. The least squares line of best fit is also shown. Data after 2 November 1975 do not fit as well.

stratification, contrary to Fig. 6a. This further confirms that the kinetic energy term of Eq. 12b is probably the main term in the eddy diffusivity in SWA.

Secondly, as a test that the rise in bottom temperature in the model is indeed due to eddy diffusion rather than to direct solar heating, a run was undertaken with $C_1 = C_2 = 0$ (no eddy diffusion), but with $E/h = 0.1$ m^{-1}; for this run, temperature at 17 m rose only 0.2°C from the winter minimum through to the following winter, *i.e.* direct heating can only account for about 0.2°C of the bottom temperature rise from August to January in Fig. 6b,c.

Finally, a run with a 15–day heat anomaly in September with the parameters as in Fig. 6b but with J_0 increased by a factor of 1.5 during the 15–day period, showed the top–to–bottom temperature difference to increase to about 3.2°C at the end of the anomaly period, but it returned to near unperturbed values in 35 days. The observed sudden onset of stratification around 11 – 17 September in Fig. 6a was probably due to such an anomaly, associated with clear, calm weather at this time, and the spikiness of the near–surface temperature trace to weather fluctuations not included in the model.

6. VERTICAL DIFFUSION OF DISSOLVED OXYGEN

The model could readily be adapted to simulate dissolved oxygen, given appropriate information on oxygen production rates. However, it is more useful to work with the bottom half of SWA only, where photosynthetic production of oxygen can be neglected (Scott, 1978b) and exchange across the air–water interface does not directly affect the oxygen content. The water below a depth z in SWA must obey the equation

$$\partial/\partial t \int_z^D O_2(z') \, A(z') \, dz' \; = \; A(z) \, (\kappa[\partial O_2/\partial z](z) - R) \tag{15}$$

where R is the rate of oxygen consumption per unit area below depth z, due both to the sediments and to detritus in the water column. We will consider the consequences of this equation for SWA, assuming the eddy diffusivity κ is given by κ_2 alone, with $C_2 = 0.025$.

6.1 Spring Warm–up Response

Fig. 8 shows measurements of dissolved oxygen concentration at 0, 10, 14 and 16 m in SWA, for the same spring warm–up period as in Fig. 6a (Vaudrey et al., 1983). Comparing Figs 8 and 6a, the dissolved oxygen difference $(O_{10} - O_{16})$ is seen to be closely correlated with the temperature difference $(T_0 - T_{16})$ from 0 to 16 m, both over the long term and to some extent also over the spring–neap cycle. A fit of the form $(O_{10} - O_{16}) = a \, (T_0 - T_{16})$ between 1 September and 22 December (Fig. 9) yields the coefficient: $a = 0.84$ litre $O_2/m^3/°C$.

If the equations of Sections 3 and 4 are assumed to apply to dissolved oxygen, then this result implies that the diffusive flux F of oxygen through a surface at about 13 m depth in SWA is roughly constant. This can be seen as follows. If only the term κ_2 is important, F should be

$$F = \kappa(t) \, [\partial O_2/\partial z](13 \text{ m})$$
$$\approx \; 1.5 \times 10^{-4} \, C_2 \, (\Delta H)^3 \, (O_{10} - O_{16}) \, / \, (T_0 - T_{16}) \text{ litre } O_2 \text{ m}^{-2} \text{ s}^{-1} \tag{16}$$

where we have used

$$\Delta\rho/\rho_0 \approx \; 2.5 \times 10^{-4} \, (T_0 - T_{16}) \tag{17}$$

Now the average value of $(\Delta H)^3$ during the period of Fig. 8 was 1.61 m³; taking $(O_{10} - O_{16}) \, / \, (T_0 - T_{16}) = a$, and $C_2 = 0.025$ yields a *constant* value for diffusive flux through a surface at 13 m, averaged over a spring–neap cycle:

$$F \approx \; 0.402 \text{ litre } O_2 \text{ m}^{-2} \text{ day}^{-1}.$$

A separate estimate, taking individual values of the right hand side of Eq. 16 and averaging them, yielded 0.439 litre O_2 m^{-2} day^{-1}.

Fig. 8 shows that the depth integral of dissolved oxygen below 13 m fell by an average of about 0.084 litres m^{-2} day^{-1} between 1 September and 22 December, so that the net consumption rate is about $0.42 + 0.08 = 0.50$ litre O_2 m^{-1} day^{-1} in the water column below 13 m.

Bulleid (1983) found a fairly constant oxygen consumption rate for the mud at the bottom of SWA of 0.39 ± 0.05 litre m^{-2} day^{-1}, and a range of 0.04–0.06 litre m^{-3} for suspended detritus. This implies a net consumption rate in the average 3.5 m average depth below 13 m of $(0.39 + (0.05 \times 3.5)) = 0.56$ litre O_2 m^{-2} day^{-1}, in good agreement with the above estimate from the sum of diffusion and oxygen depletion. It is interesting to note that during this period, diffusion brings about five times more dissolved oxygen into the water below 13 m than might be estimated from the rate of oxygen loss alone, so it is definitely incorrect to treat the bottom layer at this time as "isolated" for the purpose of estimating an oxygen budget.

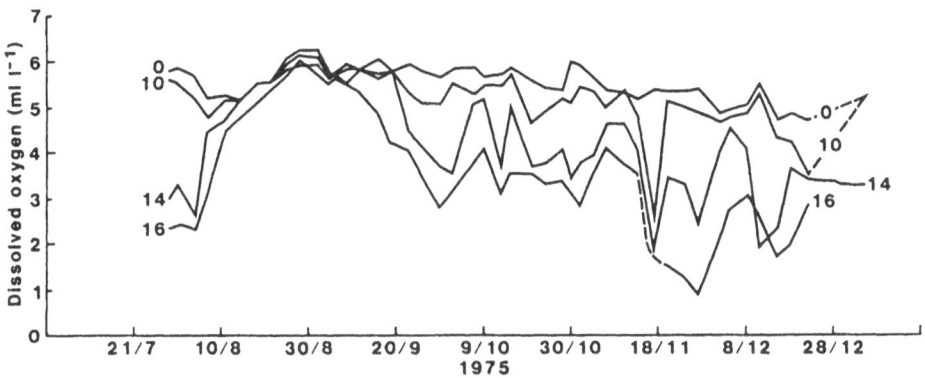

Fig. 8: Dissolved oxygen concentration in SWA at 0, 10, 14 and 16 m depths in 1975.

6.2 Response to a rainstorm

After a single rainstorm, we will again use Eq. 15 with κ estimated from Eq. 12b with $\Delta\rho/\rho_0 = 7.5 \times 10^{-4}$ $(S_0 - S_{16})$. Order of magnitude estimates with the salinities of Fig. 4 show that the diffusive term on the right of Eq. 15 will be essentially negligible until fairly close to the collapse of stratification in Fig. 4a,c. This is true even if the potential energy term κ_1 is included in the estimate of diffusivity. Thus for about 15 – 20 days after the rainstorm the oxygen content per unit area below 13m

$$\int_{13}^{20} O_2(z) \, A(z) \, dz \ / \ A(13)$$

should fall linearly at the rate $R = 0.5 - 0.6$ litre O_2 m^{-2} day^{-1}. If the water column is fully oxygenated before the rainstorm, at 5 litre m^{-3} (7.14 g m^{-3}), then the oxygen content per unit area is about 17.5 litre m^{-3}; 20 days of de–oxygenation at the above rate will reduce oxygen content to 6.5 litre m^{-2} – $i.e.$ complete de–oxygenation is approached, but not reached, before the stratification collapses. (This collapse will also re–oxygenate the bottom waters.)

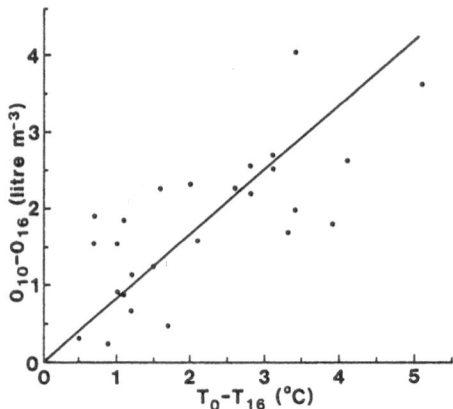

Fig. 9: Differences in oxygen concentration at 10 and 16 m depth in SWA plotted against simultaneous differences in temperature at 0 and 16 m between 1 September and 22 December 1975. A line of best fit is drawn.

However, if a second rainstorm were to occur within 14 days of the first so that the collapse of stratification were prevented, then the oxygen content would continue to decrease at roughly the rate R until complete de–oxygenation occurred. This is in agreement with Parker's (1983) observation that two successive rainstorms are needed for SWA to become completely anoxic at the bottom.

A full run of the model of Section 4 for dissolved oxygen, using observed density stratification as input for calculating eddy diffusivities, would be desirable for a more accurate test. However, this has not been done, owing to the difficulty of adequately including the effects of photosynthesis in the upper layers of SWA and of oxygen exchange across the air–water interface.

7. CONCLUSION

It has been shown that the present model gives good simulations of both rainstorm response and spring warm–up behaviour in SWA, for a choice of $C_1 = C_2 = 0.025$, $E/h = 0.1$ m^{-1}. Of these parameters, only the choice of C_2 is critical; the choice $C_1 = 0$ (corresponding to the assumption that potential energy loss is negligible for diffusion in SWA) gives essentially the same results, while variation of E/h over the observationally allowed range seems to have little effect on the results. The same choice of parameters appears to account quite well, with measured oxygen consumption rates, for the main features of the dissolved oxygen content of the bottom waters of SWA during a dry spring warm–up, and also for the fact that *two* successive rainstorms seem to be needed for complete de–oxygenation to occur (Parker, 1983).

Unfortunately, the kinetic energy influx to SWA has not been properly measured; the estimate of Eq. 11 may be in error by a factor of 3. If this estimate were too large by a factor of 2, then all the results of this chapter would be exactly reproduced by doubling C_2, *i.e.* best agreement with observation would be for $C_2 = 0.05$, the same value as found (for different energy sources) by Fischer *et al.* (1979) in a reservoir, and by Stigebrandt (1976) in a fjord. If, on the other hand, the kinetic energy estimate were too small by a factor of 3, then C_2 would need to be reduced by ⅓ to reproduce the results of this chapter: the energy conversion in SWA would then be 6 times less efficient than found by Fischer *et al.* and by Stigebrandt. Other possible defects of the present work are in the parameterization of tidal flushing, and in the neglect of the effects of internal waves inside SWA, as described by Stigebrandt (1976).

However, the fact that such a simple and inexpensive model has been shown to be useful in a freshwater, non–tidal reservoir; in a small, subtropical, fjord–like coastal lagoon (SWA), with a much greater energy level than the reservoir; and in a large true fjord, suggests that this type of modelling, with eddy diffusion estimated from the rate of energy input to the water body, may be useful in a wide range of coastal and inland waters.

ACKNOWLEDGEMENTS

The South West Arm Project was very much a team effort, and the author is indebted in one way or another to virtually everyone connected with it. However, he is particularly grateful to J. Imberger and T. Beer, whose rather austere comments on an earlier draft of this chapter led to substantial and much needed revisions.

REFERENCES

Australian National Tide Tables, 1975: Australian Hydrographic Publication 11. Canberra: Australian Government Publishing Service (1974)

Bird, E.C.F.: *Coasts*. Canberra:Australian National University Press (1968)

Bowden, K.F., Hamilton, P.: Some experiments with a numerical model of circulation and mixing in a tidal estuary. *Estuarine and Coastal Marine Science* 3, 281–302 (1975)

Bulleid, N.C.: The nutrient cycle of an intermittently stratified estuary. In: W.R. Cuff and M. Tomczak jr, eds *Synthesis and Modelling of Intermittent Estuaries*. Berlin, Heidelberg, New York: Springer (1983)

Fischer, H.B., List, E.J., Koh, R.C.Y., Imberger, J., Brooks, N.H.: *Mixing in Inland and Coastal Waters*. New York: Academic Press (1979)

Godfrey, J.S., Parslow, J.: Description and preliminary theory of circulation in Port Hacking Estuary. *CSIRO Division of Fisheries and Oceanography Report* **67** (1976)

Hansen, D.V., Rattray, M.: New dimensions in estuary classification. *Limnology and Oceanography* **11**, 319–326 (1966)

Huzzey, L.: *The Dynamics of a Topographically-controlled Estuarine Front, Port Hacking, N.S.W.* M.Sc. Thesis, The University of Sydney (1981)

Kjerfve, B., Proehl, J.A.: Velocity variability in a cross–section of a well-mixed estuary. *Journal of Marine Research* **37**, 409–418 (1979)

McEwan, A.D.: Internal mixing in stratified fluids. *Journal of Fluid Mechanics* (in press)

Ozmidov, R.V.: On the turbulent exchange in a stably–stratified ocean. *Izvestiya, Atmospheric and Oceanic Physics* **1**, 493–497 (1965)

Parker, R.R.: Some ecological effects of rainfall on the protoplankton of South West Arm. In: W.R. Cuff and M. Tomczak jr, eds *Synthesis and Modelling of Intermittent Estuaries*. Berlin, Heidelberg, New York: Springer (1983)

Parker, R.R., Rochford, D.J., Tranter, D.J.: History and organization of the Port Hacking Estuary Project. In: W.R. Cuff and M. Tomczak jr, eds *Synthesis and Modelling of Intermittent Estuaries*. Berlin, Heidelberg, New York: Springer (1983)

Pickard, G.L.: Oceanographic features of inlets in the British Columbia mainland coast. *Journal of the Fisheries Research Board of Canada* **18**, 907–999 (1961)

Pritchard, D.W.: The movement and mixing of contaminants in tidal estuaries. In: E.A. Pearson, ed. *Proceedings of the First International Conference on Waste Disposal in the Marine Environment*. New York: Pergamon Press (1960)

Scott, B.D.: Phytoplankton distribution and light attenuation in Port Hacking Estuary. *Australian Journal of Marine and Freshwater Research* **29**, 31–44 (1978a)

Scott, B.D.: Nutrient cycling and primary production in Port Hacking. *Australian Journal of Marine and Freshwater Research* **29**, 803–815 (1978b)

Scott, B.D.: Seasonal variations of phytoplankton production in an estuary in relation to coastal water movements. *Australian Journal of Marine and Freshwater Research* **30**, 449–461 (1979)

Stigebrandt, A.: Vertical diffusion driven by internal waves in a sill fjord. *Journal of Physical Oceanography* **6**, 486–495 (1976)

Thompson, R.O.R.Y.: Efficiency of conversion of kinetic energy to potential energy by a breaking internal gravity wave. *Journal of Geophysical Research* **85** (**C11**), 6631–6635 (1980)

Turner, J.S.: *Buoyancy Effects in Fluids*. Cambridge University Press (1973)

Vaudrey, D.J., Griffiths, F.B., Sinclair, R.E.: Data base for the Port Hacking Estuary Project: Parameters, monitoring procedure, and management system. In: W.R. Cuff and M. Tomczak jr, eds *Synthesis and Modelling of Intermittent Estuaries*. Berlin, Heidelberg, New York: Springer (1983)

Synthesis and Modelling of Intermittent Estuaries
(W.R. Cuff and M. Tomczak jr. eds) Berlin, Heidelberg,
New York: Springer (1983), pp. 55–75.

The Nutrient Cycle of an Intermittently Stratified Estuary

Nicholas C. Bulleid

Division of Fisheries Research
CSIRO Marine Laboratories
P.O. Box 21, Cronulla, N.S.W. 2230, Australia

Summary. The influence of stratification on the oxygen regime and nutrient concentrations was examined over four years in an estuary (South West Arm, Port Hacking, N.S.W., Australia) affected by occasional heavy freshwater run–off. The rate of detrital input to the sediment was measured using sedimentation tubes, and oxygen flux across the sediment–water interface was estimated from the oxygen consumption of intact sediment cores.

Following the onset of stratification, oxygen concentration in the bottom water decreased, frequently reaching zero, and the concentrations of nitrate, phosphate, silicate and ammonium all increased, though the occurrence of nitrate depended on the presence of oxygen. It was concluded that the major source of these nutrients was planktonic detritus recently settled from the water column, although over short periods terrestrial material (following torrential rain) and the sediment nutrient pool (during anoxic conditions) were also important.

The rate of nitrogen remineralization at the sediment surface, on a mean annual basis, was found to be 40% of that required by phytoplankton, which demonstrates the important link between benthic recycling and the productivity of the water column.

Key words: nutrient cycle, Port Hacking, South West Arm

1. INTRODUCTION

South West Arm is a 3 km long, ~300 m wide embayment of Port Hacking estuary which forms the southern boundary of the Sydney metropolitan area. The hydrology of this estuarine system was first studied by Rochford (1951) who found increased concentrations of phosphate in the bottom water of South West Arm (SWA) which, he was later able to show (Rochford, 1974), were associated with periods of de–oxygenation which accompanied stratification induced by heavy rainfall.

He suggested three potential sources of phosphate released into the bottom water, namely: the plant and animal tissues within the estuary, the detrital material introduced by freshwater run–off, and the phosphate bound to sediment particles which was mobilized as de–oxygenation became advanced. Each of these sources has received attention. Jitts (1959) found that estuarine sediment was able to bind large quantities of phosphate and suggested that this might be an important factor in estuarine productivity. More recently Scott (1978a) concluded, on the basis of light penetration measurements, that both particulate material in freshwater run–off and biodetritus originating from plankton in the water column decomposed on the bottom of SWA, remineralizing inorganic nutrients. The flux of nutrients across the sediment–water interface in SWA has been examined *in situ* using submerged chambers (Bulleid, unpublished) and the rates of nutrient release and oxygen consumption obtained are used below.

This paper examines the relative importance of these three sources to the cycling of phosphate and other inorganic nutrients in SWA and the role of stratification in nutrient recycling and distribution.

2. METHODS

2.1 Water sampling and analysis

A sampling station in SWA (Stn C, corresponding to Australian Grid Reference co–ordinates 27.1,25.2; see Fig. 1 of Vaudrey *et al.*, 1983) was occupied approximately every 14 days from February 1974 until December 1977 and from October 1979 until March 1980. Seawater samples taken from 0, 5, 10, and 15 m and the bottom (18 to 20 m depending on tide) were analysed for salinity, dissolved oxygen, nitrate, reactive orthophosphate, reactive silicate, and ammonium (the latter in 1979 and 1980 only) by methods described by Major *et al.* (1972). Additional samples were taken from June 1975 to May 1976 and analysed for particulate organic carbon (POC), dissolved organic carbon (DOC) (Airey & Hogan, 1980), and chlorophyll *a* (Jeffrey & Humphrey, 1975).

2.2 Detritus settling rate

The settling rate of particulate matter onto the sediment surface in SWA between December 1976 and October 1977 was estimated using clear acrylic sedimentation tubes (diameter 3.8 cm, height 17 cm) which were suspended in quadruplicate on a mooring 2 m above the sediment surface. These were collected approximately every 7 days and POC and particulate organic nitrogen (PON) analysed either gravimetrically (Airey & Hogan, 1980) or using a Hewlett Packard CHN Analyser, Model 185B.

2.3 Sediment oxygen uptake

Oxygen uptake by the sediment was measured on intact sediment cores (diameter 3.8 cm, length 20 cm) taken by hand by divers. Cores were returned immediately to the laboratory where the overlying water was removed by pipette and replaced with 50 ml of seawater saturated with oxygen, which was then covered with a layer of liquid paraffin. Four cores from each station were incubated at 18°C in a thermostatically controlled water bath. Samples of the supernatant were taken with a hypodermic syringe over a period of 1 – 2 h and the change in oxygen concentration measured by injecting the sample into the chamber of a Radiometer oxygen electrode Model D 616/0. The oxygen uptake was expressed as mg O_2 m^{-2} day^{-1}.

2.4 Meteorological data

Daily rainfall data were collected at Audley (2.4 km from the SWA sampling station; see Fig. 1 of Albani *et al.*, 1983) by the Bureau of Meteorology, Department of Science.

3. RESULTS

Long–term records from the southern Sydney region show a small seasonal variation in mean rainfall pattern, with precipitation varying from approximately 105 mm per month from February to June to about 70 mm from August to November and falling on an average of 12 days per month. These figures tend to disguise the irregularities of the rainfall pattern, which shows dry periods punctuated by occasional heavy rainfall.

Fig. 1 shows the daily rainfall at Audley from February 1974 to December 1977. The sporadic rainfall pattern is reflected by the salinity of SWA during the same period (Fig. 2a), which shows intense stratification near the surface associated with the freshwater run–off.

The temperature record for SWA is shown in Fig. 2b and reveals stratified conditions between October and April.

The freshwater input introduces both dissolved and particulate matter to the water column, as demonstrated by profiles taken after heavy rainstorms on 4 July 1975 and 3 March 1977 (Fig. 3a,b). Fig. 3a shows high concentrations of POC and DOC in the surface low salinity water and Fig. 3b increased dissolved silicate concentration.

The input of particulate matter is also reflected in the increase in sinking detritus caught by the sedimentation tubes during periods of heavy rainfall, *e.g.* 24 February to 9 March 1977 (Fig. 4). The material collected consisted mostly of intact phytoplankton, faecal pellets, and amorphous particulate matter. Recognizable material of terrestrial origin, *e.g.* leaf litter, was negligible.

During stratification the denser, more saline water became de–oxygenated (Fig. 2c) and occasionally anoxic.

These periods of de–oxygenation were accompanied by higher concentrations of phosphate (Fig. 2d) and silicate (Fig. 2e) in the deeper water, which correlate with the oxygen decrease. Nitrate showed a different distribution (Fig. 2f), maximum concentrations occurring either when the oxygen concentration at the bottom was low (but above zero), *e.g.* during March 1975, January 1977, and March 1977, or when water with a low oxygen concentration overlay anoxic water *e.g.* April 1974 and February/March 1976.

Nutrient distributions showed a similar pattern and relationship with oxygen during the later sampling period October 1979 to March 1980. Ammonium followed the pattern of phosphate and silicate, with a maximum in the bottom water where oxygen concentration was lowest (Fig. 5).

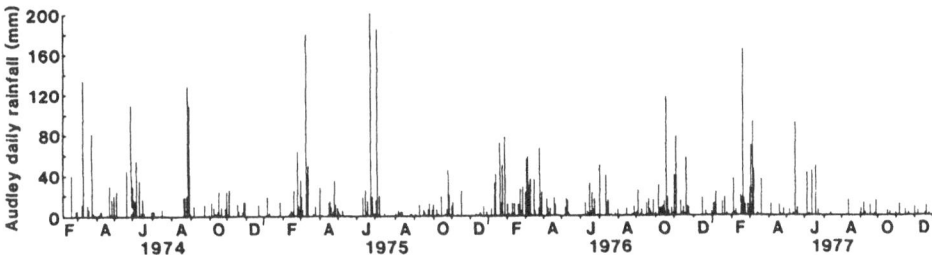

Fig. 1: Daily rainfall (mm) at the Audley station (34°04'S, 151°13'E) for the period February 1974 to December 1977. Recordings of less than 1 mm are not shown.

Oxygen uptake by the sediment at Stn C was 577 ± 70 (std. dev.) (23 March 1977), 588 ± 55 (1 June 1977) and 535 ± 25 mg O_2 m^{-2} day^{-1} (8 June 1977).

4. DISCUSSION

4.1 The influence of stratification on oxygen flux

The dominating effect of rainfall on the density structure of SWA is demonstrated by Figs 1,2a, this effect being more pronounced if it occurs during the period of temperature stratification between September and March. The hydrological conditions and circulation induced both by heavy rainfall and the topography of SWA have been described by Godfrey & Parslow (1976). Stratification reduces the oxygen input to the deeper waters both by slowing diffusion from the surface (see below and also Godfrey, 1983) and by decreasing the oxygen production by photosynthesising phytoplankton since the turbid surface layer increases light attenuation. De–oxygenation follows (Fig. 2c).

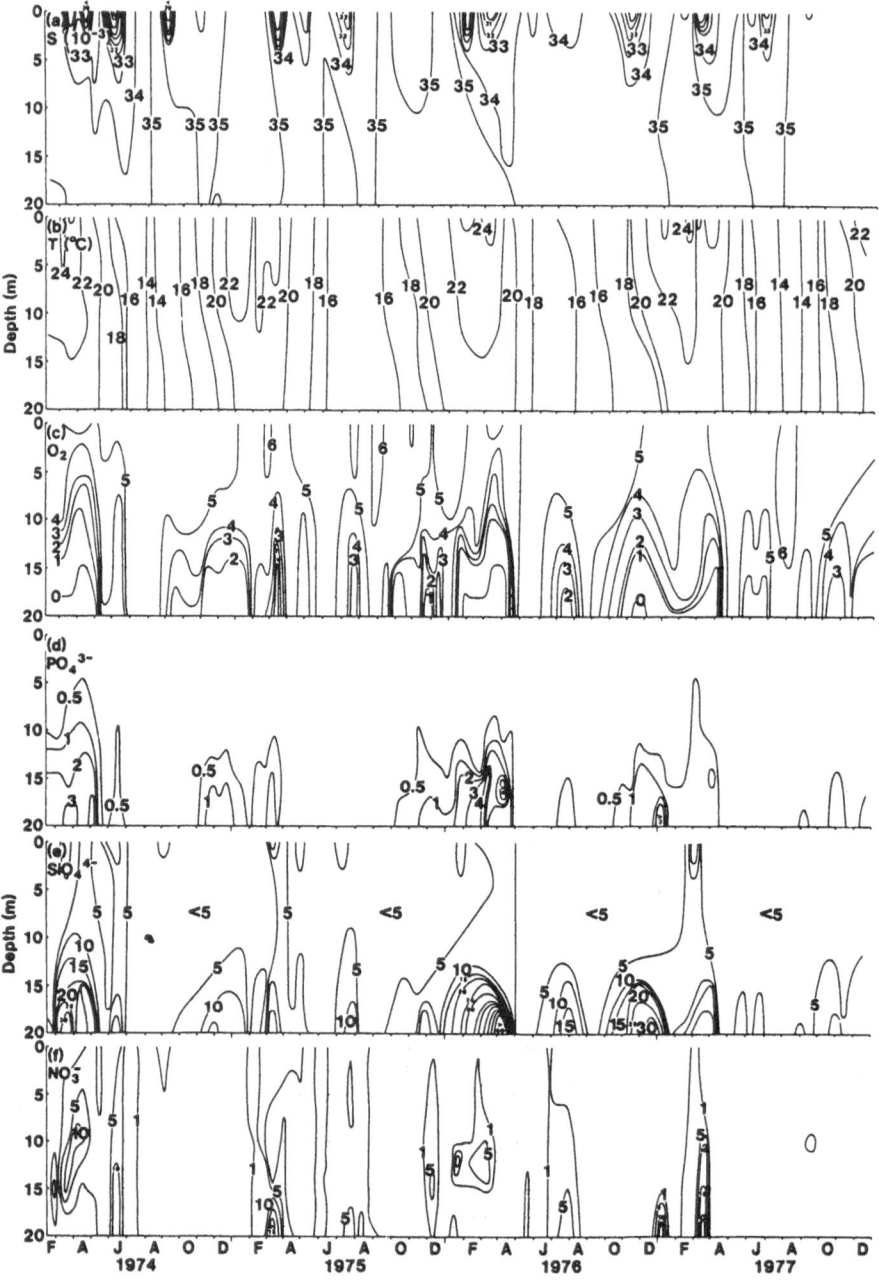

Fig. 2: Isopleths of various properties in the water column at Stn C in SWA from February 1974 to December 1977: a. salinity (10⁻³) b. temperature (°C) c. oxygen (ml l⁻¹) d. reactive orthophosphate concentration (μg–atom l⁻¹) e. reactive silicate concentration (μg–atom l⁻¹) f. nitrate concentration (μg–atom l⁻¹).

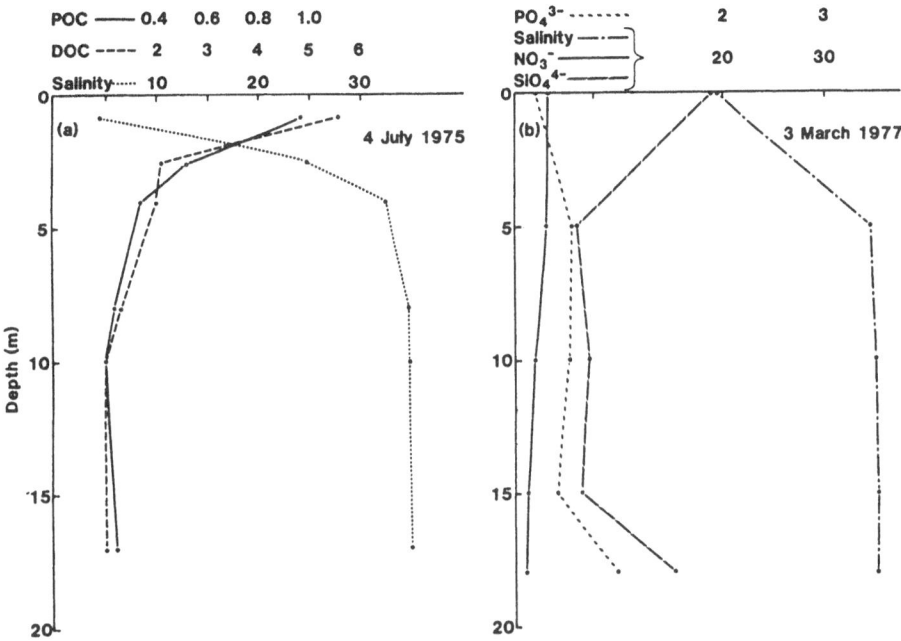

Fig. 3a: Profiles of salinity (10^{-3}), POC (mg l^{-1}), and DOC (mg l^{-1}) at Stn C in SWA on 4 July 1975.

Fig. 3b: Profiles of salinity (10^{-3}), nitrate (μg–atom l^{-1}), phosphate (μg–atom l^{-1}), and silicate (μg–atom l^{-1}) in SWA on 3 March 1977.

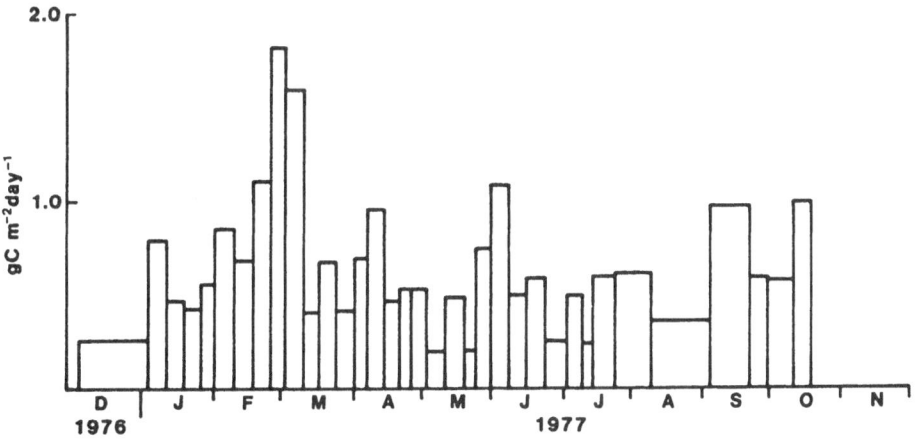

Fig. 4: Settling rate of detritus (g C m^{-2} day^{-1}) near Stn C during 1977.

The decrease in oxygen concentration in the water column below 12 m was examined over eight periods of stratification during which the salinity of the bottom water remained unchanged and when the lower water column was presumed to have remained stationary. The calculated decreases, expressed as mg O_2 m^{-2} day^{-1}, were plotted against the mean difference in density (σ_t) between the surface and bottom waters (Fig. 6). The positive relationship (r = 0.84, p < 0.01) implies a reduced influx of oxygen with increased stratification.

The decrease in oxygen concentration is the result of several processes, these being the respiration of the biota in the water column and the uptake across the sediment–water interface, as against the replenishing processes of diffusion from the surface and photosynthetic oxygen output. However, the contribution of photosynthesis below 12 m is small, owing to the small degree of light penetration (Scott, 1978a), and can be ignored.

The oxygen decrease can therefore be expressed as the equation

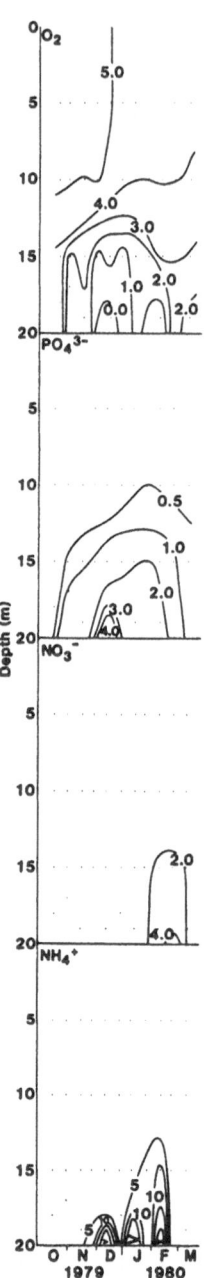

Fig. 5 Oxygen (ml l^{-1}), phosphate (μg–atom l^{-1}), nitrate (μg–atom l^{-1}), and ammonium (μg–atom l^{-1}) concentrations at Stn C from 18 October 1979 to 6 March 1980.

$$\text{Oxygen decrease} = \text{Water column respiration} + \text{Sediment uptake} - \text{Diffusion}$$

During periods of low stratification, the calculated rates of oxygen decrease ranged from 250 to 787 mg O_2 m^{-2} day^{-1} (Fig. 6). Oxygen consumption in the bottom water in SWA was measured by Bulleid (unpublished) in an *in situ* experiment as 80 mg O_2 m^{-3} day^{-1}, or 640 mg O_2 m^{-2} day^{-1} below 12 m. Mean sediment oxygen uptake measured on intact cores was 567 mg O_2 m^{-2} day^{-1}. Substituting these values into the above equation, an approximate estimate of 420–957 mg O_2 m^{-2} day^{-1} for diffusion during low stratification can be obtained, a range which overlaps that obtained by Godfrey (1983).

4.2 The influence of stratification on oxygen and nutrient concentration

The extent of de–oxygenation of a water mass can be expressed as the apparent oxygen utilization (AOU), which is the difference between the observed oxygen concentration and that at 100% oxygen saturation at the same temperature and salinity.

The de–oxygenation which follows both salinity and temperature stratification (Fig. 2a,b,c) is also accompanied by increased nutrient concentrations. The distributions of de–oxygenation and phosphate concentration are similar (Fig. 2c,d) and their relationship is further examined in Fig. 7, where phosphate concentration is plotted against AOU for each year from 1974 to 1977. Below an AOU of 4.0 ml l^{-1} the relationship is close, though at high AOU's, and with the approach of anoxic conditions, the relationship changes markedly with a greatly increased phosphate concentration per unit of oxygen consumed. There are two components of this change, the first being the increased mobility of the phosphate ion at low oxygen concentration (Mortimer, 1941, 1942; Theis & McCabe, 1978). Mortimer showed that ferric complexes and adsorbing sites in clay lattices in the surface sediment which had bound inorganic phosphate became reduced as anoxic conditions approached, liberating phosphate into the interstitial and overlying water. The second component is the failure of the AOU value to account for the additional oxygen obtained for respiratory purposes by nitrate and sulphate reduction under anoxic conditions.

The relationship between silicate concentration and AOU shows more scatter (Fig. 8) due to the effect of higher silicate concentrations in freshwater run–off and to silicate recycling via the chemical solution of silica (Nriagu, 1978). The relationship is, therefore, not directly related to oxygen consumption.

The relationship between nitrate and AOU (Fig. 9) is positive at AOU's below 4 ml l^{-1} but changes above this level, showing zero nitrate concentration at maximum AOU (zero oxygen), a trend also demonstrated in Fig. 2c,f. Nitrogen is remineralized

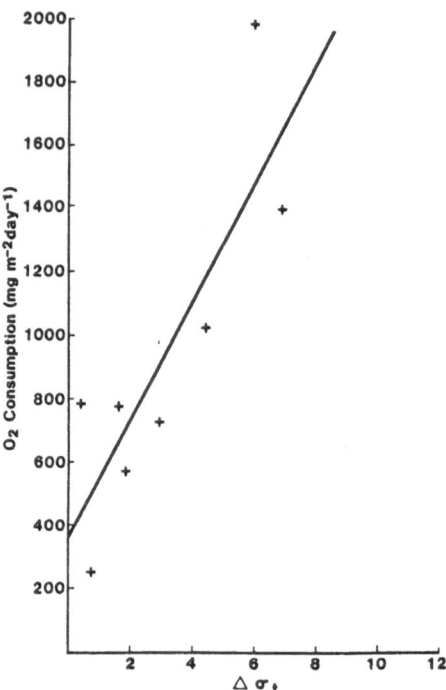

Fig. 6: The relationship between de-oxygenation (mg O_2 m^{-2} day^{-1}) in the water column below 12 m during eight stratified periods and mean σ_t difference between 20 m depth and the surface. Periods chosen were 21 August – 19 September 1974; 3 October – 27 November 1974; 26 February – 13 March 1975; 19 June – 24 July 1975; 4 September – 2 October 1975; 13 November – 27 November 1975; 23 September – 12 November 1976; and 3 March – 17 March 1977.

from organic matter principally as ammonium and the data suggest that those samples showing low oxygen concentration (high AOU) with high concentrations of phosphate and silicate also contained high ammonium concentrations, since nitrate would be absent owing to the lack of oxygen for nitrification and to the microbial denitrification of preformed nitrate. It is this complex speciation of inorganic nitrogen, furthermore, which probably accounts for the generally less distinct relationship between nitrate and AOU than that shown by phosphate.

Results from the later sampling period (1979–1980, Fig. 5) showed that increased concentrations of ammonium occurred with increased phosphate and silicate concentrations under anoxic conditions, confirming the interpretation suggested above.

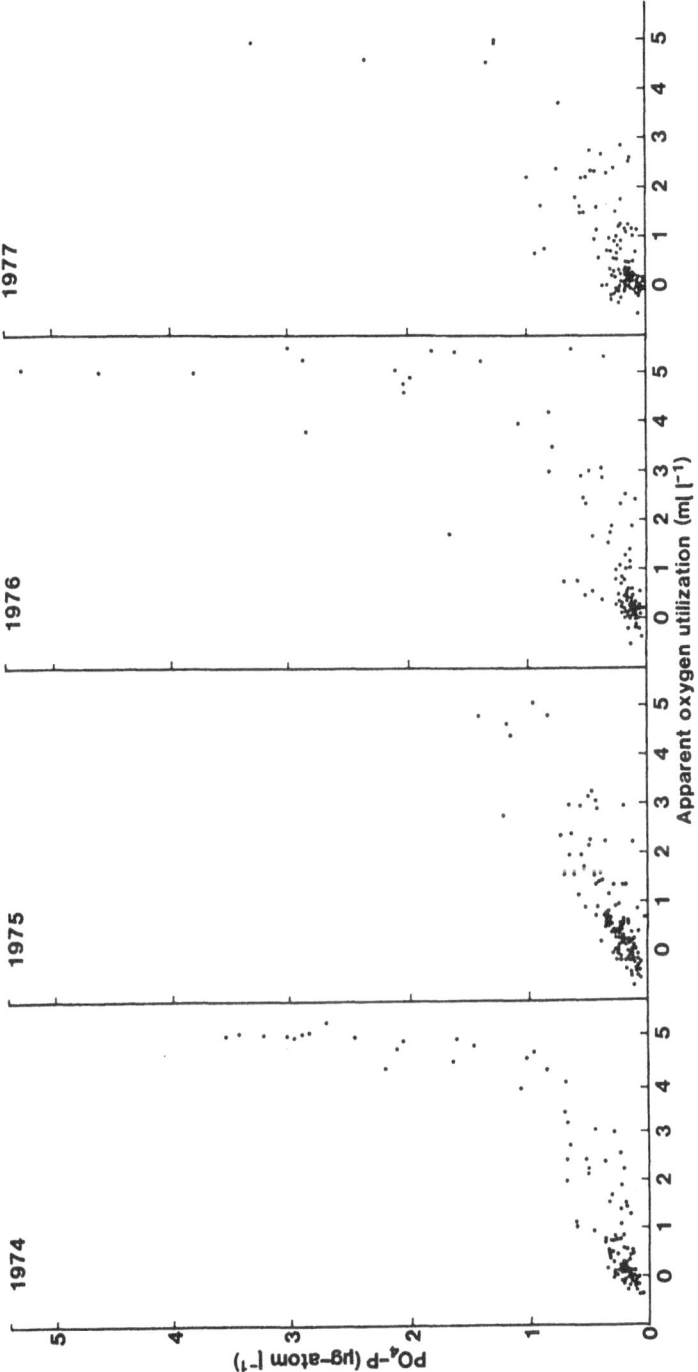

Fig. 7: The relationship between phosphate concentration (μg-atom l^{-1}) and AOU (ml l^{-1}) at all depths sampled at Stn C, 1974 – 1977.

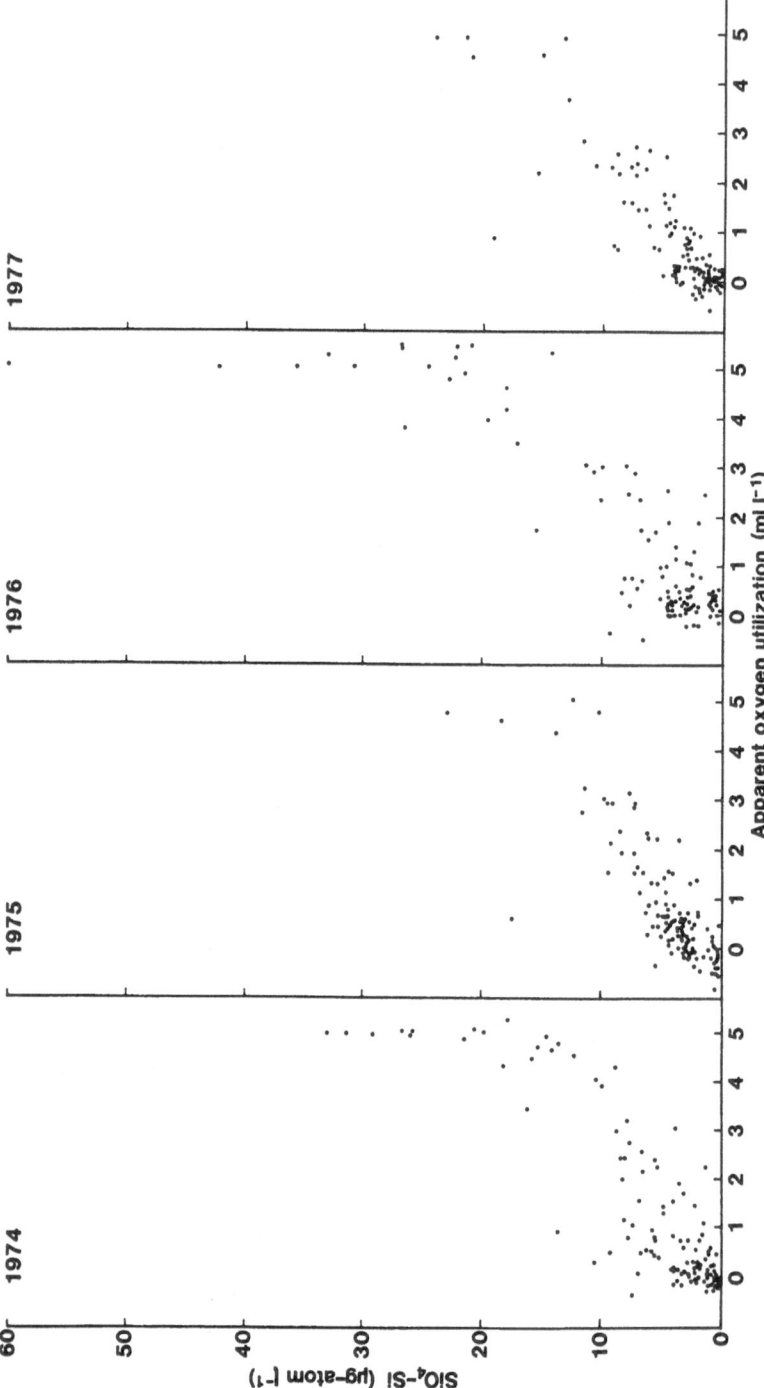

Fig. 8: The relationship between silicate concentration (μg–atom l^{-1}) and AOU (ml l^{-1}) at all depths sampled at Stn C, 1974 – 1977.

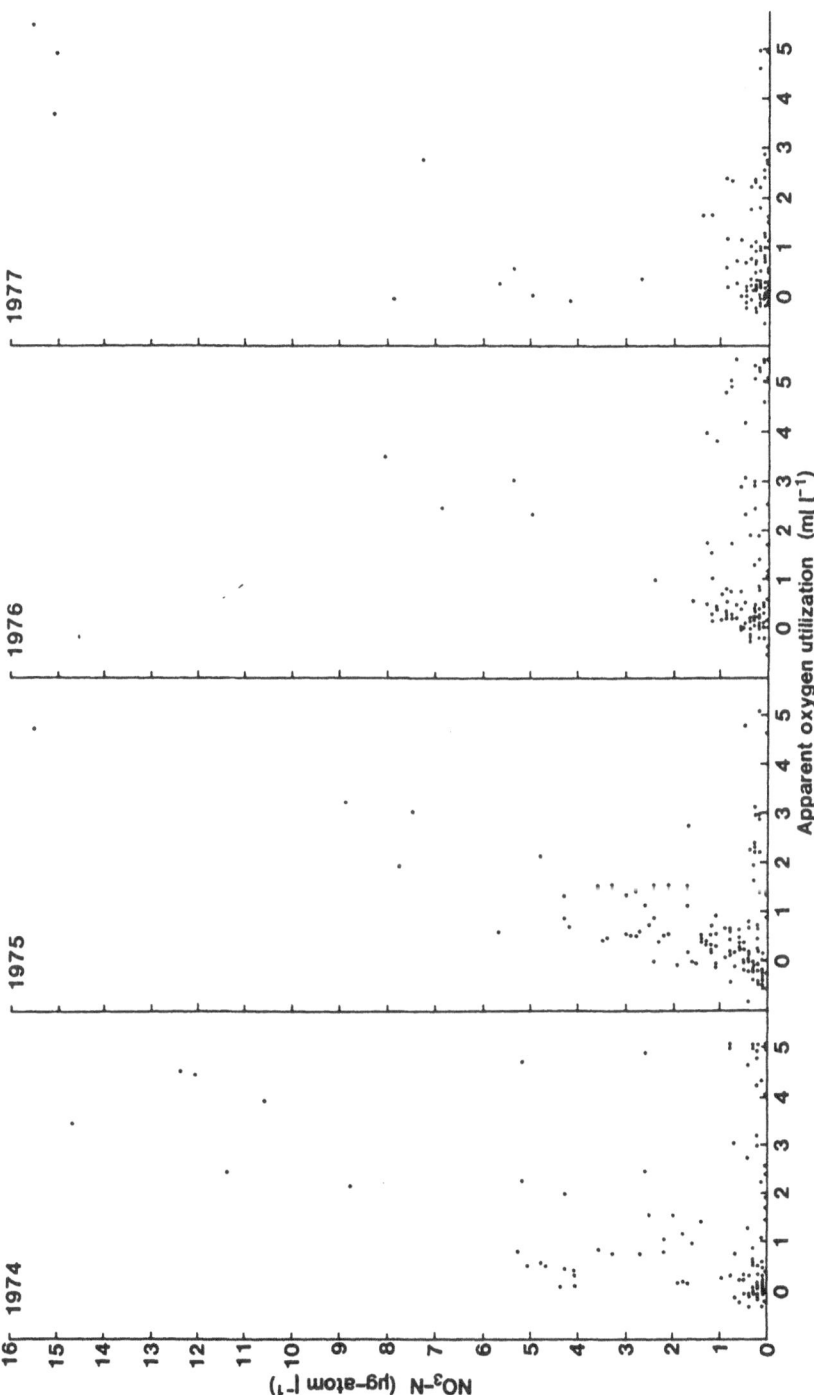

Fig. 9: The relationship between nitrate concentration (μg-atom l^{-1}) and AOU (ml l^{-1}) at all depths sampled at Stn C, 1974 – 1977.

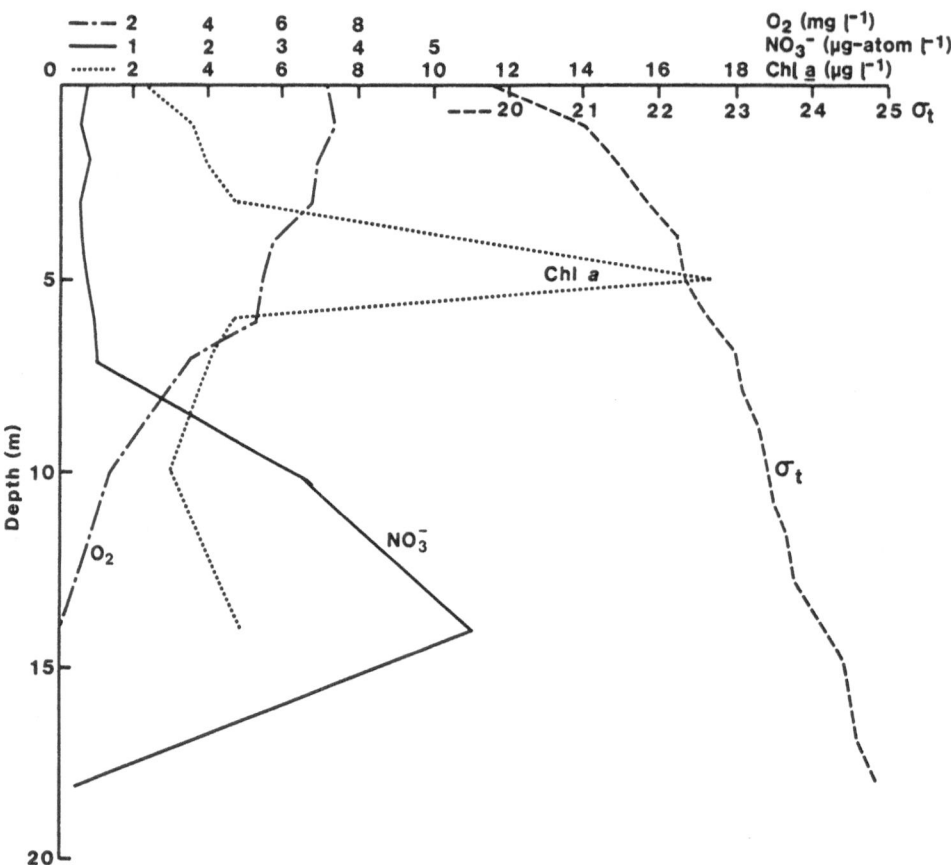

Fig. 10: Profiles of σ_t, oxygen (ml l^{-1}), nitrate (μg–atom l^{-1}), and chlorophyll a (μg l^{-1}) concentrations at Stn
        C on 15 March 1976.

The N : P ratios in the anoxic water (4.2 – 9.0 : 1) in 1979–1980 were anomalously low in comparison with the ratio of 16 : 1 expected from the remineralization of organic matter (Redfield *et al.*, 1963), a further consequence of the increased phosphate release during anoxic conditions.

The de–oxygenation of the bottom water and the increased nutrient concentrations which occur during stratification cause zonation of the water column, as demonstrated by the profiles in Fig. 10. This suggests that nutrients diffusing out of the deeper, anoxic water through the density gradient into the shallower, lighted regions are able to support the growth of phytoplankton, particularly those motile forms which are able to position themselves at a depth where optimum conditions of

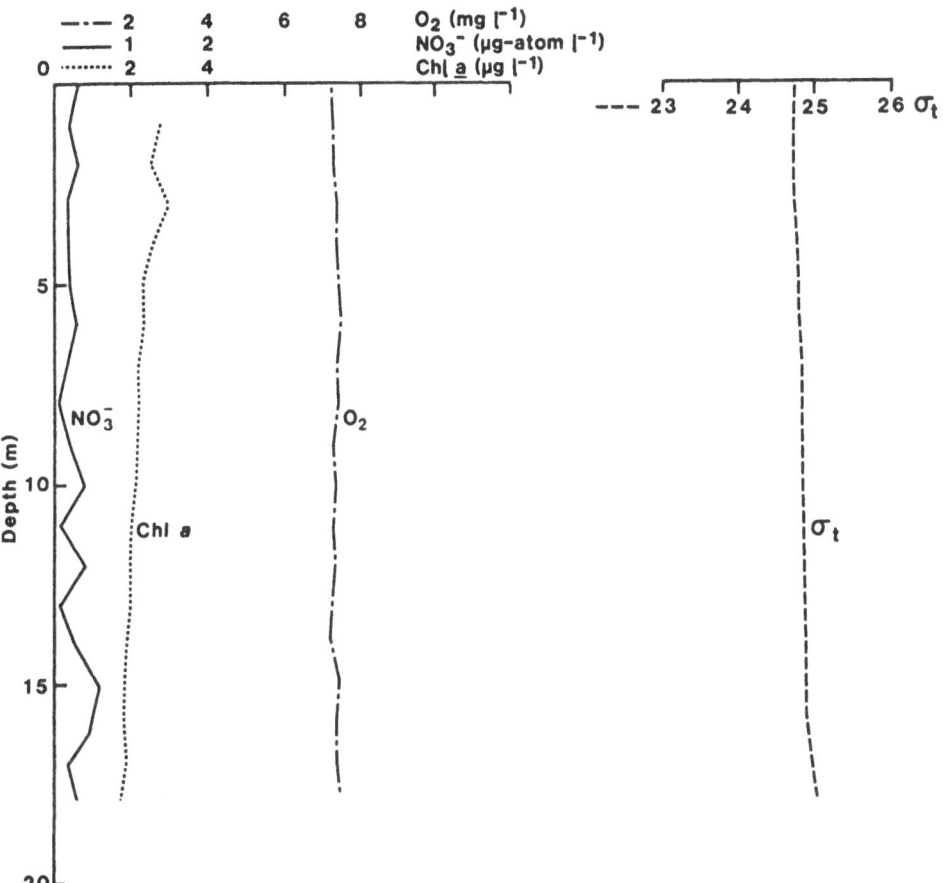

Fig. 11: Profiles of σ_t, oxygen (ml l^{-1}), nitrate (μg–atom l^{-1}), and chlorophyll a (μg l^{-1}) concentrations at Stn C on 25 May 1976.

light and nutrients prevail (Parker, 1983). Under these stratified conditions, phytoplankton distribution is frequently layered, as indicated by the chlorophyll a profile in Fig. 10. In contrast, well mixed conditions in SWA are usually accompanied by a more uniform vertical distribution of chlorophyll a (Fig. 11, see also Wood, 1964).

Stratification with the associated de–oxygenation and high nutrient concentrations usually lasts into the late summer, until either an influx of denser water from outside the estuary replaces the bottom water, as occurred in March 1975 (Scott, 1978b), or the progressive cooling of the surface water, coupled with tidal flushing of the less saline upper layers, leads to instability and turnover, as in April 1976. Such an event is demonstrated in the successive temperature and salinity profiles in Fig. 12.

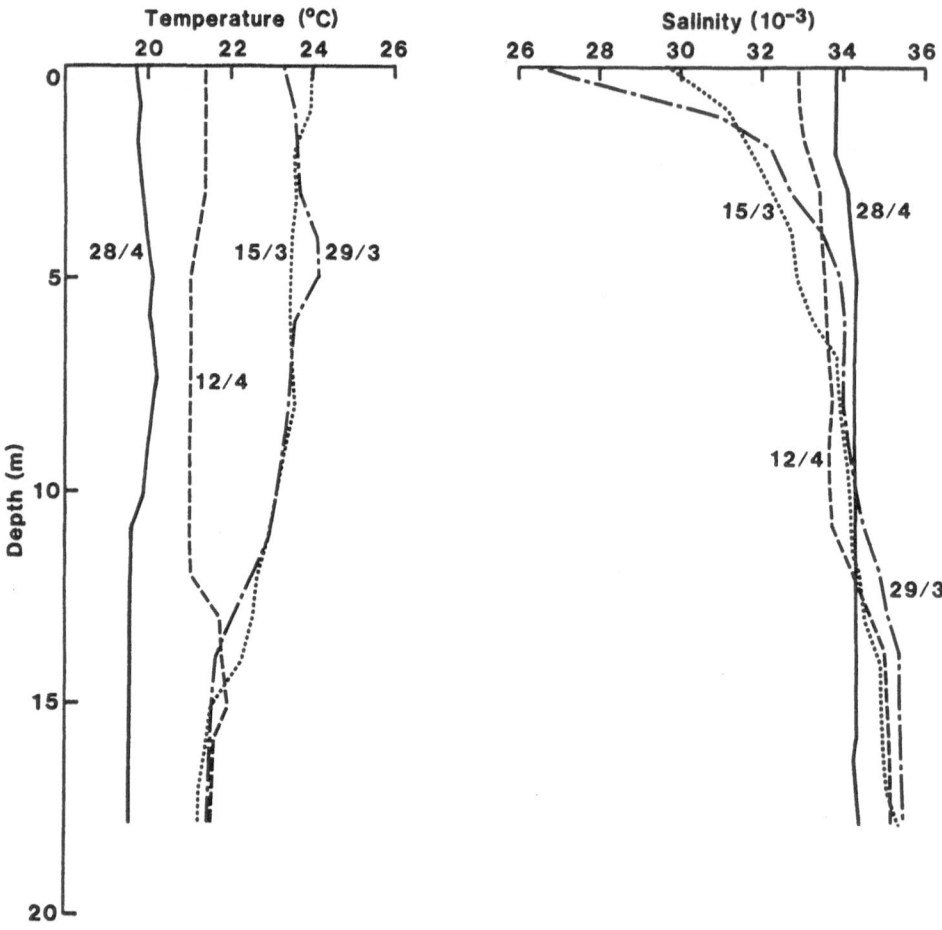

Fig. 12: Successive profiles of temperature (°C) and salinity (10^{-3}) at Stn C between 15 March and 28 April 1976.

4.3 The sources of nutrients in South West Arm bottom water

Three sources were identified above as contributing to the increased nutrient concentrations during stratification. The relative importance of settled planktonic detritus and the sediment nutrient pool was investigated by Bulleid (unpublished) during experimental studies on de–oxygenation and remineralization in SWA. It was concluded that remineralization of recently settled detritus, consisting mostly of phytoplankton and zooplankton faecal pellets, made a substantial contribution to the increase of ammonium, phosphate, and silicate concentrations and the decrease of the oxygen concentrations in the overlying water while the sediment–water interface was oxygenated. Pamatmat (1971) has reported a similar relationship in Puget Sound between seasonal variation of detrital input from the water column and oxygen consumption at the sediment surface. Only after de–oxygenation in SWA became advanced did the sediment itself contribute more significantly to the nutrient flux (expressed in terms of oxygen consumed) and then, principally, with respect to phosphate. The rate of nutrient increase (with respect to time) in the bottom water did not change markedly after the onset of anoxic conditions. The sediment pool is therefore significant as a source of nutrients during the relatively short anoxic periods, though it also is ultimately derived from settling particulate matter.

This increased flux of phosphate under anoxic conditions was also shown in the phosphate/AOU relationship in Fig. 7. Since phosphate adsorption by sediment particles is a reversible process dependent on the sediment Eh (Theis & McCabe, 1978), one can speculate that some of the phosphate becomes re–adsorbed when oxygenated conditions in the bottom water are restored, redressing the imbalance of phosphorus flux with respect to nitrogen. Thus both detrital input from the water column and the sediment nutrient pool (particularly with respect to phosphate) are important as nutrient sources, their relative magnitudes depending on the oxygen regime and therefore also on statification.

The importance of allochthonous material is more difficult to assess. Heavy rainfall and the associated particulate fallout do not appear to be a prerequisite for nutrient increase and de–oxygenation, for example the dry periods November/December 1974 and December 1976/January 1977. However, when freshwater run–off does occur, the large input of organic matter, as recorded by the sedimentation tubes (Fig. 4), will increase the substrate available for microbial breakdown, in particular in the upper reaches of SWA, where the sedimentation following rainstorms has been found to exceed that downstream by up to an order of magnitude (Bulleid, unpublished data).

However, periods of heavy freshwater run–off are relatively short–lived and the effect of their detrital load averaged over a year would be comparatively small (Fig. 4). Anoxic periods, during which the sediment nutrient pool may be mobilized, are similarly relatively infrequent. It may therefore be concluded that recent planktonic detritus provides the major substrate for remineralization at the sediment–water interface.

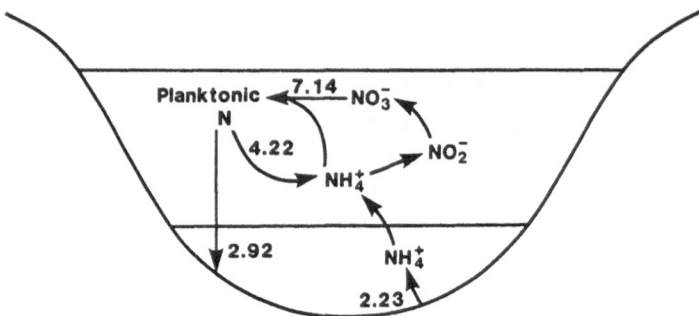

Fig. 13: Schematic representation of some nitrogen fluxes in SWA. The rates (mg–atom N m⁻² day⁻¹) used
are taken from Table 1.

4.4 The relative importance of water column and benthic remineralization

Since a proportion of the particulate matter generated in the water column sinks
to, and is remineralized at, the sediment surface, it is instructive to speculate on the
magnitude of this in comparison with recycling in the water column. Fig. 13 and
Table 1 show some major nitrogen flows within South West Arm.

Table 1.
Some rates of nitrogen flux in SWA (see Fig. 13).
Units mg–atom m⁻² day⁻¹.

Nature of N flux	Source of Data	Value
N uptake by phytoplankton	Mean annual primary production (Cuff *et al.*, 1983) x N/C ratio (Strickland, 1960) 50 μg–atom C m⁻² day⁻¹ x 1/7	= 7.14
Settling rate	Sedimentation tubes, annual mean.	= 2.92
Water column recycling rate	By difference 7.14 – 2.92	= 4.22
Recycling rate from sediment	By *in situ* measurement (Bulleid, unpublished)	= 2.23

Of the 7.14 mg–atom m^{-2} day^{-1} of nitrogen incorporated as ammonium and nitrate by phytoplankton in the top 12 m of the water column (on a mean annual basis), 2.92 mg–atom m^{-2} day^{-1} settles to the sediment surface as detritus. Assuming that there is no long–term increase of nitrogen in the SWA sediment, this should equal the annual average remineralization rate of nitrogen. This rate is comparable with the measured (short–term) nitrogen flux from the sediment to the water column of 2.23 mg–atom m^{-2} day^{-1}. The balance of the nitrogen assimilated by the phytoplankton (by difference), 4.22 mg–atom m^{-2} day^{-1}, is recycled as ammonium in the water column, mostly through zooplankton grazing. Thus approximately 40% (2.92 / 7.14) of the nitrogen turnover in SWA takes place at the sediment surface, the balance, 60%, being recycled in the water column.

5. CONCLUSIONS

Stratification has a marked effect on the distribution of dissolved oxygen, nitrate, ammonium, phosphate, and silicate. It achieves this through the restriction of circulation in the deep water, the reduction of nutrient uptake by phytoplankton in the deeper, shaded layers, the development of anaerobic processes at the sediment–water interface, and the liberation of adsorbed nutrients locked within the sediment. It thereby creates a range of environments determined by varying oxygen and nutrient concentrations and light, and each of these will provide a niche for a community of organisms which are positively advantaged by the conditions encountered.

The major source of nutrients recycled at the sediment–water interface is planktonic detritus which has recently settled from the water column, though the nutrient pool within the sediment and terrestrial material introduced with freshwater run–off are also important over short periods. The benthic remineralization rate of nitrogen is approximately 40% of the total nitrogen assimilated by phytoplankton in the water column.

ACKNOWLEDGEMENTS

A substantial part of the sampling program was performed by the Data Services Group at these Laboratories. P. Blomkamp performed the carbon analyses and R. Campbell assisted with the sediment–oxygen consumption measurements. N. Charlesworth drafted the figures. The manuscript was read by S. Jeffrey, B. Newell and L. Thompson, whose valuable suggestions have significantly improved the paper. The assistance of all the above is gratefully acknowledged.

REFERENCES

Airey, D., Hogan, M.: A manual for the determination of organic carbon in marine samples. *CSIRO Division of Fisheries and Oceanography Report* 127 (1980)

Albani, A.D., Rickwood, P.C., Tayton, J.W., Johnson, B.D.: Geological aspects of the Port Hacking estuary. In: W.R. Cuff and M. Tomczak jr, eds *Synthesis and Modelling of Intermittent Estuaries*. Berlin, Heidelberg, New York: Springer (1983)

Cuff, W.R., Sinclair, R.E., Parker, R.R., Tranter, D.J., Bulleid, N.C., Giles, M.S., Godfrey, J.S., Griffiths, F.B., Higgins, H.W., Kirkman, H., Rainer, S.F., Scott, B.D.: A carbon budget for South West Arm, Port Hacking. In: W.R. Cuff and M. Tomczak jr, eds *Synthesis and Modelling of Intermittent Estuaries*. Berlin, Heidelberg, New York: Springer (1983)

Godfrey, J.S.: Tidal flushing and vertical diffusion in South West Arm, Port Hacking. In: W.R. Cuff and M. Tomczak jr, eds *Synthesis and Modelling of Intermittent Estuaries*. Berlin, Heidelberg, New York: Springer (1983)

Godfrey, J.S., Parslow, J.: Description and preliminary theory of circulation in Port Hacking estuary. *CSIRO Division of Fisheries and Oceanography Report* 67 (1976)

Jeffrey, S.W., Humphrey, G.F.: New spectrophotometric equations for determining chlorophylls a, b, c_1 and c_2 in higher plants, algae and natural phytoplankton. *Biochemie und Physiologie der Pflanzen* 166, 191–194 (1975)

Jitts, H.R.: The adsorption of phosphate by estuarine bottom deposits. *Australian Journal of Marine and Freshwater Research* 10, 7–21 (1959)

Major, G.A., Dal Pont, G., Klye, J., Newell, B.S.: Laboratory techniques in marine chemistry: a manual. *CSIRO Division of Fisheries and Oceanography Report* 51 (1972)

Mortimer, C.H.: The exchange of dissolved substances between mud and water in lakes. I and II. *Journal of Ecology* 29, 280–329 (1941)

Mortimer, C.H.: The exchange of dissolved substances between mud and water in lakes. III and IV. *Journal of Ecology* 30, 147–201 (1942)

Nriagu, J.O.: Dissolved silica in pore waters of Lakes Ontario, Erie and Superior sediments. *Limnology and Oceanography* 23, 53–67 (1978)

Pamatmat, M.M.: Oxygen consumption by the seabed. VI. Seasonal cycle of chemical oxidation and respiration in Puget Sound. *Internationale Revue der Gesamten Hydrobiologie* 56, 769–793 (1971)

Parker, R.R.: Some ecological effects of rainfall on the protoplankton of South West Arm. In: W.R. Cuff and M. Tomczak jr, eds *Synthesis and Modelling of Intermittent Estuaries*. Berlin, Heidelberg, New York: Springer (1983)

Redfield, A.C., Ketchum, B.H., Richards, F.A.: The influence of organisms on the composition of seawater. In: M.N. Hill, ed. *The Sea* Vol.2, 26–77. New York: Interscience (1963)

Rochford, D.J.: Studies in Australian estuarine hydrology. I. Introductory and comparative features. *Australian Journal of Marine and Freshwater Research* 2, 1–116 (1951)

Rochford, D.J.: Sediment trapping of nutrients in Australian estuaries. *CSIRO Division of Fisheries and Oceanography Report* 61 (1974)

Scott, B.D.: Phytoplankton distribution and light attenuation in Port Hacking estuary. *Australian Journal of Marine and Freshwater Research* 29, 31–44 (1978a)

Scott, B.D.: Nutrient cycling and primary production in Port Hacking Estuary. *Australian Journal of Marine and Freshwater Research* 29, 803–815 (1978b)

Strickland, J.D.H.: Measuring the production of marine phytoplankton. *Fisheries Research Board of Canada Bulletin* 122, 1–172 (1960)

Theis, T.L., McCabe, P.J.: Phosphorus dynamics in hypereutrophic lake sediments. *Water Research* 12, 677–685 (1978)

Vaudrey, D.J., Griffiths, F.B., Sinclair, R.E.: Data base for the Port Hacking Estuary Project: Parameters, monitoring procedure, and management system. In: W.R. Cuff and M. Tomczak jr, eds *Synthesis and Modelling of Intermittent Estuaries*. Berlin, Heidelberg, New York: Springer (1983)

Wood, E.J.F.: Studies in microbial ecology of the Australasian region. I-VII. *Nova Hedwigia* 8, 5-54, 453-568 (1964)

Synthesis and Modelling of Intermittent Estuaries
(W.R. Cuff and M. Tomczak jr. eds) Berlin, Heidelberg,
New York: Springer (1983), pp. 77–89.

Phytoplankton Distribution and Production in Port Hacking Estuary, and an Empirical Model for Estimating Daily Primary Production

Barry D. Scott

Division of Oceanography

CSIRO Marine Laboratories

P.O. Box 21, Cronulla, N.S.W. 2230, Australia

Summary. The variations of phytoplankton biomass with both time and space are described for Port Hacking, New South Wales, a marine–dominated estuary. Frequent short–term increases in phytoplankton biomass and production were caused by estuarine hydrological events resulting in the release of regenerated nutrients, and by coastal hydrological events where slope–water intrusions enriched the coastal waters and were introduced into the estuary by tides. These frequent changes prevented any prediction of primary production. A simple empirical model was devised to estimate daily primary production by phytoplankton from measurements of phytoplankton biomass, total daily solar irradiance, and light attenuation by the water column.

Key words: phytoplankton, primary production, empirical model, Port Hacking, South West Arm

1. INTRODUCTION

This contribution describes the distribution and production of phytoplankton in Port Hacking, a small estuary on the Australian east coast, during 1975. The only previous study of phytoplankton in Port Hacking was that of Wood (1964), who studied changes in the species composition of diatoms and dinoflagellates. These studies showed that there was a general annual pattern with frequent dinoflagellate blooms during the warmer months from September to February, and a few blooms during the remainder of the year. The intrusion of some of the dinoflagellate blooms from the ocean was indicated by the dominance of typical marine species. Wood also suggested that the phytoplankton blooms during the remainder of the year were due to hydrological events occurring wholly within the estuary.

The climatology of Port Hacking and the hydrological events within the estuary which could produce phytoplankton blooms have been described by Rochford (1951, 1959, 1974). Rochford showed that either thermal or saline stratification of the estuarine water column resulted in an accumulation of regenerated nutrients in the lower water column, which were later released to the surface waters as the stratification decreased. Godfrey & Parslow (1976) described the physiography and water circulation in the estuary and the exchanges with the ocean.

Previous studies of phytoplankton in the coastal waters adjacent to Port Hacking are more numerous and diverse. Seasonal variations of phytoplankton have been described by Dakin & Colefax (1933). They found that there was a smooth annual variation in abundance with peaks in February and September. This pattern was partly confirmed later by Jeffrey & Carpenter (1974), who reported a September bloom of diatoms, but no data were reported to indicate the relative magnitude of these blooms. Humphrey (1960, 1963) found that during the 3 years from 1958 to 1960 the phytoplankton pigment concentrations showed maxima of short duration during the months from August to December with occasional maxima in February or March. These phytoplankton maxima were associated with intrusions of nutrient–rich slope water onto the coastal shelf. The occurrence and frequency of these intrusions was also discussed by Castillejo (1966). Newell (1966) obtained correlations between the amount of nitrate contained in the intrusions of slope water and the nitrogen contained in the plankton biomass. Later, in the spring of 1971, an intensive study of primary production and nutrients in the coastal waters near Port Hacking confirmed the earlier conclusions that these intrusions caused the algal blooms in the coastal region. The onset of these coastal blooms and their dissipation by sinking and by coastal currents have been described by Bulleid & Carpenter (1973), Newell & Bulleid (1975), and Carpenter & Carpenter (1976).

In the study of 1975 which is reported here, the distribution of plankton was determined using *in vivo* fluorescence and ^{14}C uptake at constant irradiance, while daily production by phytoplankton was determined by ^{14}C intake *in situ*. Results are reported in Scott (1978a, 1978b, 1979a), but a summary of results is included here. A simple empirical model is described for estimating daily production of phytoplankton from measurements of phytoplankton biomass and solar irradiance.

2. PHYTOPLANKTON DISTRIBUTION IN PORT HACKING

During the first three months of the study period in 1975 phytoplankton sampling was restricted to South West Arm (SWA), which is a small arm of Port Hacking estuary. Sampling positions and Station numbers are shown in the reference map of Vaudrey *et al.* (1983). Samples were collected at intervals of 2 – 3 days during the period at depths of 0, 1, 2, 3, 4, 6, 10, and 14 m, from Stns 3, 5, and 7. The measurements of phytoplankton biomass showed that at any one depth the phytoplankton was distributed almost uniformly between the three stations, although there were frequent and large variations with time, and also maxima at depths of 2 – 6 m (Scott, 1978a). Some of the variations with time were related to hydrological events within the estuary. Stratification of the water column caused by heating or dilution of the upper water column trapped regenerated nutrients in the lower water column, to be released later by destratification, causing phytoplankton blooms. Some of the variations of phytoplankton biomass and production could not be explained by events in the estuary, and it was not until the second phase of the work that these could be explained.

During the next 6 months of the study period phytoplankton samples were collected as before in SWA at Stn 7. It was not considered necessary to continue detailed sampling at Stns 3 and 5 due to their similarity in both phytoplankton biomass and light attenuation to Stn 7. This Station was selected since it was also used for routine measurements of temperature, salinity, oxygen, and nutrients. This reduction of effort allowed the investigation of the horizontal distribution of phytoplankton to be extended to more stations within SWA (Stns 1 – 11), and in Port Hacking Basin (Stns 12 – 14), and three more stations across the mouth of the estuary (Stns 15 – 17). These stations were sampled at two–week intervals.

The phytoplankton distribution showed that the stations could be divided into four groups which corresponded with their locality. Stns 1–7 within SWA formed one group with individual rates of photosynthesis at 1 m usually being within 20% of the group mean on each occasion. It was of interest that Stn 1 could be included in this group, since on falling tides the water at this point had come from the shallows on the extensive sand flats at the southern extremity of SWA. The second group of stations was that in Port Hacking Basin (Stns 12 – 14), where the average rate of photosynthesis was 40% greater than that within SWA at 1 m depth. Other data from the routine water sampling program indicated that the greater phytoplankton biomass in Port Hacking Basin could be explained by increased nutrient flux from the lower water column and sediments, induced by the relatively greater tidal flow in Port Hacking Basin. The third group of stations was that at the eastern or oceanic end of Port Hacking (Stns 15–17) where the rates of photosynthesis were similar to those at the other stations, but on two occasions in both August and September the rates of photosynthesis were almost twice that at any other station within the estuary. An examination of data from the coastal station, 6 km east of Port Hacking, showed that these occasions of high phytoplankton biomass had coincided with intrusions of nutrient–rich slope water onto the coastal shelf. The fourth group of stations was that near and over the shallow sand flat at the entrance of SWA (Stns 8, 9, 11) where the rates of photosynthesis were always within the range of the three groups described above, indicating that this water was of mixed origin.

Within SWA, samples were collected at 6 m depth from Stns 2 – 7, and 10, during the period of the distribution study. At Stns 3 and 4 the phytoplankton biomass was sometimes greater than at other points in this arm of the estuary. This greater biomass was possibly caused by increased nutrient flux from the sediments due to the constriction and change of direction of the tidal flow at this point. On a few occasions the rates of photosynthesis were greater at 6 m at Stns 7,10. These occasions coincided with nutrient enrichment on the coastal shelf, and the maximum at 6 m is explained by the circulation theory of Godfrey & Parslow (1976), where denser marine water enters the estuary on a rising tide at the surface, and on reaching the deeper basins sinks to an intermediate depth.

Thus the distribution of phytoplankton in Port Hacking is dependent mainly on the tides which provide energy to bring regenerated nutrients from the sediments and deeper water into the euphotic surface zone, and which transport enriched water or phytoplankton blooms into the estuary from the coastal waters.

3. CHANGES IN PHYTOPLANKTON BIOMASS AND PRODUCTION WITH TIME

As noted above, the phytoplankton biomass in SWA exhibited large and frequent variations with time. Maximum values recorded during 1975 at Stn 7 were: photosynthesis at optimum irradiance 56 mg C m^{-3} h^{-1}, and 320 mg C m^{-2} h^{-1} for the water column; chlorophyll a 12 mg m^{-3}, and 55 mg m^{-2} for the water column; photosynthesis *in situ* 334 mg C m^{-3} day^{-1}, and 1732 mg C m^{-2} day^{-1} for the water column (Scott, 1979a). More than twenty maxima of various magnitudes were detected during the year, seven of which could be explained by hydrological events within the estuary, while the remainder coincided with intrusions of nutrient–rich slope waters on to the coastal shelf. These slope–water intrusions seemed to occur regularly through the year with a period of about 15 days, but this apparent period may be an artifact of the weekly sampling period.

Despite the frequent maxima, primary production showed a seasonal trend with a maximum at the end of December. This seasonal trend was shown to be related more to the seasonal variation of solar irradiance than to the seasonal variation of water temperature in the estuary, which reached a maximum at the end of January (Scott, 1979a). The relationship of primary production to solar irradiance can be caused by a combination of increased day length during the summer and an increase in the depth of the euphotic zone. Seasonal changes in temperature will influence primary production, since in winter when the estuarine water is colder and more dense than the ocean water, any intrusion of nutrient–rich water entering the estuary on a rising tide will be removed again on the falling tide, whereas in summer the denser ocean water will be trapped in the estuary. This means that in summer there would be a net transport of nutrients into the estuary from the ocean in both inorganic and biologically combined forms, while in winter the net transport would be reversed. This seasonal change in net transport will therefore alter the size of the nutrient pool in the estuary and also the nutrient flux available for phytoplankton photosynthesis.

4. PREDICTION AND ESTIMATION OF PRIMARY PRODUCTION

4.1 Previous methods

The measurement of primary production in any large body of water presents several problems. One of these is the restriction that the *in situ* incubation method has on the number of measurements that can be made each day. This restriction is due to time, since if the samples are incubated from noon to sunset, then on any one day there would be only enough time for one worker to take samples from various depths at one position, innoculate with ^{14}C, and re–suspend them at the original depths. Even in very shallow estuaries the number of positions would be restricted by the time and distance between them. In the study reported here an attempt to overcome this problem was made by selecting one position which appeared to be representative of SWA, then making all measurements of primary production at this position. The primary production at other points in the estuary was assumed to be proportional to the phytoplankton biomass, as estimated by ^{14}C uptake at constant irradiance or by *in vivo* fluorescence. These methods were convenient since an adequate number of samples could be taken to determine the distribution of biomass. It is also assumed that the turbidity of the water is approximately the same throughout the estuary, and that the cloud cover over the estuary is distributed uniformly on any one day.

In oceanic studies of primary production, *in situ* incubations are an even greater problem since it may not be convenient or economical for a ship to wait for half a day while the samples are incubated.

While *in situ* measurements of primary production are desirable, the logistic problems associated with them have caused many workers in this field to find more convenient ways to estimate primary production. In some cases the *in situ* incubation has been avoided by using sunlit incubators to simulate the underwater irradiance. However, it is very difficult to simulate the underwater irradiance, since the irradiance field is directional near the surface and becomes diffuse with depth. Changes in spectral composition of the underwater irradiance must also be considered in the design of simulated *in situ* incubators. The changes in the underwater irradiance field also lead to problems in measuring the irradiance available for photosynthesis, both underwater and in any incubator. The control of temperature would also add to the complexity of any simulated *in situ* incubator, since temperature usually decreases with depth within the euphotic zone of the water column.

Primary production has also been estimated from a number of mathematical relationships which use measurements of phytoplankton biomass, solar irradiance, and underwater attenuation of solar irradiance. The evolution and general description of these models has been reviewed by Vollenweider (1970). Usually these mathematical relationships are simple and do not account for changes in the relationships of biomass, irradiance, and production with time of day. The diurnal changes in the photosynthetic rates of phytoplankton were first noted by Doty & Oguri (1957), and more recently the diurnal changes in the photosynthesis–irradiance relationship has been described by Taguchi (1976), MacCaull & Platt (1977), and Scott (1979b), but the most recent mathematical approach (Engqvist & Sjoberg, 1980) to the estimation of primary production fails to consider this problem. The mathematical approach also rarely considers the effect of temperature on the photosynthesis–irradiance relationship, which has been described by Nielsen & Hansen (1959) and Saijo & Ichimura (1962).

A more direct ("calculative") approach to the problem has been used by Fee (1973) and Jitts *et al.* (1976), who calculated daily production from measurements of photosynthesis–irradiance (P *v.* I) curves, solar irradiance, and light attenuation by the water column. This method suffers from the disadvantage that a large number of photosynthetic–rate measurements are necessary to establish the P *v.* I curves on each occasion, and that a specialized artificial light incubator is needed for these measurements. Jitts *et al.* (1976) also proposed that the P *v.* I curves could be used to calculate primary production on succeeding days using measurements of solar irradiance on those days. It is doubtful that this could be done in Port Hacking since the day–to–day variations in phytoplankton biomass are large, and therefore the change in the magnitude of the P *v.* I curves will be equally large. These changes can be illustrated by the change in the rate of photosynthesis at constant and optimum irradiance (P_c or P_{max}) at the principal station in SWA. During 1975 the mean photosynthetic rates at this position and depths of 1, 2, 3, 4, 5, 10, and 14 m were 6.6, 8.0, 8.0, 8.2, 8.1, 6.6, and 5.7 mg C m^{-3} h^{-1}, and the mean daily change of the rates were 1.2, 1.5, 1.4, 1.4, 1.4, 1.2, and 1.1 mg C m^{-3} h^{-1} respectively (Scott, 1979b). At times of nutrient enrichment, daily changes of up to 8 mg C m^{-3} h^{-1} were common, and since they were due to mixing of the estuarine water column or intrusions of enriched coastal water they occurred at unpredictable intervals. Due to subsurface phytoplankton maxima at depths of 2 to 6 m it would have been necessary to measure at least four P *v.* I curves from various depths to calculate the daily primary production. Since the number of measurements required for the P *v.* I curves is large and the daily changes in the estuary are large, the calculative method is not a satisfactory replacement for the conventional *in situ* incubation method. Since the *in situ* method is only useful for measuring primary production on a particular day, a method is required for estimating primary production on succeeding days.

4.2 An integral model based on P v. I curves

The usefulness of any model for predicting primary production in a variable estuarine environment like Port Hacking will depend not so much on its precision in estimating primary production at any one point, but on producing reasonable estimates of primary production at a large number of positions and depths. To do this the measurements must be simple, and preferably the amount of data processing should be small. The simplest measurement of photosynthesis is the measurement of the maximum photosynthetic rate (P_{max}), or the photosynthetic rate at constant and near optimum irradiance (P_c), while the simplest method of measuring underwater irradiance is to measure the attenuation coefficient of the water column, and monitor the solar irradiance above the water. The measurements of P_c are best made near the middle of the day when there is the least diurnal variation, and the measurements of attenuation coefficient can also be made at this time.

Since P v. I curves are of the same general shape for any particular level of phytoplankton biomass and ambient irradiance, a curve could be drawn with reasonable confidence using any point on the curve near to the optimum irradiance (Scott, 1979b). If the total daily irradiance and the P v. I curve are known it is possible to estimate the total daily photosynthesis. Therefore it is possible that the integral daily photosynthesis could be estimated as a function of total daily irradiance and P_c.

The use of the integrated total daily irradiance for estimating the integral daily production introduces an approximation since it assumes a smooth sinusoidal variation of irradiance during the day, but this may be justified by the simplification it produces.

If any one P v. I curve was selected and used to construct a series of P v. I curves with different values of P_c or P_{max} these could be used with typical daily graphs of underwater irradiance to calculate an array of values for daily primary production (P_{calc}). This array of P_{calc} values could then be plotted as a function of P_c and the total daily irradiance (I_{tot}) as shown in Fig. 1. This array would be more accurate in the region close to the value of P_c and I_{tot} corresponding to that for the original P v. I curve, and less accurate at the extremes of P_c and I_{tot}. The array could be extended and made more precise by subdividing the area of the array and using typical P v. I curves for each P_c and I_{tot} combination. The completed array could then be used to estimate daily primary production from measurements of P_c and I_{tot}.

4.3 An empirical model for estimating daily primary production

The construction of an array as described above presupposes that suitable equipment is available for measuring P v. I curves and that time is available to measure a large variety of curves, so that the array of primary production values will be large enough to cover the expected range. Often, time and equipment are not available and immediate estimates of primary production are required. In this case a simpler empirical model can be constructed as the study progresses, using photosynthetic rates measured by both the *in situ* and constant artificial irradiance methods, together with estimates of underwater irradiance calculated from measurements of solar irradiance and the attenuation coefficients of the water column.

The empirical model is constructed by plotting the values of primary production measured by the *in situ* method (P_{is}) in an array similar to that in Fig. 1. After sufficient numbers of P_{is} values have been plotted these can be contoured, and as the number of P_{is} values grow the model contours will take the same general shape as those shown in Fig. 1.

A model was constructed in this manner using the values of P_c, P_{is} and I_{tot} found at depths of 1 and 2 m in SWA during 1975. The values of I_{tot} were calculated from the half–day total solar irradiance measured at the Cronulla laboratory (3.8 km east of the station in SWA) and the attenuation coefficients measured in SWA. Due to the variability of the P_{is} values in the array, the contours could not be drawn easily and a preliminary smoothing of the data was necessary. The P_{is} values were smoothed by dividing the graph area into smaller rectangular areas, and the mean of the P_{is} values within each of these small areas was calculated and plotted at the centre of each small area. This process was repeated three more times on the same P_{is} data using rectangular grids which overlapped the other grids by half the rectangular area. The size and overlap of these rectangular areas are shown in Fig. 2. The P_{is} contours were then drawn using the grid of mean P_{is} values.

Although the contours resemble those in Fig. 1, there are several differences. At high irradiance values there seems to be no inhibition of photosynthesis, particularly at high values of P_c , and at low values of P_c and I_{tot} the values of P_{is} do not decrease until very low values of irradiance are reached. Both these differences are due to the adaptation of the phytoplankton to the environmental conditions of both irradiance and nutrient flux and the consequent alteration of the shape of the P v. I curve.

The simple empirical model was tested by comparing the actual measured values of P_{is} with the values obtained by interpolation between the contour lines (P_{est}). The relationship between P_{is} and P_{est} is shown in Fig. 3, together with the regression line and the 95% confidence limits, where:

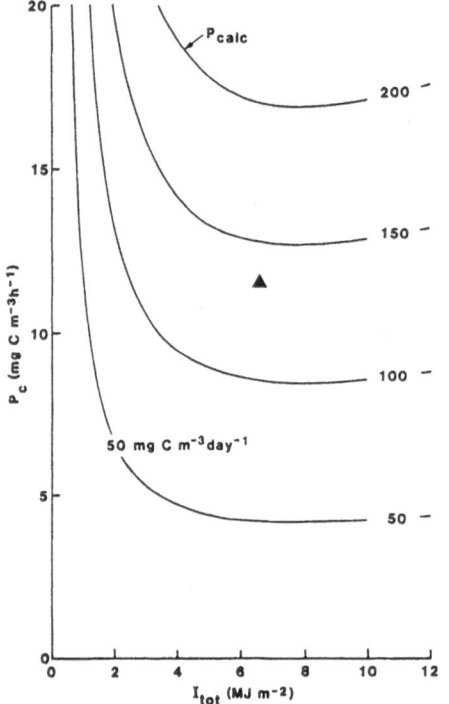

Fig- 1: A model of daily primary production calculated from a series of P v. I curves of varying amplitude. The P v. I curve used to produce this series of curves was one of many of similar shape measured during the SWA studies, and the P_c of this curve and the I_{tot} for the day are shown by the marked point (▲). The model shows the relationship of the calculated primary production (P_{calc}, expressed as mg C m^{-3} day^{-1}) to the photosynthetic rate at constant irradiance (P_c) and the total daily irradiance available for photosynthesis (I_{tot}).

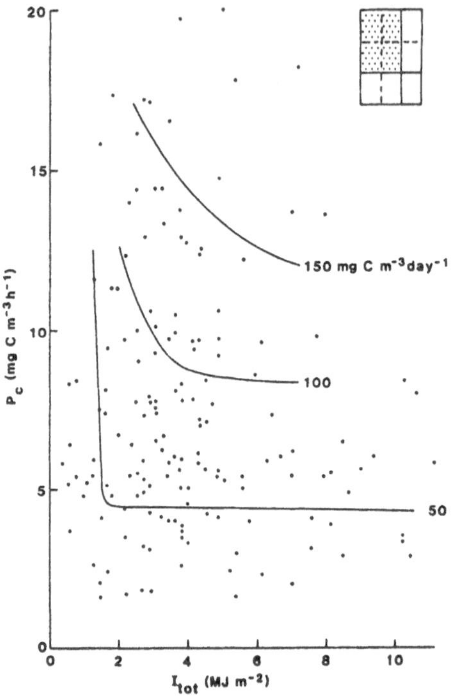

Fig. 2: An empirical model constructed from measured values of daily primary production *in situ* (P_{is}), photosynthetic rate at constant irradiance (P_c), and total daily irradiance available for photosynthesis (I_{tot}). The P_{is} contours were drawn using mean values within equal overlapping areas as shown on the upper right. The rates of photosynthesis are for samples from 1- and 2-m depth at Stn 7 in SWA.

$$P_{is} = 1.02 \ P_{est} - 0.6 \quad mg \ C \ m^{-3} \ day^{-1} \quad (r = 0.90, n = 155)$$

Some of the scatter around the regression line may be due to errors in measuring the solar irradiance, since the solar radiation was measured at the Cronulla laboratory and not directly in SWA where the primary production was measured.

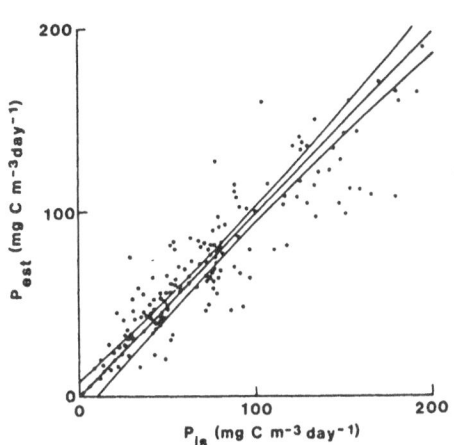

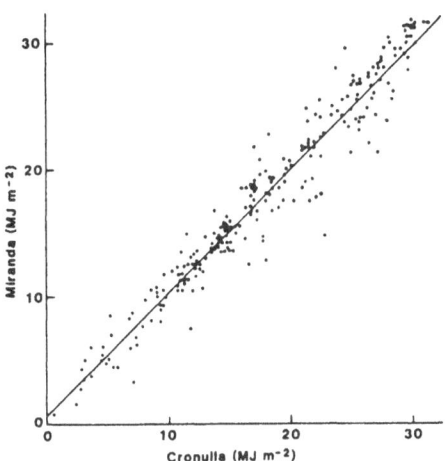

Fig. 3: A comparison of daily primary production measured *in situ* (P_{is}), with that estimated from the empirical model shown in Fig. 2 (P_{est}). The linear regression line and 95% confidence limits are shown.

Fig. 4: A comparison of the daily solar irradiance at the CSIRO laboratory on the coast at Cronulla, and 6 km inland at Miranda. The linear regression line is shown, where: Miranda = 0.97 Cronulla + 0.7 MJ m^{-2}.

During several periods in 1976 and 1977 the solar irradiance was measured at the Cronulla Laboratory and at a site in Miranda 4.3 km to the north of the SWA position, using identical pyroheliometers. These measurements show that over a long period the total solar radiation was the same at the two sites, but the daily measurements could be different by a factor of up to 1.5 (Fig. 4). Since the values of I_{tot} in Fig. 2 were calculated from the half–day solar irradiance measurements, the errors could be slightly greater than those indicated in Fig. 4 for daily irradiance values. Despite the errors possibly caused by the solar irradiance measurements the empirical model appears to give satisfactory estimates of the daily primary production, and are comparable to those found for the calculative method of Jitts et al. (1976).

For greater depths in the water column separate models should be constructed, due to the more diffuse irradiance field and the change in spectral composition with depth. These models will have different ranges of values for I_{tot} due to the decrease in irradiance with increasing depth. The construction of separate models for each range of light conditions will ensure that the value of P_{is} is correctly predicted from the measurement of P_c and I_{tot}, even though the underwater irradiance may be underestimated by the use of a directional irradiance sensor. Models for use in estuaries should be constructed for percentage ranges of surface irradiance, since scattering of light by particles is the main cause of attenuation. Oceanic models could be constructed for depth ranges since the attenuation is due mainly to selective absorption of certain wavelengths of light by the water.

Both oceanic and estuarine models could be constructed using alternative measurements of phytoplankton biomass instead of P_c, such as *in vivo* fluorescence, which has been shown to be closely related to P_c (Scott, 1978a). Integral models constructed using P_c values could not be directly converted to use *in vivo* fluorescence measurements, since the relationship of P_c to fluorescence depends on the physiological state of the phytoplankton.

The empirical model has the interesting feature that it could be used in a general way to show the ecological zones in which other workers have concentrated their studies of phytoplankton production and ecology. For instance, the surface zone of the ocean is represented by the area of high I_{tot} and low P_c, where P_{is} depends mainly on nutrient flux and very little on irradiance. Deeper zones of the ocean are represented by areas of low P_c and low I_{tot}, where P_{is} depends on both nutrient flux and irradiance. The highly productive shallow ecosystems are represented by the area of high I_{tot} and high P_c, where P_{is} appears to have reached a maximum value. Similarly the highly turbid and eutrophic conditions encountered in some freshwater locations are represented by the area of high P_c and low I_{tot}, where P_{is} depends mainly on irradiance. This feature of the model may make it useful as a teaching aid.

REFERENCES

Bulleid, N.C., Carpenter, D.J.: A multidisciplinary programme to investigate the sources and consequences of enrichment of nutrients off the New South Wales coast. In: R. Fraser, ed. *Oceanography of the South Pacific 1972.* Wellington: N.Z. National Commission for UNESCO (1973)

Carpenter, D.J., Carpenter, S.M.: Numerical analysis of primary production and associated data from waters off the east coast of Australia. *Australian Journal of Marine and Freshwater Research* **27**, 431–439 (1976)

Castillejo, F.F. de: Non–seasonal variations in the hydrological environment off Port Hacking, Sydney, *CSIRO Division of Fisheries and Oceanography Technical Paper* **21** (1966)

Dakin, W.J., Colefax, A.: The marine plankton of the coastal waters of New South Wales, I. The chief planktonic forms and their seasonal distribution. *Proceedings of the Linnaean Society of N.S.W.* **58**, 186–222 (1933)

Doty, M.S., Oguri, M.: Evidence for a photosynthetic daily periodicity. *Limnology and Oceanography* **2**, 37–40 (1957)

Engqvist, A., Sjoberg, S.: An analytical integration method of computing diurnal primary production from Steele's response curve. *Ecological Modelling* **8**, 219–232 (1980)

Fee, E.J.: A numerical model for determining integral primary production and its application to Lake Michigan. *Journal of the Fisheries Research Board of Canada* **30**, 1447–1468 (1973)

Godfrey, J.S., Parslow, J.: Description and preliminary theory of circulation in Port Hacking estuary. *CSIRO Division of Fisheries and Oceanography Report* **67** (1976)

Humphrey, G.F.: The concentration of plankton pigments in Australian waters. *CSIRO Division of Fisheries and Oceanography Technical Paper* **9** (1960)

Humphrey, G.F.: Seasonal variation in plankton pigments in waters off Sydney. *Australian Journal of Marine and Freshwater Research* **14**, 24–36 (1963)

Jeffrey, S.W., Carpenter, S.M.: Seasonal succession of phytoplankton at a coastal station off Sydney. *Australian Journal of Marine and Freshwater Research* **25**, 361–369 (1974)

Jitts, H.R., Morel A., Saijo, Y.: The relation of oceanic primary production to available photosynthetic irradiance. *Australian Journal of Marine and Freshwater Research* **27**, 441–454 (1976)

MacCaull, W.A., Platt, T.: Diel variations in the photosynthetic parameters of coastal marine phytoplankton. *Limnology and Oceanography* **22**, 723–731 (1977)

Newell, B.S.: Seasonal changes in the hydrological and biological environments off Port Hacking, Sydney. *Australian Journal of Marine and Freshwater Research* **17**, 77–91 (1966)

Newell, B.S., Bulleid. N.C.: A schema of the nitrogen cycle off Port Hacking, New South Wales. *Australian Journal of Marine and Freshwater Research* **26**, 375–388 (1975)

Nielsen, E.S., Hansen, V.K.: Light adaptation in marine phytoplankton populations and its interrelation with temperature. *Physiologia Plantarum* **12**, 353–370 (1959)

Rochford, D.J.: Studies in Australian estuarine hydrology. 1. Introduction and comparative features. *Australian Journal of Marine and Freshwater Research* **2**, 1–116 (1951)

Rochford, D.J.: Classification of Australian estuarine systems. *Archivio di Oceanografia e Limnologia* **11**, 171–177 (1959)

Rochford, D.J.: Sediment trapping of nutrients in Australian estuaries. *CSIRO Division of Fisheries and Oceanography Report* **61** (1974)

Saijo, Y., Ichimura, S.: Some considerations on photosynthesis of phytoplankton from the point of view of productivity measurement. *Journal of the Oceanographical Society of Japan* **20**, 687–693 (1962)

Scott, B.D.: Phytoplankton distribution and light attenuation in Port Hacking estuary. *Australian Journal of Marine and Freshwater Research* **29**, 31–44 (1978a)

Scott, B.D.: Nutrient cycling and primary production in Port Hacking. *Australian Journal of Marine and Freshwater Research* **29**, 803–815 (1978b)

Scott, B.D.: Seasonal variations of phytoplankton production in an estuary in relation to coastal water movements. *Australian Journal of Marine and Freshwater Research* **30**, 449–461 (1979a)

Scott, B.D.: *The Relation of Irradiance to Photosynthesis in Marine Phytoplankton.* M.Sc. Thesis, University of New South Wales (1979b)

Taguchi, S.: Short-term variability of photosynthesis in natural marine phytoplankton populations. *Marine Biology (Berlin)* **37**, 197–207 (1976)

Vaudrey, D.J., Griffiths, F.B., Sinclair, R.E.: Data base for the Port Hacking Estuary Project: Parameters, monitoring procedure, and management system. In: W.R. Cuff and M. Tomczak jr, eds *Synthesis and Modelling of Intermittent Estuaries.* Berlin, Heidelberg, New York: Springer (1983)

Vollenweider, R.A.: Models for calculating integral photosynthesis and some implications regarding structural properties of the community metabolism in aquatic systems. In: I. Setlik, ed. *Prediction and Measurement of Photosynthetic Productivity. Proceedings of the IBP/PP Technical Meeting, Trebon.* Wageningen Centre for Agricultural Publishing and Documentation (1970)

Wood, E.J.: Studies in microbial ecology of the Australasian region. I–VII. *Nova Hedwigia* **8**, 5–54, 453–568 (1964)

Synthesis and Modelling of Intermittent Estuaries
(W.R. Cuff and M. Tomczak jr. eds) Berlin, Heidelberg,
New York: Springer (1983), pp. 91–107.

Zooplankton Community Structure and Succession in South West Arm, Port Hacking

F. Brian Griffiths

Division of Fisheries Research
CSIRO Marine Laboratories
P.O. Box 21, Cronulla, N.S.W. 2230, Australia

Summary. A one–year study of the zooplankton and its community structure in South West Arm, Port Hacking was carried out between June 1975 and July 1976. Multivariable classification was used to delineate groups of zooplankton samples with similar species complements and abundance patterns. The classification yielded two groups (marine and estuarine) consistent enough to be considered communities. The marine community had significantly smaller biomass and fewer individuals per sample but significantly higher species richness, diversity and equitability than the estuarine community. A large proportion of these differences is due to variations in the abundance of the *Oithona* species group (Copepoda, Cyclopodia). These differences in community structure are attributed to the effects of different environmental structure caused by salinity stratification. The marine community had a more complex structure in a simpler, unstratified environment, while the estuarine community was structurally simple in a more complex environment. Seasonal succession occurred in the marine community, and was responsible for the change in species composition leading to the estuarine community. Seasonal succession was suppressed during the estuarine period.

1. INTRODUCTION

This study of the changes in zooplankton abundance and community structure in a marine embayment over a one–year period was carried out as part of a larger study aimed at understanding the principles of the structure and dynamics of Australian estuarine systems (Allen, 1983). Communities were identified and their relationship with the environment determined.

Dakin & Colefax (1935, 1940) have discussed the seasonal abundance of a number of neritic zooplankton species in Australian temperate waters. Kott (1955) and Hodgson (1979) reported on seasonal changes in Lake Macquarie. There is little other published information about changes in zooplankton biomass and species composition in warm temperate estuarine areas in Australia.

Changes in the seasonal abundance and succession of zooplankton have been shown to be related to temperature (Deevey, 1948; Woodmansee, 1958); seasonal wind and rainfall patterns (Wickstead, 1957; Youngbluth, 1976, 1980); predation (Cronin *et al.*, 1962; Sage & Herman, 1972; Rippingale & Hodgkin, 1974); phytoplankton availability (Woodmansee, 1958; Reeve, 1964, 1970) and to surface currents (Frolander *et al.*, 1973).

This study describes the changes in species abundance and biomass throughout a one–year period in South West Arm, Port Hacking. These results were used to test the hypothesis that a persistent change in the physical environment would have no effect on the community structure. The alternative hypothesis adopted was that a persistent alteration in the physical environment (rainfall causing stratification of the water column) would result in the development of a community unique to that environment.

2. DESCRIPTION OF SURVEY AREA

Cuff *et al.* (1983) give a detailed description of the sampling area; a map showing all sampling locations is given by Vaudrey *et al.* (1983). Briefly, South West Arm (SWA) is a drowned river valley of about 78 ha with a maximum depth of 20 m. SWA is about 4 km from the ocean, and is connected to it by a shallow channel. Exchange with the ocean is by tidal mixing (Godfrey & Parslow, 1976).

3. METHODS

Salinity and temperature were measured at various depths between the surface and bottom at Stn C (see Fig. 1 of Vaudrey *et al.*, 1983) in SWA approximately weekly between 2 June 1975 and 2 July 1976. Rainfall was measured at South Cronulla, about 4 km from SWA.

Zooplankton sampling for biomass and species abundance estimates was carried out at approximately weekly intervals by vertically hauling a 0.25m² mouth area, 100μm mesh–aperture net from 2m above the bottom to the surface. No flowmeter was used as experience showed that clogging did not occur even under phytoplankton bloom conditions in the 16–18 m vertical haul. Single hauls were taken at each of five additional locations (Vaudrey *et al.*, 1983) and mixed together. The composite sample was returned alive to the laboratory, filtered onto a preweighed 100μm–mesh metal screen, rinsed briefly with distilled water to remove salt and dried at 60°C to a constant weight. Ctenophores and medusae were separated from the samples in the field, combined together as a unit, and dried to a constant weight in preweighed glass dishes. An additional sample from Stn C was collected and preserved for counting. Subsamples obtained using the Folsom splitter were counted using procedures described by Griffiths *et al.* (unpublished). Copepods were identified using keys in Tafe (1979) and other taxa using Dakin & Colefax (1940) and Gosner (1971). Some specimens could not be identified to species; higher taxa were used instead. *Oithona similis* could be separated easily from a group of seven other *Oithona* species, of which *O. simplex* was the most abundant. The other species were *O. brevicornis, O. plumifera, O. atlantica, O. fallax, O. nana,* and *O. rigida.* Very few adult chaetognaths were found; they were identified as *Sagitta* near *neglecta* . Other abundant taxa that could not be subdivided were the polychaete larvae, larval medusae, bivalve veligers, gastropod veligers and copepod nauplii.

The polythetic agglomerative program Centperc (Dale *et al.*, 1971) available from the Taxon library of the CSIRO Division of Computing Research, Canberra, employing the information statistic ($\triangle$I) was used to classify or "group" the 52 samples. No transformations were applied to the numerical data before analysis. Species contributing most to the differences between groupings resulting from the analysis were determined using the diagnostic program Grouper (Lance *et al.*, 1968). The sample groupings were considered to represent communities if they fulfilled the criteria of Mills (1969) *viz.* "a group of organisms occurring together in a particular environment, presumably interacting with one another, and with the environment, and separable by means of ecological surveys from other groups". These groupings were further examined using the following community measures: Shannon–Weaver diversity (Shannon & Weaver, 1963); equitability (Pielou, 1966), species richness, fidelity and dominance (Fager, 1963); biomass and comparative abundance. Differences in these

elements of community structure were tested for significance using non–parametric
Mann–Whitney tests (Conover, 1971).

4. RESULTS

4.1 Environment

Marine and estuarine periods were designated using water column mean salinity
and density stratification at Stn C. SWA was considered marine when water column
mean salinities were $34.8 \cdot 10^{-3}$ or greater, and the density difference between surface
and bottom was less than 2 σ_t units, and estuarine when the water column mean
salinity was less than $34.8 \cdot 10^{-3}$, and the density difference between surface and
bottom exceeded 2 σ_t units. Light, infrequent rainfall had little effect beyond a
transient lowering of the surface salinity. More than 20 mm of rain in a day plus
follow–up rain (5–10 mm day^{-1}) resulted in a marked reduction in salinity and the
formation of a persistent halocline. SWA was in a marine state between July and
December 1975, and between May and June 1976, and estuarine periods were found
in June 1975, and between January and April 1976 (Fig. 1).

There was a regular seasonal temperature cycle in SWA, with mean water column
temperatures ranging from 14.2°C in August to 23.0°C in February (Fig. 1). These
follow the same seasonal pattern as nearby ocean temperatures, but are about 2°C
colder in winter and 2°C warmer in summer than surface water temperatures at the
CSIRO 50 m reference station 6 km east of Cronulla (CSIRO, unpublished data). The
seasonal temperature changes contributed less to stratification in SWA than did the
unpredictable rainfall.

4.2 Community differentiation

The dendrogram resulting from the Cenperc analysis of the 52 samples over the
one–year period showed three main sample groupings (Fig. 2). There were only three
samples in Group 99; detailed inspection of the species composition showed that these
three samples contained a combination of rare taxa *(Gladioferens pectinatus, Isais
uncipes,* penaeid larvae, thalassinid larvae and acantharians) that were not present in
other samples. These samples were assigned to one of the other two main sample
Groups (101 or 100, Fig. 2) using similarities of the dominant species, diversity and
species richness values.

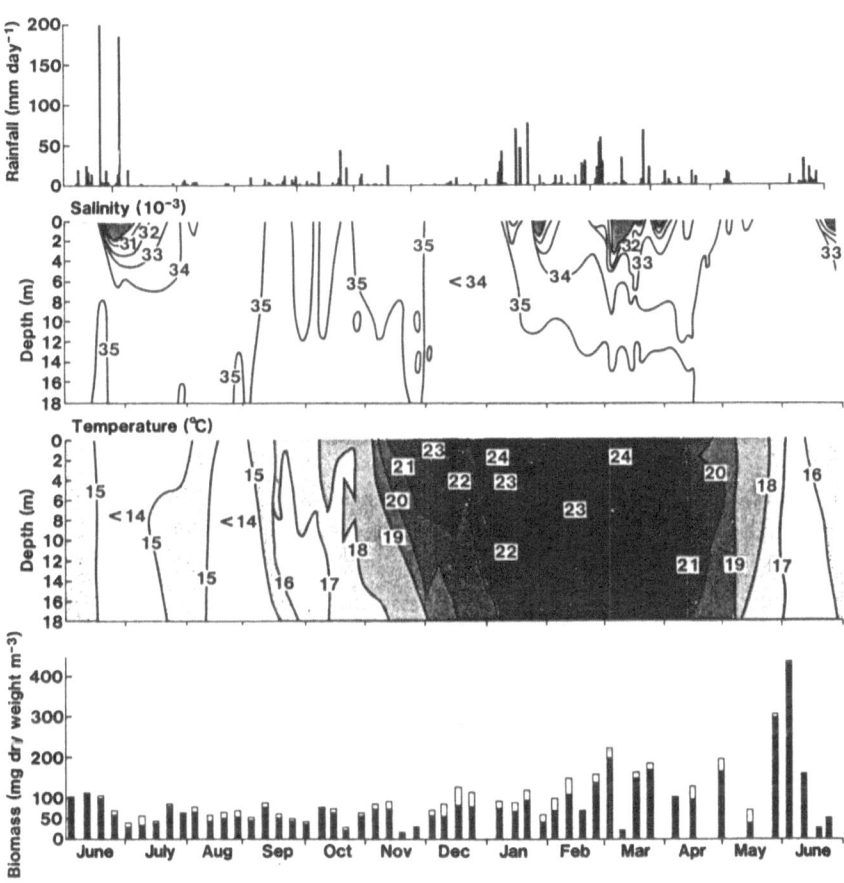

Fig. 1: Rainfall (mm day⁻¹), salinity (10⁻³), temperature (°C), zooplankton (solid, mg m⁻³), and ctenophore/medusae dry weight (open, mg m⁻³) in South West Arm between 1 June 1975 and 2 July 1976. Rainfall is plotted on a daily basis, while salinity and temperature profiles were made at approximately weekly intervals on the same day as biomass samples were taken.

The samples in Group 101, plus sample 11 from Group 99, fall mainly in the winter–spring–summer period, and when SWA was a marine embayment. This group of samples is called the "marine affiliated group". The samples in Group 100, plus samples 34 and 42 from Group 99, occur mainly in the summer–autumn–early winter period, and mainly when SWA was an estuary; it has been called the "estuarine affiliated group". The sample numbers and sampling dates for samples included in the marine affiliated group and the estuarine affiliated group as well as the total abundance, taxon richness, diversity, equitability, and dominant taxa for each sample are given in Tables 1a and 1b. The total abundance per sample was significantly lower; but the taxon richness, diversity, and equitability were significantly higher in the marine affiliated group than in the estuarine affiliated group (Mann–Whitney test, p<0.001).

In the marine affiliated group, 19 samples were consistent enough in time to fit the definition of a community adopted earlier. This community, comprising samples 8–26, occupied SWA between 21 July and 24 November 1975, when SWA was a marine embayment, and has been designated the marine community. Twenty one samples in the estuarine affiliated group (1–4, 35–39, and 42–52, taken 2 June to 23 June 1975, 2 February to 1 March 1976, and 22 March to 2 July 1976, respectively) are considered to represent a functional "estuarine community". There was significantly lower abundance, but significantly higher taxon richness, diversity, and equitability in the marine community than in the estuarine community. The marine community had 2–5 taxa co–dominant, and the taxon order was not consistent amongst the dominants, while in the estuarine community *Oithona* spp. was the dominant taxon in all but three samples, and co–dominance occured in only six samples.

A taxon list is given in Table 2, and the species found in SWA are arranged in descending order of their contribution to the difference between the two communities. The mean abundance for each taxon in each period is given, along with the community affiliation of species contributing more than 3% to the difference between communities. The first 15 taxa contributed 77.4% of the difference between the two communities. The mean abundances of 9 of these 15 taxa differ by a factor of two, and 4 taxa show large differences in fidelity between two communities. The Table emphasizes that few taxa were restricted to one time period throughout the year, or to one community, and highlights the importance of abundance differences in distinguishing the communities.

Samples 5, 29, 31 and 33 in the marine affiliated group, and samples 6, 7, 27, 28, 30, 32 and 34 in the estuarine affiliated group are considered to be representative of transitional periods between the marine and estuarine communities because of the alternation of samples between groups due to the varying abundance of marine affiliated species which were persisting in SWA. Samples 40 and 41 were classified with the marine community because they contained *Pseudodiaptomus colefaxi* and

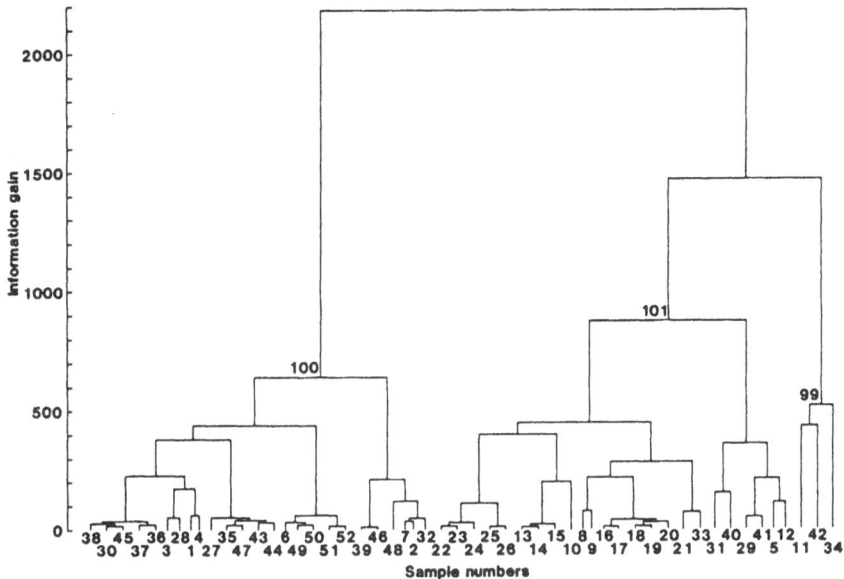

Fig. 2: Dendrogram resulting from the Centperc analysis of the 52 zooplankton samples taken between 1 June 1975 and 2 July 1976. The sample numbers are given on the x-axis, and sampling dates can be obtained by reference to Table 1a and 1b. Sample group 100 represents the estuarine-affiliated group, 101 the marine-affiliated group; and 99 represents three samples containing rare species.

pluteus larvae, two taxa normally diagnostic of the marine community. The total abundance, species richness, diversity, equitability and dominant species results suggest very strong estuarine affiliations, and these two samples may be misclassified.

4.3 Species abundance

Thirty five taxa were identified in the samples (Table 2). *Oithona* spp., with a yearly mean abundance of 23 862 m^{-3} were about four times more abundant than the second most abundant species, *Penilia avirostris,* and about an order of magnitude more abundant than *Temora turbinata* (3218 m^{-3}).

The abundance patterns of the species most characteristic of the marine and of the estuarine groups are shown in Fig. 3. These patterns show that the marine species are largely restricted to the spring period, while species diagnostic of the estuarine community have comparatively reduced abundances during the marine period, but they are generally present throughout the year.

Table 1.

Details of the two communities separated in the classification analysis.

The dominant species in each sample were determined by ordering each species in each sample by abundance and summing the percentage composition of each sample starting from the most abundant species, until the number of individuals exceeded 50% of the total in the sample. Species included in the group were called the dominants. The taxa are indicated by the following codes: B *Bestiola similis*, E *Evadne nordmanni*, N nauplii, O *Oithona* spp. (mainly *O.*

a: Marine community

Sample Number	Date	Abun[s]	Spec[1]	Div[t]	Equ[i]	Dominant Species
5	30.6.75	15 949	18	1.82	0.43	O(68)
8	21.7.75	13 426	16	2.41	0.60	O(49) T(16)
9	28.7.75	25 460	18	3.18	0.76	O(20) N(17) T(14) P(13) B(13)
10	4.8.75	32 960	18	2.90	0.70	O(29) T(24)
11	11.8.75	28 240	17	2.90	0.71	T(22) O(21) B(18)
12	18.8.75	17 216	16	2.62	0.66	B(31) T(28)
13	25.8.75	19 840	18	3.27	0.78	O(20) N(19) B(15)
14	1.9.75	14 656	21	3.38	0.77	T(16) Oi(16) O(15) B(12)
15	8.9.75	18 112	20	3.52	0.81	T(14) O(14) N(14) P(12)
16	15.9.75	20 106	20	3.62	0.84	Oi(15) N(13) P(12) E(12)
17	22.9.75	15 094	20	3.54	0.82	B(23) Oi(16) P(11) E(9)
18	29.9.75	9 952	18	3.37	0.81	P(19) B(19) Oi(15) Pn(15)
19	7.10.75	15 168	20	3.25	0.75	Pn(27) Oi(17) P(11)
20	13.10.75	16 192	22	3.41	0.77	Pn(26) P(14) Oi(12)
21	20.10.75	24 382	17	3.33	0.82	Bi(17) Oi(17) Pn(15) O(11)
22	27.10.75	17 408	18	2.82	0.68	Pn(46) Oi(10)
23	3.11.75	28 392	19	3.22	0.76	Pn(22) Bi(18) Oi(17)
24	10.11.75	13 824	18	3.40	0.82	Pn(26) Oi(12) Bi(11) Po(11)
25	17.11.75	13 312	18	3.20	0.77	O(31) P(16) Pn(15)
26	24.11.75	6 336	19	3.36	0.79	O(28) Pn(16) Bi(12)
29	15.12.75	93 400	16	2.25	0.56	O(53)
31	5.1.76	50 944	15	1.34	0.34	O(79)
33	19.1.76	74 880	18	1.76	0.42	O(70)
40	8.3.76	33 536	14	1.74	0.46	O(74)
41	15.3.76	90 240	17	2.57	0.63	O(46) P(22)

[s] Total Abundance; [1] Species Richness ; [t] Shannon–Weaver Diversity Index; [i] Pielou Equitability Index

simplex), Oi *Oikopleura dioica*, P *Paracalanus crassirostris*, Pn *Penilia avirostris*, Po *Podon polyphemoides*, T *Temora turbinata*. The percentage abundance of each dominant species is given in parentheses. Samples 5 – 8 are considered transitional between the estuarine and marine communities in July 1975, and samples 27 – 33 transitional between the marine and estuarine communities (December 1975 – January 1976).

b: Estuarine community

Sample Number	Date	Abun[s]	Spec[s]	Div[t]	Equ[i]	Dominant Species
1	2.6.75	67 840	14	1.71	0.45	O(69)
2	9.6.75	55 936	14	1.79	0.47	O(68)
3	16.6.75	51 520	17	1.93	0.47	O(63)
4	23.6.75	28 674	13	1.76	0.48	O(67)
6	7.7.75	30 062	15	1.27	0.33	O(82)
7	14.7.75	12 767	16	2.30	0.58	O(55)
27	2.12.75	48 600	19	3.22	0.76	O(25) Pn(21) B(12)
28	8.12.75	60 400	17	2.50	0.61	O(52)
30	22.12.75	33 984	12	1.88	0.52	O(67)
32	13.1.76	34 048	15	2.21	0.57	O(59)
34	27.1.76	53 248	13	2.06	0.56	O(64)
35	2.2.76	80 374	15	1.83	0.47	O(69)
36	9.2.76	132 389	15	1.65	0.42	O(75)
37	16.2.76	134 784	16	2.05	0.51	O(63)
38	23.2.76	84 352	14	2.43	0.64	O(52)
39	1.3.76	75 246	15	1.75	0.45	Pn(57)
42	22.3.76	66 048	15	2.41	0.62	O(54)
43	4.4.76	54 400	17	2.34	0.57	O(56)
44	13.4.76	47 920	17	1.99	0.49	O(66)
45	28.4.76	94 080	14	2.06	0.54	O(58)
46	5.5.76	96 704	15	1.74	0.45	O(58)
47	26.5.76	53 888	15	2.30	0.59	O(45) P(28)
48	1.6.76	100 736	16	2.41	0.60	O(58)
48	10.6.76	38 720	15	2.75	0.70	O(36) T(23)
50	18.6.76	24 832	13	2.62	0.71	O(49) P(10)
51	23.6.76	8 512	14	2.66	0.70	N(33) O(29)
52	2.7.76	11 325	13	2.95	0.80	P(21) O(21) N(21)

[s] Total Abundance; [s] Species Richness; [t] Shannon–Weaver Diversity Index; [i] Pielou Equitability Index

Table 2.
Taxon list for animals found in South West Arm, Port Hacking.

Taxa are listed in order of their contribution to the differences between the marine affiliated (mar.) and estuarine affiliated (est.) groups of samples as determined by the diagnostic program Grouper (see text).

Taxon	% ctb§	aff¶	mean abundance no m⁻³ mar.	est.	fidelity mar.	est.
Oithona spp.	8.8	e	12312	34857	1.0	1.0
Penilia avirostris Dana	7.3	e	2006	8060	1.0	0.93
Pluteus larvae	6.1	m	107	57	0.58	0.09
Bivalve veligers	6.1	e	1156	1235	0.96	1.0
Paracalanus crassirostris Dahl	5.7	e	2604	3471	1.0	1.0
Pseudodiaptomus colefaxi Bayly	5.6	m	148	0	0.58	0
Temora turbinata Dana	5.3	e	2599	3786	0.89	1.0
Oithona similis Claus	4.8	e	396	853	0.95	0.91
Gastropod veligers	4.7	e	310	972	0.89	1.0
Oikopleura dioica Fol	4.4	m	1440	558	1.0	0.88
Podon polyphemoides Leuckart	4.2	m	165	0	0.48	0
Polychaete larvae	4.0	m	539	475	0.89	0.91
Sagitta near *neglecta*	3.7	e	235	410	0.95	0.97
Evadne nordmanni Lovén	3.4	m	358	14	0.44	0.15
Medusa larvae	3.3	e	79	120	0.84	0.61
Harpacticoids	2.9		51	79	0.60	0.52
Tortanus barbatus	2.8		64	14	0.56	0.11
Copepod nauplii	2.5		1694	1964	1.0	0.93
Bestiola similis (Sewell)	2.2		2010	2043	0.96	0.96
Bipinnaria larvae	1.5		11	58	0.16	0.19
Fish eggs	1.5		76	83	0.56	0.63
Acartia tranteri Bradford	1.4		807	980	0.84	0.78
Carid larvae	1.3		42	17	0.36	0.19
Zoea	0.9		3	9	0.04	0.07
Oncaea spp.	0.9		17	21	0.12	0.07
Gladioferens pectinatus Brady	0.9		207	33	0.08	0.04
Isias uncipes Bayly	0.9		3	24	0.08	0.04
Ostracods	0.3		0	2	0	0.04
Lucifer hanseni Nobili	0.3		0	1	0	0.04
Gippslandia estuarina Bayly & Arnott	0.3		0	5	0	0.04
Isopods	0.3		3	0	0.04	0
Penaeid larvae	0.2		0	9	0	0.04
Thalassinid larvae	0.3		20	0	0.04	0
Acantharians	0.3		0	9	0	0.04
Pilidium larvae	0.3		28	0	0.08	0
Copepodites unidentified	0.3		5	0	0.04	0

§ % contribution to difference between communities; ¶ Affinity; e = estuarine; m = marine

4.4 Biomass

The zooplankton and ctenophore/medusae biomass results are shown in Fig. 1. Zooplankton biomass in the estuarine affiliated group was significantly greater than that in the marine affiliated group (105.5 mg m^{-3} v. 55.7 mg m^{-3}, Mann–Whitney test, $p<0.001$). The coefficient of variation in the estuarine affiliated group was higher than that in the marine affiliated group. There was no significant difference in ctenophore/medusae biomass between the estuarine and marine affiliated groups (15.5 mg m^{-3} and 10.3 mg m^{-3} respectively).

5. DISCUSSION

The three main limitations to this study of zooplankton communities in South West Arm are that it was of a single year's duration, the sampling plan adopted, and the use of taxons instead of the identification of all animals to species. The repeatability of seasonal succession patterns of different species in SWA cannot be established on a single year's data, particularly as rainfall is unpredictable. Program constraints limited sampling to a single vertical net haul from one station. Although the sampling was generally representative of species richness in SWA, abundance estimates were less representative (Griffiths, unpublished data), probably because the vertical haul would not have integrated any horizontal patchiness in SWA. Good community ecology is dependent on good taxonomy, and our inability to routinely separate complexes like *Oithona* spp. detracts somewhat from the results. It was too time consuming to dissect free and examine the diagnostic characters of nearly equal sized *Oithona* species, and the differences between juveniles of the different species are not known. No keys were available to separate larval medusae, bivalve and gastropod veligers, all crustacean larvae, polychaete larvae and pluteus larvae. In spite of the limitations, the patterns in the data are sufficiently reproducible to allow an examination of factors which may influence community structure.

One feature of the multivariable analysis is that taxa were diagnostic of a community by virtue of their abundance patterns, and a reduction in a taxon's abundance (e.g. *Oithona* spp. in September/October) is as diagnostic of a community as the presence or absence of a taxon (*e.g. Podon polyphemoides* in summer). The diagnostic taxa are not necessarily the dominants in the community. For example, the dominant taxa in the marine community (Table 1a) are often taxa that are much more abundant during the estuarine period (Table 2), but are not dominant in that period because of the abundance of *Oithona* spp. The marine community is a group of samples, consecutive in time, with similar taxon composition, abundance patterns and community measures. The estuarine community has a slightly different taxon

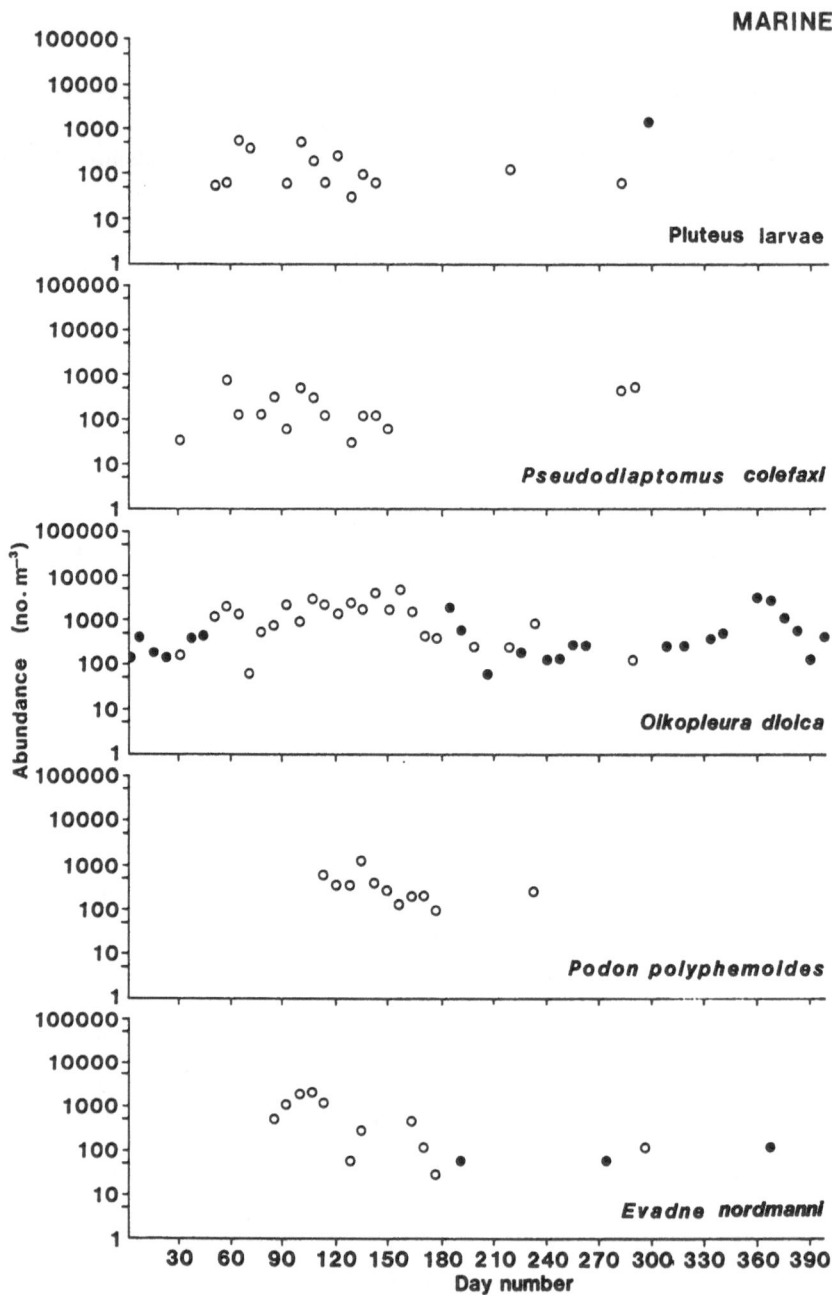

Fig. 3: Abundance patterns of the five species most characteristic of the marine and estuarine communities. Solid symbols show those samples grouped with the marine community and open symbols those grouped with the estuarine community.

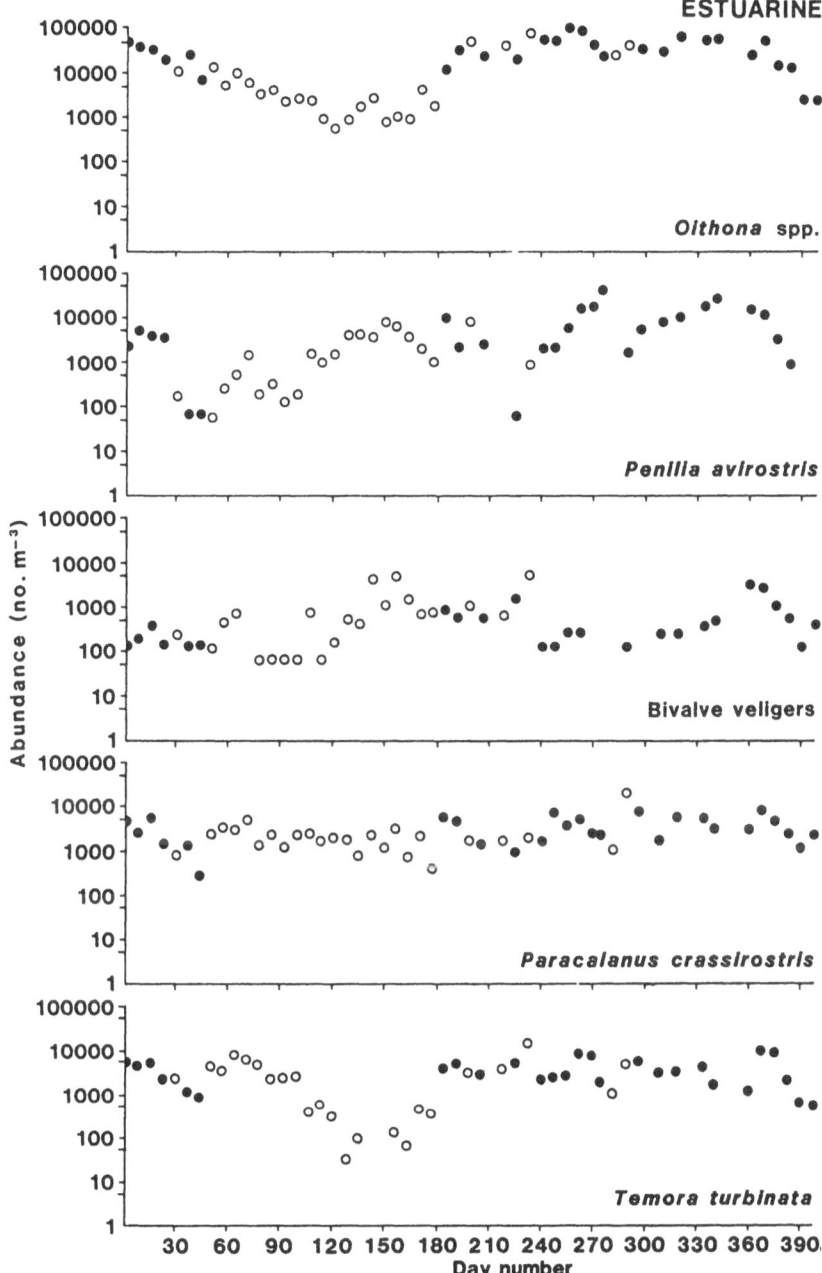

composition, but similar abundance patterns and community measures. There are significant differences in the abundance patterns and community measures between the two communities.

Succession occurs at two levels: seasonal succession, describing the seasonal abundance patterns of particular taxa, and community succession, which is the orderly, progressive sequence of replacement of communities. There was obvious seasonal succession in the appearance and disappearance of *Podon polyphemoides,* and seasonal abundance changes in *Oithona* spp., *Penilia avirostris,* bivalve veligers, *Paracalanus crassirostris, Pseudodiaptomus colefaxi, Temora turbinata, Oikopleura dioica* and others (see Fig. 3). These succession patterns are similar to those found by Kott (1955), Harper (1972) and Hodgson (1979) in Lake Macquarie, although several differences were found. Kott (1955) and Hodgson (1979) found that *Oikopleura dioica* was primarily a summer species in Lake Macquarie, while in SWA it was a spring, marine species. *Temora turbinata* was abundant from October to May in Lake Macquarie while in SWA it was most abundant between January and May, in spite of it being one of the dominant species in the marine community. Hodgson found that bivalve veligers were abundant only in December, and concluded that spawning by bivalves in Lake Macquarie was erratic. In SWA, bivalve veligers were present all year, but abundant only in January–March. Dakin *et al.* (1952) suggested that bivalves spawn at any time of the year, with a peak between December and April. The successional patterns seen in SWA in the winter–spring–early summer period, between late July and December, closely resemble those found by Dakin & Colefax (1940) for nearby neritic areas, but the pattern in SWA during the estuarine period differs with the lack of clear–cut succession patterns. This estuarine period in SWA also differs from the pattern of continuous succession seen in Lake Macquarie by Hodgson (1979). In SWA, then, during the marine period the co–dominance, the succession of the dominant species and the species abundance patterns are similar to those seen in the nearby ocean, and in Lake Macquarie. By contrast, in the estuarine period in SWA, there is little co–dominance, and the abundance patterns of a number of species show a broad plateau, suggesting that the community present is able to exploit the fluctuating environment without succession occuring. The patterns of species succession, co–dominance, high diversity and high species richness suggest that the marine community is more complex than the estuarine community. The estuarine community contains euryhaline species, and these species form a background fabric of species in the marine community. The marine community has stenohaline species as well that apparently cannot survive in SWA during estuarine periods.

Parker (1983) has shown that salinity stratification of the water column can lead to a change in the species composition and particle size distribution of the protoplankton. A monospecific bloom of *Ceratium furca* persists as long as a double halocline and de–oxygenation of the bottom water is present. This is in addition to the normal succession from a diatom dominated phytoplankton community in the spring to a dinoflagellate dominated community in the summer (Dakin & Colefax, 1940). Tafe (1979) considers *Oithona* spp. to be raptorial herbivores, adapted to feeding on large particles such as dinoflagellates. It can be postulated that zooplankton succession in SWA is occuring in the spring due to a changing food resource, while the lack of succession seen in the summer estuarine conditions is due to a relatively constant food resource, with the community best able to exploit this resource being present.

Predation can also affect community structure (Cronin *et al.*, 1962; Sage & Herman, 1972; Rippingale & Hodgkin, 1974). Hodgson (1979) considers planktonic Cladocera as predators, with *Evadne nordmanni* and *Podon polyphemoides* preying on copepodites, nauplii and other small zooplankters. These species, present in the spring, may be limiting the abundance of *Oithona* spp., *Temora turbinata* and *Acartia tranteri* in the spring, marine period, as these latter species increase in abundance once the Cladocera decrease in abundance. Predation by ctenophores can affect community biomass, but as ctenophore biomass was not significantly different in the two periods, predation effects of this group (mainly *Mnemiopsis* sp.) should be relatively constant through a year.

In summary, the zooplankton in SWA during 1975–1976 forms two communities, one coincident with a marine environment, and a second coincident with an estuarine environment. Seasonal succession occurs in individual taxa during the marine period, but is suppressed during the estuarine period. The marine community may be structured by predation and seasonal succession, while the estuarine community may be structured by food availability and salinity structure of the water column.

ACKNOWLEDGEMENTS

I thank D. Tafe and C. Lanzing for help in counting the zooplankton samples. Special thanks are due to G. Arnott and B. Hodgson for critically reviewing an early draft of the manuscript, and to P. Young for help with computing.

REFERENCES

Allen, K.R.: Introduction to the Port Hacking Estuary Project. In: W.R. Cuff and M. Tomczak jr, eds *Synthesis and Modelling of Intermittent Estuaries.* Berlin, Heidelberg, New York: Springer (1983)

Conover, W.J.: *Practical Nonparametric Statistics.* New York: Wiley (1971)

Cronin, L.E., Daiber, J.C., Hulbert, E.M.: Quantitative seasonal aspects of zooplankton in the Delaware River estuary. *Chesapeake Science* 3, 63–93 (1962)

Cuff, W.R., Sinclair, R.E., Parker, R.R., Tranter, D.J., Bulleid, N.C., Giles, M.S., Godfrey, J.S., Griffiths, F.B., Higgins, H.W., Kirkman, H., Rainer, S.F., Scott, B.D.: A carbon budget for South West Arm, Port Hacking. In: W.R. Cuff and M. Tomczak jr, eds *Synthesis and Modelling of Intermittent Estuaries*. Berlin, Heidelberg, New York: Springer (1983)

Dakin, W.J., Colefax, A.N.: Observations on the seasonal changes in temperature, salinity, phosphates, and nitrate–nitrogen and oxygen of the ocean waters on the continental shelf off New South Wales and the relationship to plankton production. *Proceedings of the Linnean Society of N.S.W.* **60**, 303–314 (1935)

Dakin, W.J. Colefax, A.N.: The plankton of the Australian coastal waters off New South Wales. *Publication of the University of Sydney Department of Zoology Monograph* 1 (1940)

Dakin, W.J., Bennett, I., Pope, E.: *Australian Seashores*. Sydney: Angus and Robertson (1952)

Dale, M.B., Lance, G.N., Albrecht, L.: Extension of information analysis. *Australian Computer Journal* **3**, 29–34 (1971)

Deevey, G.B.: The zooplankton of Tisbury Great Pond. *Bulletin of the Bingham Oceanographic Collection Yale University* **12**, 1–44 (1948)

Fager, E.W.: Communities of organisms. In: M.N. Hill, ed. *The Sea, Vol. 2*. New York: Interscience (1963)

Frolander, H.F., Miller, C.B., Flynn, M.J., Meyers, S.C., Zimmerman, S.T.: Seasonal cycles of abundance in zooplankton populations of Yaquina Bay Oregon. *Marine Biology (Berlin)* **21**, 277–288 (1973)

Godfrey, J.S., Parslow, J.: Description and preliminary theory of circulation of Port Hacking estuary. *CSIRO Division of Fisheries and Oceanography Report* **67** (1976)

Gosner, K.L.: *Guide to Identification of Marine and Estuarine Invertebrates*. New York: Interscience (1971)

Harper, S.: *The Hydrology and some aspects of the Ecology of selected Zooplankton Organisms of Lake Macquarie and the Tuggarah Lakes System*. Honours Thesis, School of Zoology, University of New South Wales (1972)

Hodgson, B.R.: *The Hydrology and Zooplankton Ecology of Lake Macquarie and the Tuggarah Lakes, New South Wales*. Ph.D. thesis, University of New South Wales (1979)

Kott, P.: The zooplankton of Lake Macquarie, 1953–1954. *Australian Journal of Marine and Freshwater Research* **6**, 429–442 (1955)

Lance, G.N., Milne, P.W., Williams, W.I.: Mixed data classificatory programs III: Diagnostic systems. *Australian Computer Journal* **1**, 178–181 (1968)

Mills, E.L.: The community concept in marine zoology, with comments on continuation and instability in some marine communities. *Journal of the Fisheries Research Board of Canada* **26**, 1415–1428 (1969)

Parker, R.R.: Some ecological effects of rainfall on the protoplankton of South West Arm. In: W.R. Cuff and M. Tomczak jr, eds *Synthesis and Modelling of Intermittent Estuaries*. Berlin, Heidelberg, New York: Springer (1983)

Pielou, E.C.: The measurement of diversity in different types of biological collections. *Journal of Theoretical Biology* **13**, 131–144 (1966)

Reeve, M.R.: Feeding of zooplankton, with special reference to some experiments with *Sagitta*. *Nature (London)* **201**, 211–213 (1964)

Reeve, M.R.: Seasonal changes in the zooplankton of south Biscayne Bay and some problems of assessing the effects on the zooplankton of natural and artifical thermal and other fluctuations. *Bulletin of Marine Science* **20**, 894–921 (1970)

Rippingale, R.J., Hodgkin, E.P.: Predation effects on the distribution of a copepod. *Australian Journal of Marine and Freshwater Research* **25**, 81–91 (1974)

Sage, L.E., Herman, S.S.: Zooplankton of the Sandy Hook Bay area, N.J. *Chesapeake Science* **13**, 29–39 (1972)

Shannon, C.E., Weaver, W.: *The Mathematical Theory of Communication*. Urbana: University of Illinois Press (1963)

Tafe, D.J.: *Copepod Feeding in Port Hacking Estuary*. M.Sc. Thesis, The University of Sydney (1979)

Vaudrey, D.J., Griffiths, F.B., Sinclair, R.E.: Data base for the Port Hacking Estuary Project: Parameters, monitoring procedure, and management system. In: W.R. Cuff and M. Tomczak jr, eds *Synthesis and Modelling of Intermittent Estuaries*. Berlin, Heidelberg, New York: Springer (1983)

Wickstead, J.H.: A survey of the larger zooplankton of Singapore Straits. *Journal du Conseil International pour l'Exploration de la Mer* **23**, 340–353 (1957)

Woodmansee, R.A.: The seasonal distribution of zooplankton of Chicken Key in Biscayne Bay, Florida. *Ecology* **39**, 247–262 (1958)

Youngbluth, M.J.: Zooplankton populations in a polluted, tropical, embayment. *Estuarine and Coastal Marine Science* **4**, 481–496 (1976)

Youngbluth, M.J.: Daily, seasonal and annual fluctuations among zooplankton populations in an unpolluted tropical embayment. *Estuarine and Coastal Marine Science* **10**, 265–287 (1980)

Synthesis and Modelling of Intermittent Estuaries
(W.R. Cuff and M. Tomczak jr. eds) Berlin, Heidelberg,
New York: Springer (1983), pp. 109–133.

Seasonal Abundance, Geographical Distribution and Feeding Types of the Copepod Species Dominant in Port Hacking, New South Wales

Dennis J. Tafe, F. Brian Griffiths

Division of Fisheries Research
CSIRO Marine Laboratories
P.O. Box 21, Cronulla, N.S.W. 2230, Australia

Summary. Zooplankton samples were collected over a one–year period by net haul and light trap at four locations in Port Hacking Estuary. Copepods accounted for approximately three quarters of the total zooplankton numbers taken on a yearly basis and 94% of this fraction was composed of eleven copepod species. The seasonal abundance and geographical distribution of these eleven species were recorded and their gut and faecal pellet contents examined. *Acartia tranteri, Bestiola similis, Gladioferens pectinatus, Oithona brevicornis, O. plumifera, O. simplex, Paracalanus crassirostris, P. indicus* and *Temora turbinata* were classified as herbivores, *Acartia bispinosa* and *Tortanus barbatus* as omnivores.

Key words: copepods, seasonal abundance, distribution, feeding, taxonomy, Port Hacking, South West Arm

INTRODUCTION

This study was carried out in order to provide specific information to the Port Hacking Estuary Project (Parker *et al.*, 1983) on the copepod fraction of the zooplankton. Since copepods constitute a large fraction of the total zooplankton numbers of the area (Griffiths, 1983), such information was considered to provide a useful link in the understanding of the estuarine foodweb. The seasonal abundance and geographical distribution of marine and estuarine copepods of Australian waters are not well known, the most extensive work having been done by Dakin & Colefax (1940) on plankton samples taken off the N.S.W. coast. Other work dealing with the distribution and abundance of Australian neritic and estuarine copepods includes Farran (1913, 1936, 1949), Russell & Coleman (1934), Nicholls (1944), Kott (1955, 1957), Tranter (1962), Bayly (1965, 1973), Arnott (1974), Nyan Taw (1975, 1978), Greenwood (1976, 1977, 1978, 1979, 1980, 1982a, 1982b), Alldredge & King (1977), Nyan Taw & Ritz (1978, 1979), Kennedy (1978) and Hodgson (1979). Australian work which relates to copepod feeding is limited to a few recent papers by Hodgkin & Rippingale (1971), Arnott (1974), Rippingale & Hodgkin (1974, 1977), Smith *et al.* (1979), Griffiths & Caperon (1979), Hodgson (1979), Tafe (1979) and Tranter *et al.* (1981). Previous knowledge of copepod feeding types was based mostly on feeding experiments carried out in other parts of the world. A comprehensive review of copepod feeding, outlining the major worldwide contributions, has been compiled by Marshall (1973). The bulk of the work contained in the review relates to only a few species, the most common of which is *Calanus finmarchicus*. Many other species are assumed to have feeding habits similar to those of the species which have been studied but without supporting evidence there remains a large element of uncertainty. Some of the general assumptions concerning feeding in copepods now appears very doubtful in the light of recent work by Alcaraz *et al.* (1980).

Analysis of the gut contents of preserved animals shows what has been ingested prior to capture and often the prey items can be identified to species. Beklemishev (1954), Marshall & Orr (1962) and Mullin (1966) were able to identify species of diatoms in the guts of calanoid copepods. Beklemishev (1954) found that diatoms, radiolarian fragments, and to a lesser extent crustacean appendages constituted the major portion of the recognizable food remains. However most copepods had empty guts when examined, either because they had voided their gut contents or because they had not fed recently. Even when food is found in the gut many types of food organisms, such as naked flagellates, leave no recognizable remains. Likewise faecal pellet analysis can be employed usefully only where identifiable remains are present. Honjo & Roman (1978) collected faecal pellets from laboratory cultures of *Calanus* and *Acartia* and were able to identify diatoms and coccolithophores. Marshall & Orr (1955) have used the sizes of identifiable remains in faecal pellets to estimate the sizes of food particles consumed.

An artifact which should be taken into account when associating the gut contents with particles ingested *in situ* is the possibility of ingestion taking place between capture and fixation of the sample. At this time the plankton is much more highly concentrated than usual and so the availability of some prey is subject to change. Material in the faecal pellets is not as likely to be influenced by the artificial situation which is set up during capture and therefore a comparision of gut and faecal contents is useful.

In the following examination of gut and faecal pellet contents the recognizable remains of material ingested in the natural environment by individuals of selected copepod species were identified and measured. Eleven species of copepods were classified as omnivore or herbivore on the basis of the constituents of their guts and faecal pellets. The seasonal abundance of these species was determined from a series of vertical net hauls taken regularly at two stations over a one–year period and their world wide geographical distribution was recorded from the available literature.

2. METHODS

2.1 Seasonal abundance patterns

Plankton samples were collected at approximately weekly intervals at Stn C in South West Arm (SWA) of Port Hacking (see Fig. 1 in Vaudrey *et al.*, 1983) between 2 June 1975 and 26 May 1976 by vertically hauling a 100–µm mesh net of 0.25 m² mouth area, from 1 m above the bottom (16 – 18 m depth) to the surface. The samples were fixed in Steedmans's Fixative (Steedman, 1976), subsampled using a Folsom Splitter (F.B. Griffiths, unpublished) and counted under a low–power stereomicroscope fitted with a dark–field condenser (Heron, 1969) in order to estimate total zooplankton and copepod abundance.

2.2 Gut and faecal pellet analysis

Freshly caught specimens were used for gut and faecal pellet analysis. They were collected by light trap (Tranter *et al.*, 1981) at Stn Gg (Vaudrey *et al.*, 1983), at a station about 200 m northwest of Stn Gg (in Gunnamatta Bay) and from the CSIRO Marine Laboratories' pool, between 1800 and 2100 hr in September 1978, and narcotized by adding drops of 96% ethyl alcohol or by bubbling carbon dioxide gas through the water. The guts and faecal pellets were dissected from adult females of the selected species, macerated on microscope slides, examined and photographed using interference contrast microscopy at magnifications of 600x and 1000x. The number of guts and faecal pellets examined for each species was recorded, along with the time of sampling.

Water samples were collected using Jitts bottles at the same time and location as zooplankton sampling in order to determine the species of phytoplankton present. The most abundant species were photographed for comparison with the cellular remains of phytoplankton found in the guts and faecal pellets.

Table 1.
List of species.

1	*Acartia bispinosa* Carl	b	*Oithona brevicornis* Giesbrecht *f.* new
2	*Acartia tranteri* Bradford	c	*Oithona brevicornis* Giesbrecht *f.* new
3	*Bestiola similis* (Sewell)	24	*Oithona fallax* Farran
4	*Calanopia elliptica* Dana	25	*Oithona nana* Giesbrecht
5	*Centropages furcatus* Dana	26	*Oithona plumifera* Baird
6	*Centropages orsinii* Giesbrecht	27	*Oithona rigida* Giesbrecht
7	*Clausocalanus arcuicornis* Dana	28	*Oithona similis* Claus
8	*Clytemnestra scutellata* Dana	29	*Oithona simplex* Farran
9	*Corycaeus (Ditrichocorycaeus) anglicus* Lubbock	30	*Oncaea media* Giesbrecht
		31	*Oncaea venusta* Phillippi
10	*Corycaeus (Corycaeus) clausi* F. Dahl	32	*Paracalanus aculoeatus* Giesbrecht *f. major* Sewell
11	*Corycaeus (Onychocorycaeus) giesbrechti* F. Dahl	a	*Paracalanus aculoeatus* Giesbrecht *f. minor* Sewell
12	*Corychaeus (Urocorycaeus) lautus* Dana	33	*Paracalanus crassirostris* F. Dahl *f. cochinensis* Wellershaus
13	*Euterpina acutifrons* Dana	a	*Paracalanus crassirostris* F. Dahl *f. typica* Fruchtl
14	*Gippslandia estuarina* Bayly & Arnott	34	*Paracalanus indicus* Wolfenden
15	*Gladioferens pectinatus* (Brady)	35	*Pseudodiaptomus colefaxi* Bayly
16	*Labidocera acutum* Dana	36	*Pseudodiaptomus* sp. 1
17	*Labidocera detruncatum* Dana Sydney variety Dakin & Colefax	37	*Pseudodiaptomus* sp. 2
18	*Macrosetella gracilis* Dana	38	*Sapphirina gemma* Dana
19	*Mecynocera clausi* Thompson	39	*Sapphirina auronitens* Claus
20	*Microsetella rosea* Dana	40	*Temora discaudata* Giesbrecht
21	*Oithona atlantica* Farran	41	*Temora turbinata* Dana
22	*Oithona attenuata* Farran	42	*Thaumaleus longisinosus* Bourne
23	*Oithona brevicornis* Giesbrecht *f. aruensis* Fruchtl	43	*Tortanus barbatus* (Brady)
a	*Oithona brevicornis* Giesbrecht *f. typic* Wellerhaus	44	*Mesocalanus vulgaris* Dana

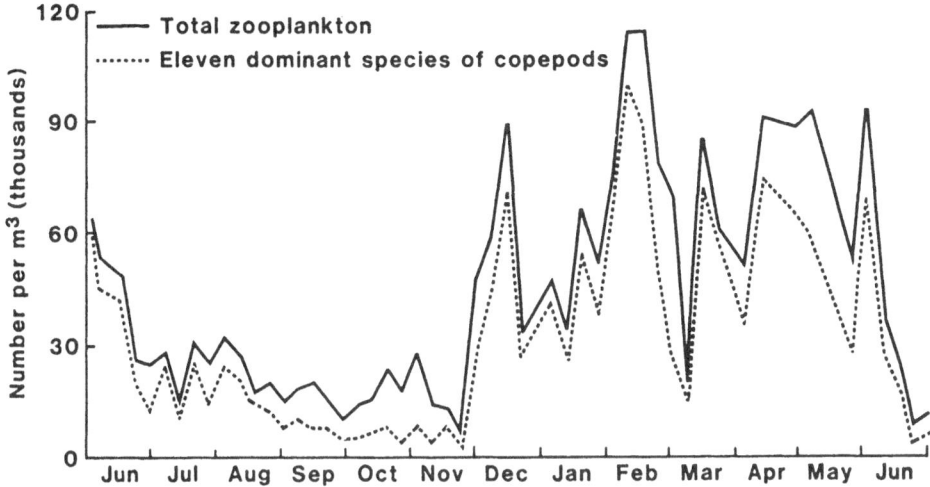

Fig. 1: Abundance of total zooplankton and of the eleven dominant species of copepods at Station C, Port Hacking Estuary between June 1975 and June 1976.

Voucher specimens of all species recorded during the period of the study are lodged in the Copepod Reference Collection of the CSIRO Marine Laboratories.

3. RESULTS

The results are divided into three Sections: a seasonal abundance summary, a feeding types summary and a section giving details on the seasonal abundance, geographical distribution and gut and faecal pellet analysis of the 11 most abundant copepod species found in Port Hacking.

3.1 Seasonal abundance

In all, 45 species of copepods were found in samples collected by net and light trap in Port Hacking (Table 1). The 11 species listed in Table 2 accounted for 94.3% of the total copepod numbers caught at Stn C. The total zooplankton abundance, and the abundance of these 11 species at approximately weekly intervals between June 1975 and June 1976 is shown in Fig. 1. There was a gradual decline in zooplankton numbers through winter and spring to a minimum in November. Zooplankton abundance increased sharply in December and remained high although variable until June 1976, when the winter decrease recommenced.

Variations in the abundance of *Oithona* spp. were responsible for most of the variation in total zooplankton numbers. Eight species of *Oithona* were recorded from Port Hacking, namely *Oithona simplex, O. brevicornis, O. plumifera, O. similis, O. fallax, O. rigida, O. attenuata* and *O. atlantica*. However, only the first two species were found in very large numbers and it is these which constitute the bulk of the *Oithona* count. Samples caught by light trap cannot be quantified for between–night comparisons, but during September *Gladioferens pectinatus* was the most abundant copepod in these samples.

Table 2.

Abundance of the dominant copepods of Port Hacking Estuary, expressed as a percentage of the total number of copepods collected over one year.

	Species	Abundance (%)
1	*Acartia bispinosa*	2.5
2	*Acartia tranteri*	
3	*Bestiola similis*	6.1
4	*Gladioferens pectinatus*	0.2
5	*Oithona brevicornis*	
6	*Oithona plumifera*	66.4
7	*Oithona simplex*	
8	*Paracalanus crassirostris*	8.8
9	*Paracalanus indicus*	
10	*Temora turbinata*	10.1
11	*Tortanus barbatus*	0.2
	total	94.3

The dominant copepod genera for each month and their percentage occurrences are listed in Table 3. *Oithona (Oith.)* was by far the most commonly found genus, accounting for between 8.9% (1500 individuals/m^3) and 67.2% (70600 individuals/m^3) of the total zooplankton numbers in the months of October 1975 and February 1976 respectively. *Temora (Tem.)* was the next most commonly found genus, taken on a yearly mean, closely followed by *Paracalanus (Parac.), Bestiola (Best.)* and *Acartia*

(Acar.). During June 1975 these five genera were together responsible for 88.8% of the total zooplankton count. Copepod nauplii (c.n.) of a variety of genera were also common at certain times of the year.

Table 3.

The dominant copepod genera recorded for each month in Port Hacking Estuary between June 1975 and June 1976, and their relative abundance, expressed as a percentage of total zooplankton numbers.

Period	Dominant genera and relative abundance (%)									Total (%)
6/75	*Oith.*	67.2	*Tem.*	9.0	*Parac.*	6.8	*Best.*	5.3	*Acar.*	0.5	88.8
7/75	*Oith.*	52.6	*Tem.*	10.8	*Best.*	9.9	*Parac.*	7.8	c.n.	7.1	88.2
8/75	*Oith.*	24.0	*Tem.*	21.1	*Best.*	18.5	*Parac.*	11.5	c.n.	10.0	85.1
9/75	*Best.*	13.1	*Oith.*	12.8	*Parac.*	11.7	c.n.	8.9	*Tem.*	7.7	54.2
10/75	*Best.*	8.9	*Oith.*	8.9	*Parac.*	8.1	*Acar.*	7.0	c.n.	5.2	38.1
11/75	*Oith.*	14.3	*Parac.*	9.9	*Best.*	8.5	*Tem.*	2.8	c.n.	3.4	38.9
12/75	*Oith.*	51.4	*Tem.*	12.7	c.n.	6.4	*Parac.*	6.0	*Best.*	5.4	81.9
1/76	*Oith.*	64.8	*Tem.*	12.1	c.n.	3.7	*Parac.*	3.0	*Acar.*	2.5	86.1
2/76	*Oith.*	67.2	*Tem.*	5.1	*Parac.*	4.5	*Best.*	2.6	*Acar.*	1.7	81.1
3/76	*Oith.*	48.8	*Parac.*	11.4	*Tem.*	5.0	*Acar.*	3.7	*Best.*	2.3	71.2
4/76	*Oith.*	62.1	*Parac.*	5.2	*Tem.*	4.3	*Acar.*	3.5	*Best.*	2.3	77.4
5/76	*Oith.*	53.3	*Parac.*	9.2	c.n.	3.9	*Tem.*	2.9	*Best.*	1.7	71.0
6/76	*Oith.*	47.1	*Tem.*	12.3	*Parac.*	9.3	c.n.	9.3	*Best.*	5.0	83.0

3.2 Feeding types

The material contained in the guts and faecal pellets of the copepods examined, particularly the small herbivores, consisted predominantly of unrecognizable remains (Fig. 2), but some fragments of diatom frustules and chitinized crustacean remains were easily identified (Figs 3 – 7). A comparision of the ingested material with the range of possible prey species in the water column at the time of sampling aided the identification of gut and faecal contents. The dominant forms of phytoplankton in the water column during the sampling period were *Leptocylindricus danicus, Eucampia zoodiacus, Nitzschia seriata* and a number of *Chaetoceros* species. Other phytoplankton which were present in lesser numbers included the chain–forming diatoms *Asterionella, Thalassiosira* and *Skeletonema,* the centric diatoms *Ditylum* and *Rhizosolenia* and the pennate diatom *Thalassiothrix.* There were also numerous unidentified microflagellates in the water column. Phytoplankton species which were identified in copepod guts are noted in the next Section.

Table 4.

The food type characteristics of the dominant species of copepods as determined from a gut and faecal pellet analysis.

Species	Feeding type	Common prey items	Prey size (μm)
1 *A. bispinosa*	omnivore	unicellular diatoms, dinoflagellates, nauplii	10 – 40
2 *A. tranteri*	herbivore	unicellular diatoms, dinoflagellates	10 – 25
3 *B. similis*	herbivore	unicellular diatoms, dinoflagellates	< 20
4 *G. pectinatus*	herbivore	unicellular and chain diatoms, dinoflagellates	1.5 – 35
5 *O. brevicornis*	herbivore	chain diatoms, dinoflagellates, bacteria	< 20
6 *O. plumifera*	herbivore	diatoms, dinoflagellates	< 45
7 *O. simplex*	herbivore	chain diatoms, bacteria, microflagellates	< 15
8 *P. crassirostris*	herbivore	unicellular diatoms, dinoflagellates	2.5 – 15
9 *P. indicus*	herbivore	unicellular diatoms, dinoflagellates	< 20
10 *T. turbinata*	herbivore	unicellular diatoms, dinoflagellates	10 – 40
11 *T. barbatus*	omnivore	dinoflagellates, crustaceans	> 15

The material examined from the guts was similar to that of the faecal pellets for any one individual (Figs 6, 7), suggesting that gut contents were not being biased by ingestion of items after the copepods were captured. A summary of the gut and faecal pellet analyses, and a food type classification based on this examination, is given in Table 4.

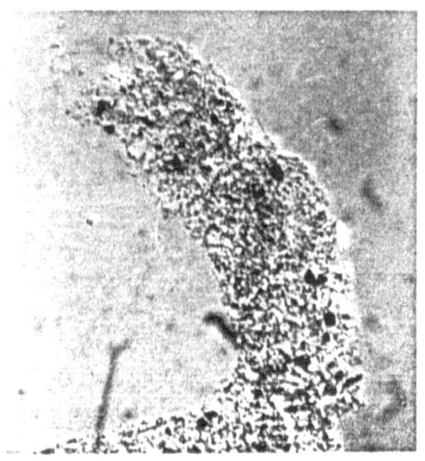

Fig. 2: Faecal pellet of *Gladioferens pectinatus* (x600).

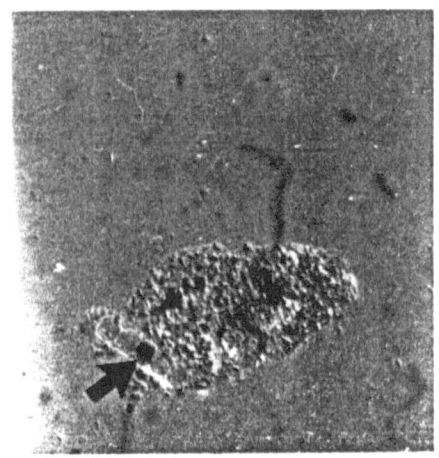

Fig. 3: Faecal pellet of *Gladioferens pectinatus* showing chain diatom (x600)

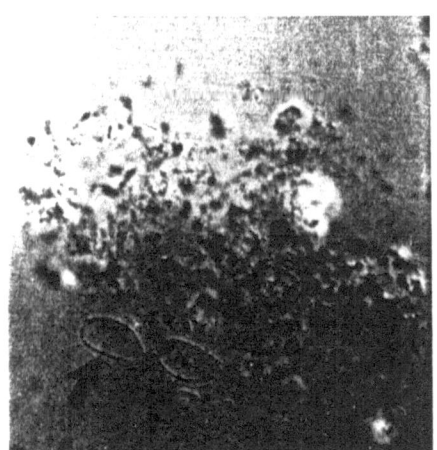

Fig. 4: Gut content of *Temora turbinata* showing two dinoflagellates of the genus *Prorocentrum* (x600)

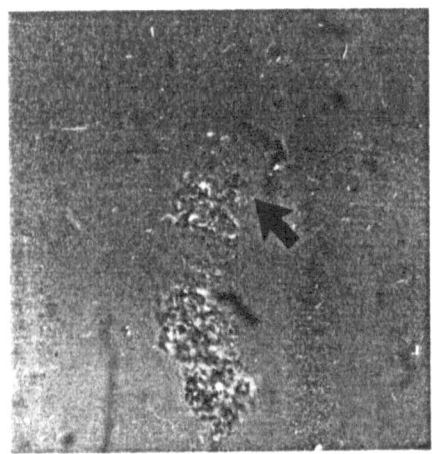

Fig. 5: Faecal pellet of *Acartia tranteri* showing centric diatom (x600).

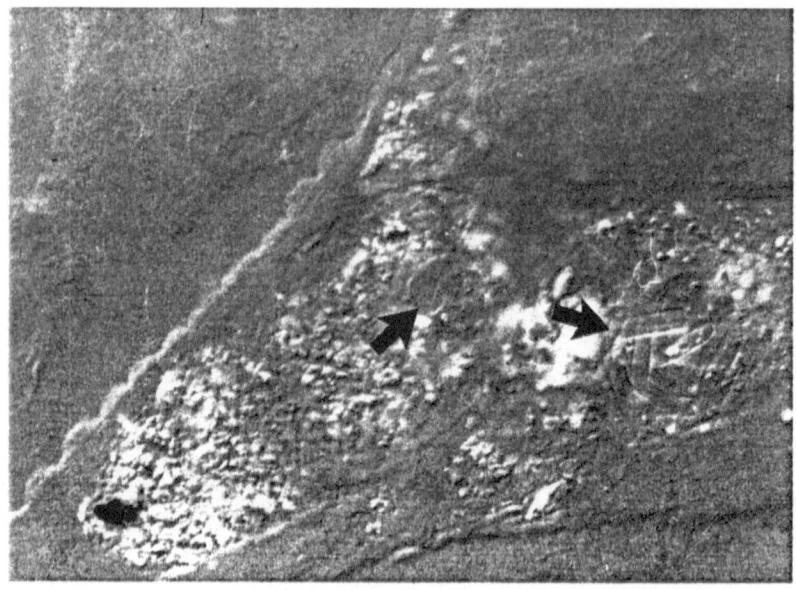

Fig. 6: Gut of *Gladioferens pectinatus* showing centric diatoms (x1000).

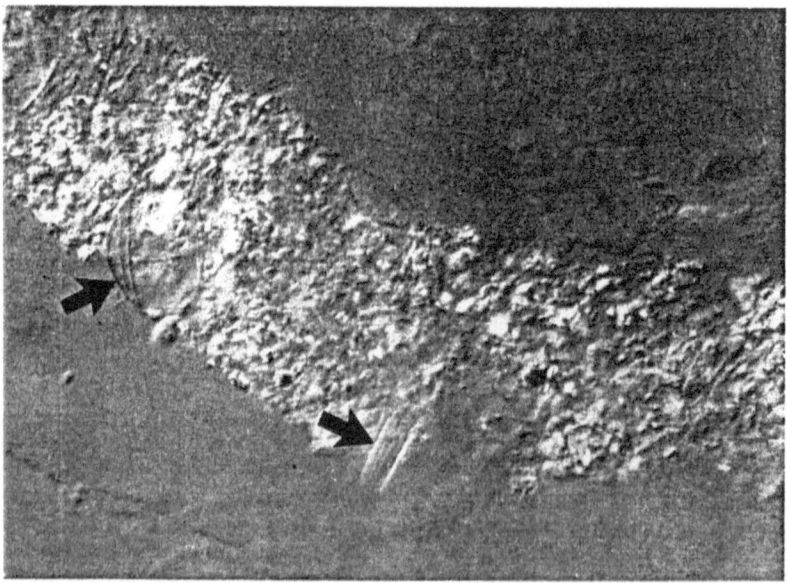

Fig. 7: Faecal pellet of *Gladioferens pectinatus* showing centric diatoms (x1000).

3.3 Seasonal abundance, geographical distribution and gut and faecal pellet analysis of the dominant species

3.3.1 *Acartia bispinosa* Carl

Acartia bispinosa Carl, 1907: 13–14, pl. 1, figs 1,2. *Acartia amboinensis* Sewell, 1914: 243, pl. 19, figs 1–7 (female). *Acartia (Odontacartia) bispinosa* Steuer, 1923: 121, figs 157–165, 22a–c, 27. *Acartia* sp. Dakin & Colefax, 1940: 106, figs 159a–e. *Acartia (Odontacartia) bispinosa* Grice, 1964: 261–262, figs 35–37.

This species was most abundant in SWA in March – April, with a peak water column abundance of 4736 m^{-3}, and was not found in June – August. In March – April it was often the most abundant copepod in night samples, a phenomenon also observed by Dakin & Colefax.

The species has also been recorded from the central coast of New South Wales (Dakin & Colefax), Malay Archipelago, Fiji Islands, Galapagos Islands (Grice), Palao Harbour, Ceylon Pearl Banks, MacPherson Strait and the neighbouring Indian waters (Sewell). These records suggest it is an estuarine and coastal Indo–Pacific species.

Three guts and three faecal pellets were examined and found to contain large centric diatoms (10 and 40 μm diameter), a variety of dinoflagellates and young naupliar stages of copepods. The species was classified here as omnivorous.

3.3.2 *Acartia tranteri* Bradford

Acartia clausi Dakin & Colefax, 1940: 106, figs 157a–c; *Acartia tranteri* Bradford, 1976: 192–194, figs 29–31, 33.

This species was common throughout the year. It has a seasonal abundance similar to *A. bispinosa* with a slight peak in early autumn and reduced numbers in winter. Hodgson (1979, p. 124) found a comparable pattern in Tuggerah Lakes, N.S.W.

Since the species was described by Bradford it has been recorded only from Port Hacking Estuary and nearby coastal regions although many of the recordings of *A. clausi* probably also include the above species. The specimens which Dakin & Colefax had previously collected from the coastal waters of N.S.W. and had called *A. clausi* were in fact *A. tranteri* according to Bradford. Other Australian recordings of *A. clausi* which may incorporate *A. tranteri* include those in the Swan River (Hodgkin & Rippingale, 1971), Port Phillip Bay and its estuaries (Arnott & Hussainy, 1972; Arnott, 1974; Neale & Bayly, 1974), Lake Macquarie and Port Stephens (Kott, 1955), estuaries of South Australia and southern Queensland (Bayly, 1973) and Moreton Bay (Greenwood, 1973).

The guts and faecal pellets of four individuals consisted of at least 90% unidentifiable green matter, the only recognizable remains being phytoplankton cells, mainly centric diatoms, which varied in diameter between 10 and 25 μm. The species appeared to be herbivorous, at least during September 1976. Fig. 5 shows the faecal pellet of a specimen with the empty frustule of a centric diatom at one end.

3.3.3 *Bestiola similis* (Sewell)

Acrocalanus similis Sewell, 1914: 211, pl. 17, figs 3–5; Wellershaus, 1969: 253, figs 10–20. *Bestiola similis* Andronov, 1972: 290–292; Greenwood, 1976: 20, fig. 7a–j; Hodgson, 1979: vol. 2, 423.

The genus name *Bestiola* was recently introduced (Andronov) to separate two groups of closely related copepods within the genus *Acrocalanus*.

The species was commonly found throughout the year with slightly increased numbers in autumn, winter and early summer. The yearly maximum (6208 m^{-3}) occurred in June and the maximum in percentage of total zooplankton (Table 3) occurred in September.

It has also been recorded from the Gulf of Mannar (Sewell), Cochin Backwater (Wellershaus), Great Barrier Reef and Moreton Bay (Greenwood) and Tuggerah Lakes (Hodgson). According to Wellershaus the species is very common in the Cochin Backwater, between 10 and 33 · 10^{-3} salinity.

The guts and faecal pellets of three specimens contained unicellular diatoms and dinoflagellates of less than 20 μm diameter, indicating that the species is basically herbivorous and may be a small–particle feeder.

3.3.4 *Gladioferens pectinatus* Brady

Gladioferens pectinatus Brady, 1899: 36, pl. 9, figs 24–27. *Gladioferens brevicornis* Henry, 1919: 35–37; Dakin & Colefax, 1940: 89–91, fig. 108. *Gladioferens pectinatus* Bayly, 1963: 201–207, figs 2, 3; 1965: 328; Arnott, 1974: 95, figs 4–14; Hodgson, 1979: vol. 1, 116–122, figs 33, 34, pl. 2; vol. 2, 426, fig. 115.

This species occurred twice in daytime hauls at Stn C but at night in light trap samples taken at Stn Gg it was found in small numbers from late winter to late spring and again in autumn, with abundance peaks in early spring and early autumn. The animal is known to exhibit a strong diurnal migratory rhythm (Hodgson). At certain times of the year it was the most abundant zooplankter collected in light traps at night, yet during the day it virtually disappeared from the water column. The light trap samples were taken in shallow water (2 – 4 m) in close proximity to seagrass beds whereas the daytime net hauls were taken well away from seagrass areas in deep water (18 – 20 m). Kirkman (pers. comm.) has found that *Gladioferens* was a common resident of seagrass beds (mainly *Posidonia australis* and *Zostera capricorni*) by day in Port Hacking.

The species has been widely recorded from the coastal, estuarine and lake areas in eastern and southern Australia, including Tasmania (Bayly; Hodgson; Dakin & Colefax). Dakin & Colefax found it to be the most common copepod in the southern coastal lakes of New South Wales and abundant in salinities below 28 · 10^{-3}. It has also been recorded from fresh water at Cumbalum, New South Wales (Bayly) and from New Zealand (Brady). Bayly (1965) believes it is an estuarine species which extends its range into marine conditions during the cooler months. Hodgson found that it was "by far the dominant copepod" in the brackish waters of Tuggerah Lakes, N.S.W. during the periods of late spring to summer and late autumn to winter. At other times of the year he found "the density was generally low".

The guts and faecal pellets of five specimens consisted predominantly of green matter. One faecal pellet (Fig. 3) contained a chain diatom of five cells with overall dimensions of 7.0 x 35.0 μm. The third cell appeared to be alive and had apparently passed through the gut unharmed. It is well known that some phytoplankters pass through the host unharmed (Porter, 1973) and may actually obtain nutriment but the phenomenon has not previously been recorded in association with *G. pectinatus*.

Recognizable material in the other specimens included cells of *Chaetoceros* and *Eucampia*. The presence of long *Nitzschia* spines in the guts shows that the animal is able to ingest spined prey. Centric diatoms of approximately 12 to 18 μm diameter were the most commonly found prey items. All food items were in the size range of 1.5 to 35.0 μm. Diatom frustules are clearly visible in photographs of *G. pectinatus* (Figs 6, 7).

The gut and faecal pellet analysis indicated that the species is herbivorous.

3.3.5 *Oithona brevicornis* Giesbrecht *forma typica* Wellershaus

Oithona brevicornis Giesbrecht, 1892: 546, pl. 34, figs 6, 7; Rosendorn, 1917: 34, fig. 19a–g; Sewell, 1924: 792, pl. 46, fig. 3; Kiefer, 1929: 8; Rose, 1933: 280, fig. 353; Sewell, 1934: 82; Dakin & Colefax, 1940: 116, fig. 203a; Grice, 1960: 487, figs 1–6; Kasturirangan, 1963: 76, 77, figs 82, 83; Cheng *et al.*, 1974: 36, pl. 4, figs 1–8. *Oithona brevicornis f. typica* Wellershaus, 1969: 279, figs 103–106, 109–119; Nishida *et al.*, 1977: 129, figs 4, 5.

This species ranked with *O. simplex* as the most commonly found copepod. It was abundant throughout the year with peaks in summer and autumn. Four forms were collected, the most common being *forma arnensis* which was closely followed by *forma typica*. The other two forms are new records, both being closely related to *forma typica*. *Form 2* is not uncommon but *form 3* is represented by only three specimens (Tafe, 1979).

O. brevicornis is abundant in neritic waters and waters of the warm regions of the world. It has been recorded from the Pacific Ocean (Sewelll), tropical Atlantic Ocean (Sewell), Adriatic Sea, Arabian Sea (Sewel), East China Sea, Yellow Sea (Cheng *et al.*), Hong Kong (Giesbrecht), Japan (Nishida *et al.*), Malay Archipelago (Sewell), brackish waters around India (Wellershaus), Aru Islands of Indonesia, the Congo River Estuary (Rosendorn), Tisbury Great Pond, Chesapeake Bay and the Gulf of Mexico (Grice).

The guts and faecal pellets of five specimens consisted predominantly of very small fragments, the bulk of which were unrecognizable. Two centric diatoms of diameter 5 and 17 μm were identified in the gut of one specimen, along with 15 rod–shaped bacteria of dimensions 0.7 x 2.0 μm. Another gut contained two *Leptocylindricus* cells, a dinoflagellate and a chain diatom of four cells, identified as *Thalassiosira*. The faecal pellet of a third specimen contained a *Thalassiothrix* cell of dimensions 1.0 x 10.0 μm, a chain diatom, *Chaetoceros*, of dimensions 9.0 x 17.0 μm, a dinoflagellate, *Gymnodinium*, of dimensions 7.0 x 10.0 μm and three rod–shaped bacteria. The most commonly found food items were small chain diatoms of the genus *Chaetoceros* which indicated an herbivorous diet, at least during the time of the year when *Chaetoceros* dominated the phytoplankton.

3.3.6 *Oithona plumifera* Baird

Oithona plumifera Baird, 1843: 59–60, fig.b p.60; Giesbrecht, 1892: 541, pl. 4, fig. 10, pl. 34, figs 12,13, 22, 25, 27–29, 32, 33, pl. 44, figs 1, 7, 12–15; Scott, 1909: 194; Farran, 1913: 183; Rosendorn, 1917: 10–12, fig. 1a–d; Steuer, 1923: 64, fig. 43; Gurney, 1927: 158; Kiefer, 1929: 4–5, fig. 1; Rose, 1933: 282–283, fig. 358; Mori, 1937: 109–110, pl. 60, figs 3–15; Dakin & Colefax, 1940: 115, fig. 200a–c; Grice, 1960: 487; Tanaka, 1960: 61, pl. 26, figs 11–13; Kasturirangan, 1963: 74–75, fig. 81b,c; Wellershaus, 1970: 470; Cheng *et al.*, 1974: 36–37, pl. 4, figs 9–14; Nishida *et al.*, 1977: figs 14,15.

This species was found throughout the year with a peak abundance in summer.

It is tolerant of brackish water and has even been collected from fresh water (Tanaka). It is distributed widely in the temperate and tropical regions of the Indian, Pacific and Atlantic Oceans (Farran). It has also been recorded from the Arctic Ocean, North Sea (Rosendorn), southwest coast of Ireland, Gulf of Guinea, Cape of Good Hope, Mediterranean Sea (Farran), Red Sea (Rosendorn), Arabian Sea (Farran), Naples (Giesbrecht), Yellow Sea, East China Sea (Cheng *et al.*), Japan (Tanaka), Malay Archipelago, Christmas Island, Fiji Islands (Farran), Florida west coast, California coast, Chesapeake Bay, Woods Hole and Martha's Vineyard, Gulf of Maine, Narragansett Bay, Gulf of Mexico and Great Barrier Reef (Grice).

The guts and faecal pellets of three specimens consisted of centric diatoms, dinoflagellates and chain diatoms up to 45 μm in length. It is classified as herbivorous.

3.3.7 *Oithona simplex* Farran

Oithona simplex Farran, 1913: pl. 29, figs 10–14, pl. 30, figs 1, 2; Rosendorn, 1917: 44, fig. 26a–f; Steuer, 1923: 73; Gurney, 1927: 160; Kiefer, 1929: 9; Grice, 1960: 488, figs 12–18; Tanaka, 1960: 64, pl. 28, figs 1–6; Gonzalez & Bowman, 1965: fig. 21f–i; Wellershaus, 1970: 464; Cheng *et al.*, 1974: 34, pl. 3, figs 4–9; Nishida *et al.*, 1977: 151, fig. 23.

This species ranked with *O. brevicornis* as the most abundant copepod and had a similar seasonal abundance, with peaks in summer and autumn.

It has also been recorded from the Pacific Ocean (Tanaka), Indian Ocean (Rosendorn), Suez Canal (Gurney), Aru Archipelago (Steuer), Christmas Island (Farran), Bay of Biscay (Rosendorn), Japan (Nishida *et al.*), East and South China Seas Florida west coast (Grice), Puerto Rico and Gulf of Mexico (Gonzalez & Bowman), mouth of the Amazon River and in the Atlantic Ocean south of 55°N (Rosendorn).

Like *O. brevicornis* the contents of the guts and faecal pellets of five specimens consisted predominantly of finely granulated, unrecognizable green material. A small diatom (2.5 x 3.0 μm) with a dividing nucleus was identified in the gut of one specimen and a large number of bacteria were present in the faecal pellet of another. The species was classified as herbivorous.

3.3.8 *Paracalanus crassirostris* Dahl *forma typica* Fruchtl

Paracalanus crassirostris Dahl, 1894: 21, pl. i, figs 27, 28; Sewell, 1924: 780; Gurney, 1927: 144–147, figs 16b–d, 17a–c; Sewell, 1929: 72–76, figs 27a, b; 1933: 25; Tanaka, 1960: 23, pl. 6, figs 1–7; Bowman, 1971: 18. *Paracalanus crassirostris f. typiça* Fruchtl, 1923: 456; 1924: 36–39; fig. 12, table 5; Gonzalez & Bowman, 1965: 244, figs 2f–n, 3a; Wellershaus, 1969: 250, table 1; Greenwood, 1976: 22, fig. 8.

This species ranked third in abundance after the *Oithona* group and *Temora turbinata*. It was found commonly throughout the year with abundance peaks in late summer and autumn. Another form, *forma cochinensis*, was also collected but it accounted for less than 5% of the catch of this species. According to Sewell (1933) there are several local forms in different parts of the world. The species inhabits brackish, estuarine and coastal waters.

The species has a wide distribution in tropical and sub–tropical waters. It has been recorded from the Indian Ocean, Cochin Backwater, Sea of Penang, coast of Sri Lanka, Chilka Lake, littoral waters of Nicobar Islands (Sewell, 1929), Suez Canal (Gurney), mouth of the Congo (Dahl), Gulf of Guinea, Cape of Good Hope (Tanaka), Aru Archipelago (Fruchtl), Florida coast, Chesapeake Bay, Tisbury Great Pond, Martha's Vineyard, Long Island Sound, Delaware Bay, Raritan Bay, Louisiana, coast of Brazil and Puerto Rico (Gonzalez & Bowman).

According to Gonzalez & Bowman the species is frequently found in brackish water and is widespread in the coastal waters of the Indian Ocean but it has not been found in the coastal waters of Japan despite thorough investigation. It has also never been reported from the Mediterranean Sea or eastern Atlantic Ocean.

The guts and faecal pellets of five specimens contained diatoms and dinoflagellates in the size range of 2.5 to 15 μm which indicated that the species is herbivorous.

3.3.9 *Paracalanus indicus* Wolfenden

Paracalanus parvus var. *indicus* Wolfenden, 1905: 998, pl. 96, figs 7, 9 11. *Paracalanus indicus* Bowman, 1971: 24, 28, figs 22b–m; 23a.

This species was found in small numbers throughout the year.

It has also been recorded from the Indian Ocean, Gulf of Naples, Nankauri Harbour and West Indies (Bowman). According to Bowman the species is not as numerous as a number of other species of the genus *Paracalanus* in coastal areas and may be more oceanic in its distribution.

The guts and faecal pellets of three specimens contained unicellular diatoms and dinoflagellates up to 20 μm in diameter. The species was classified as herbivorous.

3.3.10 *Temora turbinata* Dana

Temora turbinata Dana, 1849: 12; Giesbrecht, 1892: 329, 336, 338, pl. 17, figs 14, 17, 18, 21, pl. 38, fig. 27; Scott, 1909: 118; Fruchtl, 1924: 54; Mori, 1937: 64, pl. 32, figs 3–8; Sewell, 1947: 165; Krishnaswamy, 1953: 124; Tanaka, 1960: 49; Kasturirangan, 1963: 40, fig. 36; Vervoort, 1965: 99–100; Bradford, 1977: 131–144, figs 1–7; Dakin & Colefax, 1940: 93, fig. 118a–d.

This species ranked second in abundance after the *Oithona* group and was found in estuarine samples throughout the year except in late October, with an abundance peak in early summer. Other Australian records include Lake Macquarie, Port

Stephens and Port Hacking (Kott, 1955), Tuggerah Lakes (Hodgson, 1979), Moreton Bay (Greenwood, 1973), Port Phillip Bay (Arnott, 1974) and N.S.W. coastline (Dakin & Colefax).

The species is widely distributed throughout the Indian Ocean (Sewell) and tropical waters of the Atlantic and Pacific Oceans (Tanaka). It has also been recorded from the Sulu Sea (Dana), Arabian Sea, Persian Gulf, Gulf of Maine, Sri Lanka (Sewell), Malay Archipelago (Scott), Hong Kong (Giesbrecht), New Zealand (Bradford) and Woods Hole (Sewell). According to Bradford the species attains very large numbers only in enclosed or semi–enclosed bodies of water. She suggests that threshold nutritional conditions occur beneath which *T. turbinata* will not breed; this threshold is consistently achieved only in coastal and estuarine waters.

In Australian and New Zealand offshore waters the distribution of *T. turbinata* is closely associated with sub–tropical currents. Its presence in the offshore waters off Eden, N.S.W., has been taken to indicate the presence of northern coastal water (Kott in Bradford). The species is tolerant of a wide range of temperatures and salinities but the observations of worldwide distribution (Vervoort) suggest that there is a lower temperature limit to survival. In the Indian and Pacific Oceans *T. turbinata* is found as far south as 35°S and in the Pacific Ocean as far north as 40°N.

The guts and faecal pellets of four specimens contained a large proportion of green material and some irregular shaped brown fragments. One specimen contained two dinoflagellates (13.0 x 25.0 μm) identified as *Prorocentrum micans* (Fig. 4) while a second gut contained diatoms of the genera *Thalassiothrix* and *Leptocylindricus*. Large phytoplankton dominated the gut material and the presence of two large flagellates indicated that motile prey are ingested. The species was classified as herbivorous.

3.3.11 *Tortanus barbatus* Brady

Corynura barbata Brady, 1883: 71, pl. 31, figs 10–12. *Corynura denticulata* Giesbrecht, 1889: 26; 1892: 525, pl. 31, 42. *Corynura barbata* Giesbrecht, 1892: 525. *Tortanus denticulatus* Giesbrecht & Schmeil, 1898: 158. *Tortanus barbatus* Giesbrecht & Schmeil, 1898: 158; Scott, 1909: 189, pl. 40, figs 16–18; Fruchtl, 1924: 37–41, figs 6–8, tables 2, 3; Dakin & Colefax, 1940: 106, fig. 160a,b.

This species was only found in estuarine samples in winter and spring, and then only in small numbers (< 320 individuals/m³). Other Australian recordings include N.S.W. coastline (Dakin & Colefax), the Gippsland Lakes (Arnott, 1968), the estuaries of Port Phillip Bay (Neale & Bayly, 1974; Arnott & Hussainy, 1972), Tuggerah Lakes and Lake Macquarie (Hodgson, 1979). It has also been recorded from the Philippines (Brady), Red Sea (Giesbrecht), Malay Archipelago, coast of Burma, Patanic Bay and Thailand (Scott).

The gut and faecal pellet material of three specimens included copepod legs, dinoflagellates (*Ceratium fusus*) and a chain diatom (15 x 18 μm). The species was classified as omnivorous and appears to be an opportunistic feeder capable of ingesting copepods and other crustaceans as well as diatoms and a number of spiny dinoflagellates.

4. DISCUSSION

Thirty five of the 45 species of copepods identified in Port Hacking (Table 1) were recorded by Dakin & Colefax (1940) in the neritic waters along the New South Wales coastline while 20 of them were recorded by Tranter & Tafe (unpublished) in the offshore waters between 100 and 300 km due east of Sydney. Griffiths (1983) recorded fewer species in Port Hacking and we have attributed this to the difference in sampling method. Griffiths sampled only during the day by vertical net haul at Stn C in SWA whereas our samples included a number taken at night using light traps at 3 stations closer to the mouth of the estuary as well as net hauls at Stn C. We also examined routinely both net and light trap samples for rare species.

The species recorded in Table 2 were also recorded in coastal samples taken by Dakin & Colefax (1940), with the exception of *Oithona simplex*. They identified five species of *Oithona* and believed that there were probably others which they had not identified.

The five most abundant genera (Table 3), *Oithona, Temora, Paracalanus, Bestiola* and *Acartia* also occurred commonly in samples taken by Dakin & Colefax (1940), Hodgson (1979) and Griffiths (1983). Tranter *et al.* (1981) obtained large numbers of *Acartia tranteri, Temora turbinata, Tortanus barbatus* and *Gladioferens pectinatus* in light trap samples in Port Hacking while *Isias uncipes*, a species not recorded in Table 1, occurred in their traps at certain times, particularly on moonless nights. Griffiths (1983) considers the abundant species in SWA as euryhaline species commonly inhabiting coastal/estuarine areas. The higher abundances in summer when SWA was classified as estuarine, are thought to be due to increased amounts of food and habitats in the water column during salinity stratification, while low abundances in the spring occurred under marine conditions and no salinity stratification.

The feeding type classifications assigned to each of the 11 species (Table 4) are in accord with the findings of other workers on closely related species although *Acartia tranteri, Temora turbinata* and *Tortanus barbatus* require further qualification. *Acartia tranteri* and *Temora turbinata* have been classified as herbivores because no animal remains were found in their guts and faecal pellets. However they have mouthparts adapted to taking in both plant and animal prey (Tafe, 1979). The fact that diatoms dominated the plankton in the water column at the time of sampling and were present in the guts of these two species suggests that their food intake was largely influenced by food availability. Marshall (1973) found that food preference of a number of copepod species depends on the composition of the plankton they are feeding on and to some extent on what they have previously been eating.

Marshall (1949) and Conover (1956) examined the gut contents of *Acartia clausii*, a species closely related to *A. tranteri*, and found that most of the identifiable remains were large diatoms such as *Coscinodiscus* and moderately sized chain diatoms such as *Thalassiosira* as well as a few small diatoms such as *Skeletonema* and flagellates. Occasionally the remains of other copepods, usually of young stages, were found. While Lebour (1922) classified *A. clausii* as a diatom feeder, after observing it feeding almost wholly on diatoms, Marshall & Orr (1962) have shown that it will also ingest other copepods and *Artemia* nauplii. Although feeding studies of *A. tranteri* have not been carried out, its similarity in mouthpart structure and overall body size to *A. clausii*, the mandibular edge index (Tafe, 1979) and the present gut and faecal pellet findings indicate that it too is capable of ingesting animal prey but can survive on a strictly herbivorous diet.

The feeding type and behaviour of a number of species of *Temora* have been well documented, at least with regard to laboratory cultures. Paffenhofer & Knowles (1978) found that the adult stage of *T. turbinata* fed on the diatom *Leptocylindricus danicus*, of cell diameter 14–18 μm. They did not ascertain the species' natural diet. Lebour (1922) found that *Temora longicornis* fed in culture on diatoms and sometimes smaller copepods. Cushing (1958) and Berner (1962) successfully fed *T. longicornis* on *Skeletonema costatum* chains and Marshall & Orr (1962) maintained cultures of the same species on mixtures of flagellates and diatoms. Gauld (1951, 1953) and Raymont (1963) classified *T. longicornis* as an omnivore after culturing the animals on a number of plant and animal species. Gauld (1966) stated that *Temora* concentrates on plant food but does take animal food. Itoh (1970) calculated the mandibular edge index of *T. discaudata* and subsequently classified the species as omnivorous with a preference for phytoplankton. Tafe (1979) found that the mandible of *T. turbinata* closely resembled that of *T. discaudata* and the mandibular edge index indicated that *T. turbinata* was an omnivore. *T. turbinata* is a slightly smaller species and was commonly found in the estuary whereas *T. discaudata* was common in neritic waters (Dakin & Colefax, 1940) but only rarely found in the estuary. *T. turbinata* is therefore probably more reliant on phytoplankton and prefers the smaller cells which the estuary often provides in abundance.

Tortanus barbatus has been classified in Table 4 as an omnivore because it took both plant and animal prey. However the predominance of animal remains suggests that this species may best be classified as a facultative carnivore. The mandibular edge index indicates that the animal is a carnivore (Tafe, 1979) but further investigation into its food preferences are required to confirm this.

Previous findings in relation to the feeding behaviour of other species of *Tortanus* show they are true carnivores. *Tortanus gracilis* and *T. forcipatus* have been found, by Wickstead (1962) and Anraku & Omori (1963), to consume animal prey, particularly smaller copepods. Anraku & Omori also found that *T. discaudatus* would survive in culture on a mixed diet of *Artemia* nauplii and the small calanoid copepod

Pseudocalanus minutus. Tafe (1979) examined the guts of *Tortanus scaphus* and found them to contain crustacean remains including copepod mandibles, the chelae of a decapod larva and a shrimp pleopod. *T. scaphus*, being larger than *T. barbatus*, is able to handle a greater variety of animal prey. Itoh (1970) classified *T. forcipatus* and *T. discaudatus* as true carnivores according to his edge index, and Marshall (1973) classified the genus *Tortanus* as carnivorous.

There was a lack of recognizable remains in the guts and faecal pellets of *Oithona brevicornis* and *O. simplex* with over 95% of the ingested material being composed of unidentifiable fine granules and bacteria. The feeding type classifications assigned to these cyclopoid species (Table 4) agree with the limited findings of other authors but less is known about the feeding types of the *Oithona* group than any of the eight other species investigated. Murphy (1923) reared *Oithona* on pieces of kelp but does not say how they ate it; Gauld (1966) suggests kelp cannot make a large contribution to their diet in nature. Lebour (1922) has classified *Oithona similis* as a diatom feeder and records "green remains and bits of diatom" from its gut. Marshall & Orr (1966) have found that *O. similis* will ingest flagellates and diatoms in laboratory experiments. Many copepods will also ingest plastic beads in such experiments and therefore a distinction must be made between food types ingested in nature and those ingested in the laboratory. Gauld (1966) has suggested from a study of the mouthparts of *O. similis* that "the food of *Oithona* is necessarily largely plant food because, while relatively few animals are small enough to be captured in a net of such a size, diatom chains and individual cells of larger diatoms and dinoflagellates are all about the right size."

Fryer (1957), Gauld (1966) and Tafe (1979) have concluded from studies of mouthpart morphology that species of *Oithona* are raptorial feeders, primarily because the mouthparts lack the fine meshwork of setules seen in filter feeding calanoids. However if particles were seized and eaten in the same way described for carnivorous copepods, one would expect to see an array of diatom and dinoflagellate frustules in guts of *Oithona* but few were found. Marshall & Orr (1955) have described how diatoms can be broken open by the mandibles and the contents sucked into the animal by dilation of the oesophagus, and Cushing (1964) has seen *Pseudocalanus* take a *Odontella (Biddulphia) sinensis* cell almost as big as itself, break it open, and filter off some of the contents. Tafe (1979) found that the mouthparts of *Oithona* are well adapted to such a process of "diatom spoliation" (Cushing, 1964) as the maxillipeds and 2nd maxillae are adapted for seizing and holding prey while the 1st maxillae and mandibles are designed for piercing. The mouthparts of *O. plumifera* are structurally similar to those of the other two species but much larger, and are, therefore, probably adapted to capturing a larger size range of prey. A large numbers of diatom frustules were found in the guts of this species, indicating that phytoplankters were being ingested whole. The species may be feeding differently to *O. brevicornis* and *O. simplex*. Gauld (1966) observed *O. similis* manipulating a large centric diatom between the 1st maxillae and mandibles. The animal appeared to be scraping epiphytes or bacteria from the surface of the diatom with the mandibles as it rotated it with the 1st maxillae.

Feeding either by "bacterial scraping" or by "diatom spoliation" could explain the lack of recognizable material in the digestive tracts of *O. brevicornis* and *O. simplex*.

In summary, we should emphasize that the seasonal abundances of copepods in Port Hacking were considerably higher than in the nearby neritic waters and the species found were common euryhaline, estuarine/neritic species, most of which have a wide geographic distribution. The differences found in the food types between the species from Port Hacking and those of previous studies in other localities are due partly to interspecific differences and partly to the differential availability of prey species at the time of feeding.

ACKNOWLEDGEMENTS

We express our thanks to B. Hodgson and G. Arnott for critical revision of the first draft, to D. Tranter and M. Tomczak for constructive advice, and to G. Hallegraeff and J. Stauber for identifying phytoplankton species and bacteria in the guts of copepods.

REFERENCES

Alcaraz, M., Paffenhoffer, G.-A., Strickler, J.R.: Catching the Algae: A first account of visual observations on filter–feeding calanoids. In: W.C. Kerfoot, ed. *Evolution and Ecology of Zooplankton Communities.* Boston : University Press of New England (1980)

Alldredge, A.L., King, J.M.: Distribution, abundance, and substrate preferences of demersal reef zooplankton at Lizard Island Lagoon, Great Barrier Reef. *Marine Biology (Berlin)* **41**, 317–333 (1977)

Andronov, V.N.: Veslonogie rachki *Bestiola* gen. n. (Copepoda, Paracalanidae). *Zoologische Verhandelingen* **51**, 290–292 (1972)

Anraku, M., Omori, M.: Preliminary survey of the relationship between the feeding habits and the structure of the mouth parts of marine copepods. *Limnology and Oceanography* **8**, 116–126 (1963)

Arnott, G.H.: *An Ecological Study of the Zooplankton of the Gippsland Lakes with special reference to the Calanoida*, Hons Thesis, Monash University, Victoria (1968)

Arnott, G.H.: *Studies on the Zooplankton of Port Phillip Bay and adjacent waters with special reference to Copepoda.* Ph.D. Thesis, Monash University, Victoria (1974)

Arnott, G.H., Hussainy, S.U.: Brackish–water plankton and their environment in the Werribee River, Victoria. *Australian Journal of Marine and Freshwater Research* **23**, 85–97 (1972)

Baird, W. Notes on British Entomostraca. *Zoologist* **1**, 55–61 (1843)

Bayly, I.A.E.: A revision of the coastal water genus *Gladioferens* (Copepoda: Calanoida). *Australian Journal of Marine and Freshwater Research* **14**, 194–217 (1963)

Bayly, I.A.E.: Ecological studies on the planktonic Copepoda of the Brisbane River estuary with special reference to *Gladioferens pectinatus* (Brady) (Calanoida). *Australian Journal of Marine and Freshwater Research* **16**, 315–350 (1965)

Bayly, I.A.E.: Australian Estuaries. *Proceedings of the Ecological Society of Australia* **8**, 41–66 (1973)

Beklemishev, K.V.: The food of some common planktonic copepods in the Far Eastern Seas. (In Russian). *Zoologischeskii Zhurnal* **33**, 1210–1230 (1954)

Berner, A.: Feeding and respiration in the copepod *Temora longicornis* (Muller). *Journal of the Marine Biological Association of the Unived Kingdom* **42**, 625–640 (1962)

Bowman, T.E.: The distribution of calanoid copepods off the southeastern United States between Cape Hatteras and southern Florida. *Smithsonian Contributions to Zoology* **96**, 1–58 (1971)

Bradford, J.M.: Partial revision of the *Acartia* subgenus *Acartiura* (Copepoda: Calanoida: Acartiidae). *New Zealand Journal of Marine and Freshwater Research* **10**, 159–202 (1976)

Bradford, J.M.: Distribution of the pelagic copepod *Temora turbinata* in New Zealand coastal waters, and possible trans-Tasman population continuity. *New Zealand Journal of Marine and Freshwater Research* **11**, 131–144 (1977)

Brady, G.S.: Copepoda obtained by H.M.S. 'Challenger' during the years 1873–76. *Report on the Scientific Results of the Voyage of H.M.S. Challenger* **8**(1), 1–142 (1883)

Brady, G.S.: On the marine Copepoda of New Zealand. *Transactions of the Zoological Society of London* **15**, 31–54 (1899)

Carl, J.: Copépodes d'Amboine. *Revue Suisse de Zoologie* **15**, 7–18 (1907)

Cheng, Q., Zhang, S., Zhu, C.: On planktonic copepods of the Yellow Sea and the East China Sea. *Studia Marina Sinica* **9**: 27–76 (1974)

Conover, R.J.: Oceanography of Long Island Sound, 1952–1954. VI. Biology of *Acartia clausi* and *A. tonsa*. *Bulletin of the Bingham Oceanographic Collection Yale University* **10**, 156–233 (1956)

Cushing, D.H.: The effect of grazing in reducing the primary production: a review. *Rapports et Procès-Verbaux des Réunions Conseil International pour l'Exploration de la Mer* **166**, 149–154 (1958)

Cushing, D.H.: The work of grazing in the sea. In: D.J. Crisp, ed. *Grazing in Terrestrial and Marine Environments*. Oxford: Blackwell (1964)

Dahl, F.: Die Copepodenfauna des unteren Amazonas. *Berichte der Naturforschenden Gesellschaft zu Freiburg im Breisgau* **8**, 10–23 (1894)

Dakin, W.J., Colefax, A.N.: The plankton of the Australian coastal waters off New South Wales. Part 1. *Publications of the University of Sydney Department of Zoology* **1**, 1–211 (1940)

Dana, J.D.: Conspectus crustaceorum quae in orbis terrarum circumnavigatione, Caroli Wilkes e Classe Reipublicae Foederatae Duce, lexit et descripsit. Pars II. *Proceedings of the American Academy of Arts and Sciences* **2**, 9–61 (1849)

Farran, G.P.: Plankton from Christmas Island, Indian Ocean. II. On Copepods of the genera *Oithona* and *Paroithona*. *Proceedings of the Zoological Society of London* **1913**, 181–193 (1913)

Farran, G.P.: Copepoda. *Scientific Reports of the Great Barrier Reef Expedition* **5**(3), 73–142 (1936)

Farran, G.P.: The seasonal and vertical distribution of the Copepods. *Scientific Reports of the Great Barrier Reef Expedition* **2**(9), 291–312 (1949)

Fruchtl, F.: Notizen über die Variabilität nordadriatischer Planktoncopepoden. *Verhandlungen der Zoologisch-Botanischen Gesellschaft in Wien* **73**, 135–157 (1923)

Fruchtl, F.: Die Cladoceren und Copepoden-Fauna des Aru-Archipels. *Arbeiten aus dem Zoologishen Institut der Universitat Innsbruck* **2**(2), 25–136 (1924)

Fryer, G.: The feeding mechanism of some freshwater Cyclopoid copepods. *Proceedings of the Zoological Society of London* **129**, 1–25 (1957)

Gauld, D.T.: The grazing rate of planktonic copepods. *Journal of the Marine Biological Association of the United Kingdom* **29**, 695–706 (1951)

Gauld, D.T.: Diurnal variations in the grazing of planktonic copepods. *Journal of the Marine Biological Association of the United Kingdom* **31**, 461–474 (1953)

Gauld, D.T.: The swimming and feeding of planktonic copepods. In: H. Barnes, ed. *Some Contemporary Studies in Marine Science.* London: Allen and Unwin (1966)

Giesbrecht, W.: Elenco dei Copepodi pelagici raccolti dal tenente di vascello Gaetano Chierchia durante il viaggio della R. Corvetta "Vettor Pisani" negli anni 1882–1885, e dal tenente di vascello Francesco Orsini nel Mar Rosso, nel 1884. *Atti dell'Accademia Nazionale dei Lincei Rendiconti* **5(2)**, 24–29 (1889)

Giesbrecht, W.: Systematik und Faunistik der pelagischen Copepoden des Golfes von Neapel. *Fauna und Flora des Golfes von Neapel und der angrenzenden Meeresabschnitte* **19**, 1–831 (1892)

Giesbrecht, W., Schmeil, O.: Copepoda, I Gymnoplea. *Tierreich* **6**, 1–169 (1898)

Gonzalez, J.G., Bowman, T.E.: Planktonic copepods from the Bahia Fosforescente, Puerto Rico, and adjacent waters. *Proceedings of the United States National Museum* **117**, 241–304 (1965)

Greenwood, J.G.: *Calanoid Copepods of Moreton Bay: A Taxonomic and Ecological Account.* Ph.D. Thesis, University of Queensland (1973)

Greenwood, J.G.: Calanoid copepods of Moreton Bay (Queensland) 1. *Proceedings of the Royal Society of Queensland* **87**, 1–28 (1976)

Greenwood, J.G.: Calanoid copepods of Moreton Bay (Queensland) 2. Families Calocalanidae to Centropagidae. *Proceedings of the Royal Society of Queensland* **88**, 49–67 (1977)

Greenwood, J.C.: Calanoid copepods of Moreton Bay (Queensland) 3. Families Temoridae to Tortanidae, excluding Pontellidae. *Proceedings of the Royal Society of Queensland* **89**, 1–21 (1978)

Greenwood, J.G.: Calanoid copepods of Moreton Bay (Queensland) 4. Family Pontellidae. *Proceedings of the Royal Society of Queensland* **90**, 93–111 (1979)

Greenwood, J.G.: Composition and seasonal variations of zooplankton populations in Moreton Bay, Queensland. *Proceedings of the Royal Society of Queensland* **91**, 85–103 (1980)

Greenwood, J.G.: Calanoid copepods of Moreton Bay (Queensland) 5. Ecology of the dominant species. *Proceedings of the Royal Society of Queensland* **93**, 49–64 (1982a)

Greenwood, J.G.: Dominance, frequency and species richness patterns in occurrences of calanoid copepods in Moreton Bay, Queensland. *Hydrobiologia* **87**, 217–227 (1982b)

Grice, G.D.: Copepods of the genus *Oithona* from the Gulf of Mexico. *Bulletin of Marine Science Gulf and Caribbean* **10**, 485–490 (1960)

Grice, G.D.: Two new species of calanoid copepods from the Galapagos Islands with remarks on the identity of three other species. *Crustaceana (Leiden)* **6**, 255–264 (1964)

Griffiths, F.B.: Zooplankton community structure and succession in South West Arm, Port Hacking. In: W.R. Cuff and M. Tomczak jr, eds *Synthesis and Modelling of Intermittent Estuaries.* Berlin, Heidelberg, New York: Springer (1983)

Griffiths, F.B., Caperon, J.: Description and use of an improved method for determining estuarine zooplankton grazing rates on phytoplankton. *Marine Biology (Berlin)* **54**, 301–309 (1979)

Gurney, R.: Zoological results of the Cambridge Expedition to the Suez Canal, 1924. VIII. Report on the Crustacea; Copepoda and Cladocera of the plankton. *Transactions of the Zoological Society of London* **22**, 139–172 (1927)

Henry, M.: On some Australian freshwater Copepoda and Ostracoda. *Journal of the Royal Society of New South Wales* **53**, 29–48 (1919)

Heron, A.C.: A dark–field condenser for viewing transparent plankton under a low–power stereomicroscope. *Marine Biology (Berlin)* **2**, 321–324 (1969)

Hodgkin, E.P., Rippingale, P.J.: Interspecies conflict in estuarine copepods. *Limnology and Oceanography* **16**, 573–576 (1971)

Hodgson, B.: *The Hydrology and Zooplankton Ecology of Lake Macquarie and the Tuggerah Lakes, New South Wales.* Ph.D. Thesis, University of New South Wales (1979)

Honjo, S., Roman, M.R.; Marine copepod faecal pellets: production, preservation and sedimentation. *Journal of Marine Research* 36, 45–57 (1978)

Itoh, K.: A consideration on feeding habits of planktonic copepods in relation to the structures of their oral parts. *Bulletin of the Plankton Society of Japan* 17, 1–10 (1970)

Kasturirangan, L.R.: *A Key for the Identification of the more common Planktonic Copepoda of Indian Coastal Waters.* New Delhi: Council of Scientific and Industrial Research (1963)

Kennedy, G.R.: Plankton of the Fitzroy River estuary, Queensland. *Proceedings of the Royal Society of Queensland* 89, 29–37 (1978)

Kiefer, F.: Crustacea Copepoda 2. Cyclopoida Gnathastoma. *Tierreich* 53, 1–102 (1929)

Kott, P.: The zooplankton of Lake Macquarie, 1953–1954. *Australian Journal of Marine and Freshwater Research* 6, 429–442 (1955)

Kott, P.: Zooplankton of east Australian waters. *CSIRO Division of Fisheries and Oceanography Report* 14 (1957)

Krishnaswamy, S.: Pelagic Copepoda of the Madras coast. *Journal of the Madras University* 23, 61–75 (1953)

Lebour, M.V.: The food of plankton organisms. *Journal of the Marine Biological Association of the United Kingdom* 12, 644–677 (1922)

Marshall, S.M.; On the biology of the small copepods in Lock Striven. *Journal of the Marine Biological Association of the United Kingdom* 28, 45–122 (1949)

Marshall, S.M.: Respiration and feeding in copepods. *Advances in Marine Biology* 11, 57–120 (1973)

Marshall, S.M., Orr, A.P.: On the biology of *Calanus finmarchicus* VIII. Food uptake, assimilation and excretion in adult and stage V calanus. *Journal of the Marine Biological Association of the United Kingdom* 34, 495–529 (1955)

Marshall, S.M., Orr, A.P.: Food and feeding in copepods. *Rapports et Procès-Verbaux des Réunions Conseil International pour l'Exploration de la Mer* 153, 92–98 (1962)

Marshall, S.M., Orr, A.P.: Respiration and feeding in some small copepods. *Journal of the Marine Biological Association of the United Kingdom* 46, 513–530 (1966)

Mori, T.: *The Pelagic Copepoda from the Neighbouring Waters of Japan.* Tokyo: Yokendo (1937)

Mullin, M.M.: Selective feeding by calanoid copepods from the Indian Ocean. In: H. Barnes, Ed. *Some Contemporary Studies in Marine Science.* London: Allen and Unwin (1966)

Murphy, H.E.: The life–cycle of *Oithona nana* reared experimentally. *University of California Publications in Zoology* 22, 449–454 (1923)

Neale, I.M., Bayly, I.A.E.: Studies on the ecology of the zooplankton of four estuaries in Victoria. *Australian Journal of Marine and Freshwater Research* 25, 337–350 (1974)

Nicholls, A.G.: Littoral Copepoda from South Australia (II) Calanoida, Cyclopoida, Notodelphyoida, Monstrilloida and Caligoida. *Records of the South Australian Museum* 8, 1–62 (1944)

Nishida, S., Tanaka, O., Omori, M.: Cycloid copepods of the Family Oithonidae in Suruga Bay and adjacent waters. *Bulletin of the Plankton Society of Japan* 24, 119–158 (1977)

Nyan Taw: *Zooplankton and Hydrology of the South East Coastal Waters of Tasmania.* Ph.D. Thesis, University of Tasmania (1975)

Nyan Taw: Some common components of the plankton of the southeastern coastal waters of Tasmania. *Papers and Proceedings of the Royal Society of Tasmania* 112, 69–136 (1978)

Nyan Taw & Ritz, D.A.: Zooplankton distribution in relation to hydrology in the Derwent River estuary. *Australian Journal of Marine and Freshwater Research* 29, 763–775 (1978)

Nyan Taw & Ritz, D.A.: Influence of subantarctic and subtropical water on the zooplankton and hydrology of waters adjacent to the Derwent River estuary, south–eastern Tasmania. *Australian Journal of Marine and Freshwater Research* 30, 179–202 (1979)

Paffenhofer, G.A., Knowles, S.C.: Feeding of marine planktonic copepods on mixed phytoplankton. *Marine Biology (Berlin)* **48**, 143–152 (1978)

Parker, R.R., Rochford, D.J., Tranter, D.J.: History and organization of the Port Hacking Estuary Project. In: W.R. Cuff and M. Tomczak jr, eds *Synthesis and Modelling of Intermittent Estuaries.* Berlin, Heidelberg, New York: Springer (1983)

Porter, K.G.: Selective grazing and differential digestion of algae by zooplankton. *Nature (London)* **244**, 179–180 (1973)

Raymont, J.E.G.: *Plankton and Productivity in the Oceans,* Oxford: Pergamon Press (1963)

Rippingale, R.J., Hodgkin, E.P.: Predation effects on the distribution of a copepod. *Australian Journal of Marine and Freshwater Research* **26**, 81–91 (1974)

Rippingale, R.J., Hodgkin, E.P.: Food availability and salinity tolerance in a brackish water copepod. *Australian Journal of Marine and Freshwater Research* **23**, 1–7 (1977)

Rose, M.: Copépodes pélagiques. *Faune de France* **26**, 1–374 (1933)

Rosendorn, I.: Copepods 1. Die Gattung *Oithona. Wissenschaftliche Ergebnisse der Deutschen Tiefsee-Expedition auf dem Dampfer 'Valdivia' 1898–1899* **23**, 1–58 (1917)

Russell, F.S., Coleman, J.S.: The zooplankton 2. The composition of the zooplankton of the Barrier Reef lagoon. *Scientific Reports of the Great Barrier Reef Expedition* **2(6)**, 159–176 (1934)

Scott, A.: *The Copepoda of the Siboga Expedition.* Leiden: Brill (1909)

Sewell, R.B.S.: Notes on the surface Copepoda of the Gulf of Mannar. *Spolia Zeylanica* **9**, 191–263 (1914)

Sewell, R.B.S.; Fauna of the Chilka Lake; Crustacea Copepoda. *Memoirs of the Indian Museum* **5**, 771–852 (1924)

Sewell, R.B.S.: The Copepoda of the Indian Seas. *Memoirs of the Indian Museum* **10**, 1–221 (1929)

Sewell, R.B.S.: Notes on a small collection of marine Copepoda from the Malay states. *Bulletin of the Raffles Museum* **8**, 25–31 (1933)

Sewell, R.B.S.: A study of the fauna of the Salt Lakes, Calcutta. *Records of the Indian Museum (Calcutta)* **36**, 45–121 (1934)

Sewell, R.B.S.: The free swimming planktonic Copepoda. *John Murray Expedition 1933–34 Scientific Reports* **8**, 1–303 (1947)

Smith, D.F.N., Bulleid, N.C., Campbell, R., Higgins, H.W., Rowe, F., Tranter, D.J., Tranter, H.A.; Marine food web analysis: an experimental study of demersal zooplankton using isotopically labelled prey species. *Marine Biology (Berlin)* **54**, 49–59 (1979)

Steedman, H.F.; General and applied data on formaldehyde fixation and preservation of marine zooplankton. In: H.F. Steedman, ed. *Zooplankton Fixation and Preservation.* Paris: Unesco Press (1976)

Steuer, A.: Bausteine zu einer Monographie der Copepodengattung *Acartia. Arbeiten aus dem Zoologishen Institut der Universitat Innsbruck* **1(5)**, 89–144 (1923)

Tafe, D.J.: *A Study of the Feeding Types of the Dominant Species of Copepods from Port Hacking Estuary with an Illustrated Taxonomic Field Guide for the Identification of fifty one Species of Copepods from the Port Hacking area, N.S.W.* M.Sc. Thesis, The University of Sydney (1979)

Tanaka, O.: Pelagic Copepoda. *Biological Results of the Japanese Antarctic Research Expedition* **10**, 1–95 (1960)

Tranter, D.J.: Zooplankton abundance in Australasian waters. *Australian Journal of Marine and Freshwater Research* **13**, 106–142 (1962)

Tranter, D.J., Bulleid, N.C., Campbell, R., Higgins, H.W., Rowe, F., Tranter, H.A., Smith, D.F.: Nocturnal movements of phototactic zooplankton in shallow waters. *Marine Biology (Berlin)* **61**, 317–326 (1981)

Vaudrey, D.J., Griffiths, F.B., Sinclair, R.E.: Data base for the Port Hacking Estuary Project: Parameters, monitoring procedure, and management system. In: W.R. Cuff and M. Tomczak jr, eds *Synthesis and Modelling of Intermittent Estuaries.* Berlin, Heidelberg, New York: Springer (1983)

Vervoort, W.: Pelagic Copepoda, part 2. *Atlantide Report* **9**, 9–216 (1965)

Wellershaus, S.: On the taxonomy of planktonic Copepoda in Cochin Backwater (a south Indian estuary). *Veroffentlichungen des Instituts fur Meeresforschung in Bremerhaven* 11, 245–286 (1969)

Wellershaus, S.: On the taxonomy of some Copepoda in Cochin Backwater (a South Indian Estuary). *Veroffentlichungen des Instituts fur Meeresforschung in Bremerhaven* 12, 463–490 (1970)

Wickstead, J.H.: Food and feeding in pelagic copepods. *Proceedings of the Zoological Society of London* 139, 545–555 (1962)

Wolfenden, R.N.: Notes on the collection of Copepoda. In: J.S. Gardiner, ed. *The Fauna and Geography of the Maldive and Laccadive Archipelagoes* 2, (suppl. 1), 989–1040 (1905)

Synthesis and Modelling of Intermittent Estuaries
(W.R. Cuff and M. Tomczak jr. eds) Berlin, Heidelberg,
New York: Springer (1983), pp. 135–146.

Some Ecological Effects of Rainfall on the
Protoplankton of South West Arm

Robert R. Parker

Division of Fisheries Research
CSIRO Marine Laboratories
P.O. Box 21, Cronulla, N.S.W. 2230, Australia

Summary. Changes in ecological parameters affecting protoplankton in South West Arm of Port Hacking (Sydney region, Australia) were followed over a 94 day period during which several major rainstorms occurred. The precipitation pattern resulted in the formation of a double pycnocline and virtual isolation of the bottom water. A monospecific bloom of *Ceratium furca* developed in close association with the lower pycnocline. It is suggested that this population may have been utilizing the eutrophic water immediately below for a nutrient supply and the euphotic water above for energy. An unidentified green pigment accumulated in the deep basin during anaerobic periods. Toward the end of the 94 day period South West Arm returned to the normal marine condition with a completely mixed water column and with about the same amount and diversity of standing stock of protoplankton as at the beginning.

Key words: estuaries, protoplankton bloom, rain, stratification, anaerobic conditions, Port Hacking, South West Arm

1. INTRODUCTION

The hydrodynamic mechanisms of South West Arm (SWA), a shallow estuary with an entrance sill largely exposed at low water, have been described by Godfrey & Parslow (1976). Following an intense storm, rainwater on the surface flows out on the ebb tide, mixing in the seaward channel with marine water. The flooding stream passes the sill, and being denser than the surface water, sinks to a depth where it merges with water of similar density in the pycnocline. The succeeding ebb again removes the surface water and the cycle begins again, to continue as long as low–salinity water persists at the surface. Initially the pycnocline is intense and close to the surface, and is maintained there by continuing runoff. When the storm is past and runoff ceases, the pycnocline is progressively weakened and deepened by this circulation, until mixing occurs throughout the water column.

The perturbation of South West Arm from a well mixed marine embayment to a stratified estuarine environment must have profound effects on the populations of protoplankton (single–celled biota living in the water column; Wood, 1963) and on primary production. My purpose was to examine some of these consequences.

2. METHODS AND MATERIALS

This study has drawn on information from the data base for the Port Hacking Estuary Project (see Vaudrey et al., 1983) on file at the CSIRO Marine Laboratories, Cronulla, N.S.W. I consider a 94 day period, from 29 September to 31 December 1976, during which several rainstorms occurred within the watershed. Daily precipitation was estimated from records of the Royal National Park at Audley and a rain gauge maintained at South Cronulla by the Bureau of Meteorology. These two sites are on either side of the SWA.

All samples used herein were from a single station (Stn C; see Fig. 1 of Vaudrey et al., 1983) over the deepest basin of SWA. A submersible pump was employed to deliver water from depths (usually each metre) up to 18 m to the deck of the vessel. An expanded chamber, about 50 mm I.D. and 150 mm high, was used at the end of the hose. Water entered at the bottom and upwelled over the top, providing rapid flushing of the chamber. Salinity and temperature (Hamon conductivity meter; Lockwood, 1970) and oxygen (EIL 1520) were measured by inserting the probes in this chamber. These instruments were calibrated daily. The oxygen probe was calibrated using near–surface and near–bottom water samples (modified Winkler method; Strickland & Parsons, 1972) and assuming a linear response over the recorded range.

A "Variosens" (Früngel & Koch, 1976) *in situ* fluorometer was attached to the underwater pump and linear values (Parker & Vaudrey, 1980) of relative chlorophyll *a* fluorescence were calculated. Water samples obtained from the pumped stream were analysed for chlorophyll using an acetone extraction method (Jeffrey, 1974). Absorption was measured at 630, 647, 663 and 750 nm using a Leitz PMQ4 spectrophotometer with a 5–nm bandwidth, and trichromatic equations (Jeffrey & Humphrey, 1975) were used to estimate chlorophylls *a*, *b* and *c* concentrations.

Samples for protoplankton analysis were fixed with Lugol's iodine solution and examined by the Utermöhl method (Lund *et al.*, 1958). Biomass was estimated using micro–measurements in geometric formulae for volumes of spheres or cylinders (Kovala & Larrance, 1966). These calculations are subject to the errors discussed by SCOR Working Group 33 (1974), by Paerl (1978), and by Hasle (1969).

3. RESULTS

Available measurements of temperature, salinity, dissolved oxygen, chlorophyll *a*, fluorescence, and protoplankton biomass are listed by Vaudrey *et al.* (1983).

Rainfall during the period is shown in Fig. 1. The effect this precipitation had on water column stability (*sensu* Sverdrup *et al.*, 1942) is shown in Fig. 2a. A change of 0.2 in σ_t per metre depth (σ_t = (specific gravity–1) x 10^3 as calculated from salinity and temperature) is taken to depict the separation between a stable and an unstable water column. This value was selected after plotting isopleths connecting $\Delta\sigma_t/\Delta z$ (z = depth in metres) values from 0.05 to 0.5. It depicts the position of a pycnocline which suppressed vertical turbulence, and which persisted unless eroded by the Godfrey–Parslow mechanism described in the Introduction above.

The distribution of the stable layer ($\Delta\sigma_t/\Delta z > 0.2$) reacted quickly to changes in precipitation. Prior to Day 4 the entire water column was unstable, *i.e.* it mixed with a relatively small input of energy. The pulse of rain after the fourth day caused a stable layer of lower–salinity water at the surface. The system quickly exported its fresh water and became unstable again on Day 12. Godfrey & Parslow (1976) described a major pulse of fresh water as being essentially purged in about 7.5 days. A second rainstorm event occurred between Day 17 and Day 25. A stable layer formed and progressively deepened to about 4 m. Subsequently the stable layer dropped rapidly and by Day 30 it had progressed to about 10 m. At this time another pulse of fresh water entered the system and started a new estuarine circulation system in the

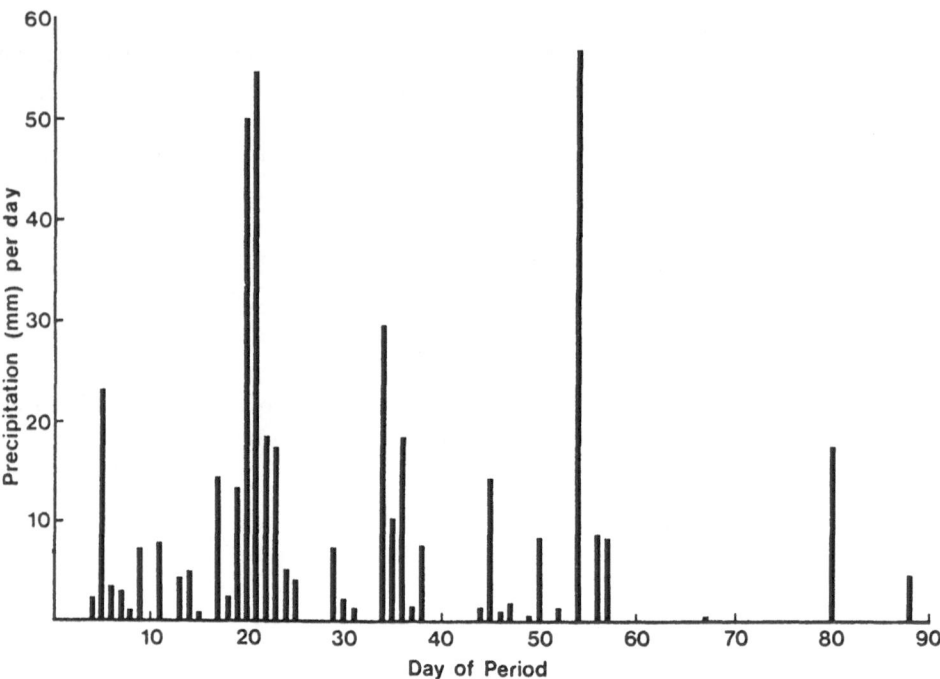

Fig. 1: Estimated daily total precipitation on South West Arm watershed during the period 29 September to
31 December 1976.

surface waters, thus effectively isolating the original pycnocline from the entering
flood–tide water. A double pycnocline system was formed in which the lower
(original) pycnocline persisted at the 8 – 12 m depth over a 23 day period while the
upper pycnocline underwent the dynamic adjustments to fresh water. When the upper
pycnocline disappeared the lower one began again the process of deepening and
eroding until it disappeared on Day 86. At this time the entire water column was once
again unstable.

The 0 and 10% oxygen saturation levels are shown in Fig. 2b. Shortly after Day 25 the 10% oxypleth moved progressively upward to reach a depth of less than 11 m on Day 54. Thereafter it coincided with the descent of the pycnocline. At and below 10% oxygen saturation, nitrate is utilized by heterotrophs as an oxygen source (Rochford, 1974). Below the primary (lowermost) pycnocline, oxygen was completely consumed (as verified by the odour of hydrogen sulphide) much of the time after Day 46. Complete de–oxygenation, however, was interrupted by short periods of non–zero values suggesting that oxygenated water occasionally penetrated below the primary pycnocline. Godfrey's (1983) model suggests a slightly higher $\Delta\sigma_t/\Delta z$ value for complete de–oxygenation and this seems appropriate in view of these observations.

The 1% light level, defining the bottom of the euphotic zone, was probably above the 10% oxygen level during the double–pycnocline period (Scott, 1978).

On Day 1 particulate chlorophyll was found at low concentrations throughout the water column except for a denser (2.5 – 5.0 μg Chl a l^{-1}) subsurface (4 – 6 m) layer (Fig. 2c). By the end of the study period chlorophyll distribution in the water column had returned to its initial distribution but at a larger standing stock: the subsurface layer contained between 5.0 and 10.0 μg Chl a l^{-1} and in the rest of the water column the concentration was generally greater than 2.5 μg l^{-1}. During the interim, a dense aggregation (sometimes exceeding 20 μg Chl a l^{-1}) had developed near the primary pycnocline and the 10% dissolved oxygen layer. De–oxygenation in the bottom water was followed by a build–up of dead particulate material containing an unidentified green pigment (UGP), misidentified as chlorophyll a & b by the trichromatic method. This pigment persisted in the bottom water until instability returned.

The presence of UGP was signalled by particulate acetone extracts with absorption maxima at 652 – 653 nm (Fig. 3), by abnormally low fluorescence emission per unit "chlorophyll a" and "chlorophyll b : a" ratios exceeding 0.5 (Table 1). The distribution of this pigment in the water column, its bright olive–green colour and its photolability suggest identification as either a chlorophyll a degradation product (not pheophytins, pheophorbides, chlorophyllides) or chlorobium chlorophyll 650 from green sulphur bacteria.

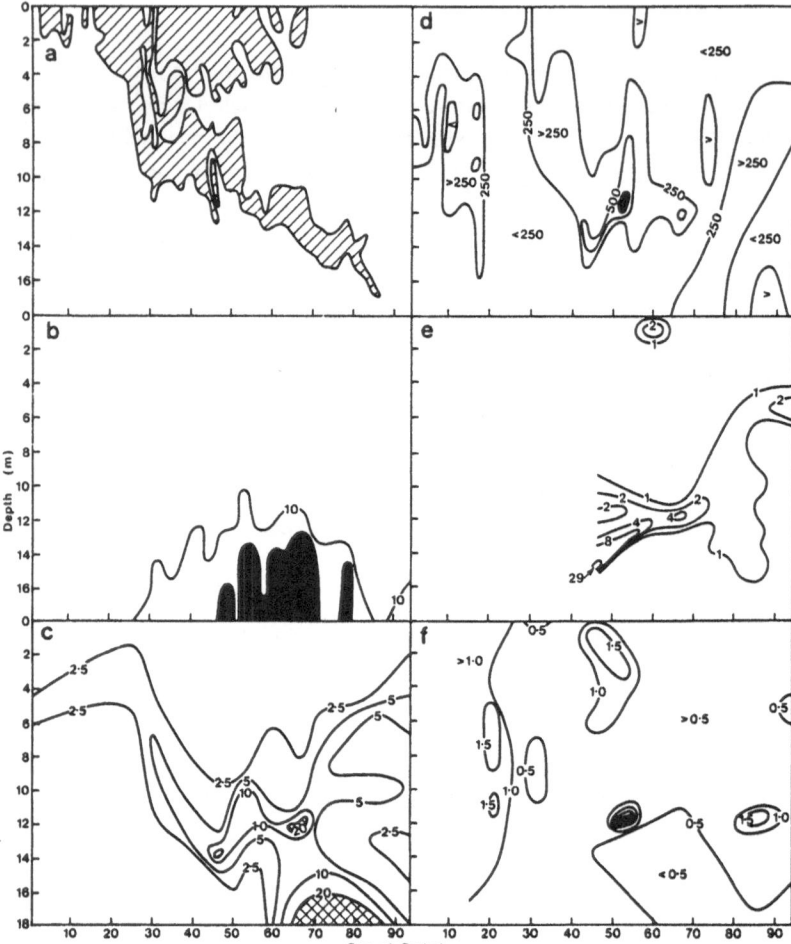

Fig. 2: Distribution of water column stability, dissolved oxygen, extractable chlorophyll *a*, *in vivo* fluorescence, biomass, and the ratio of *in vivo* fluorescence to chlorophyll *a*.

a. Stability. The shaded portion includes depths at which $\Delta\sigma_t/\Delta$ m $>$ 0.2, thus arbitrarily defining a stratum in which the water column was stable and did not mix.

b. Isopleths of 10% dissolved oxygen saturation and ~0% dissolved oxygen. The shaded portion defines the stratum in which no significant oxygen was detected.

c. Isopleths of "extractable" chlorophyll *a*. The isopleths shown are on a logarithmic or doubling scale. The shaded areas indicate levels of more than 20 μg l^{-1}. The large accumulation after Day 60 in the bottom water (z $>$ 14 m) was due to an unidentified green pigment.

d. Isopleths of *in vivo* fluorescence in arbitrary linear units. The black area centred at Day 52 indicates values exceeding 2000 units.

e. Isopleths of biomass of protoplankton, calculated from micro–measurements and geometric formula for spheres and cylinders. Isopleth values are doublings of biomass. Units are μm^3 ml^{-1}.

f. Isopleths of the ratio of *in vivo* chlorophyll *a* fluorescence and extractable chlorophyll *a* (F_v/Chl *a*) x 10^{-2}. The scale is linear, and the blackened area indicates a maximum $>$ 2.5.

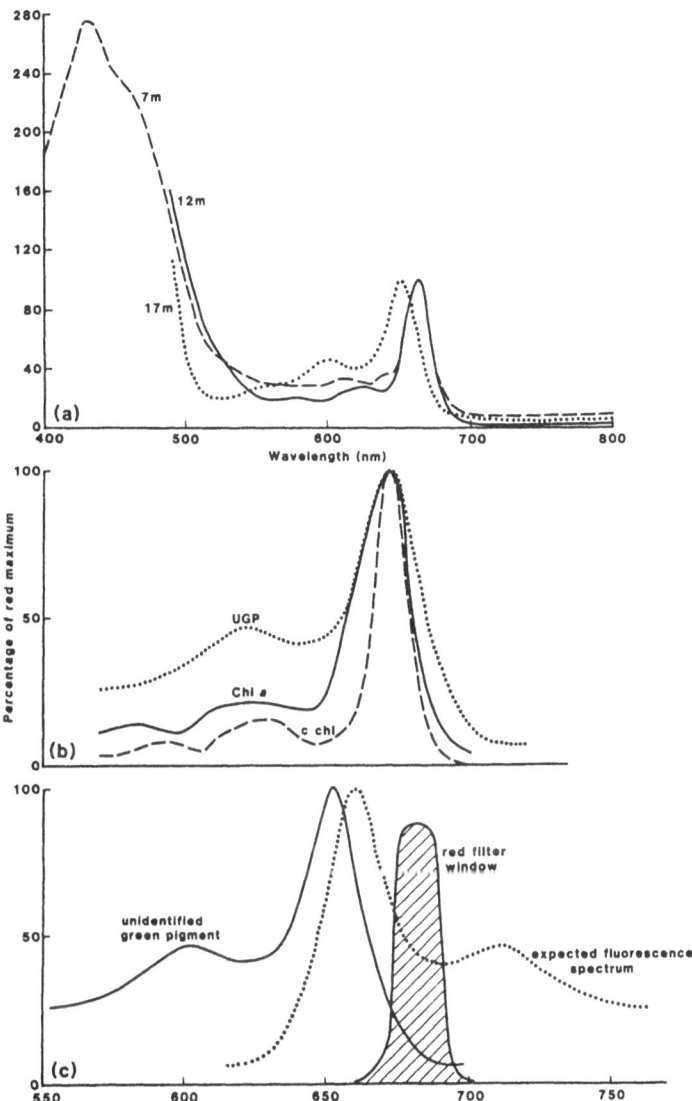

Fig. 3: a. Comparison of absorption spectra from pigments extracted from 7, 12, and 17 m samples, 3 December 1976. Pigments in 90% acetone.

b. Absorption spectra of the unidentified green pigment (UGP) (17 m) compared with those of purified chlorobium chlorophyll (c chl) in ether (after Stanier, 1969) and chlorophyll *a* in acetone (after Gaffron, 1960). Spectra are shifted to align red maxima.

c. Comparison of expected fluorescence spectrum with the observed spectrum due to the unidentified green pigment (UGP). The transmission window of the red filter used in the *in situ* fluorometer is

Table 1.

Comparison of extractable chlorophyll data with estimates of
protoplankton biomass, taxon composition, and
in vivo **fluorescence of selected samples at Stn C.**

Depth (m)	O_2 (% sat)	Chl a (μg/l)	$b : a$ (ratio)	Biomass (μm^3/ml)	Taxon Composition (%, by weight)			In vivo fluorescence	
					Diat[†]	Dino[‡]	Flag[§]	F_v	F_v/Chl a
3 Dec 76									
6	46	1.8	0.21	0.5	7	5	77	128	71
12	25	30.0	0.02	4.0	5	70	25	683	23
17	0	20.0	2.9	<0.1	36	0	64	271	14
22 Dec 76									
7	99	3.9	0.24	0.5	1	55	43	371	95
12	77	5.2	0.42	0.8	15	52	33	236	45
17	2	7.4	2.3	0.6	22	62	16	473	64

† Diatoms
‡ Dinoflagellates
§ Flagellates

In vivo fluorescence isopleths are shown in Fig. 2d. During the anaerobic period (Day 45 – 65) the maximum fluorescence usually occurred 1 m above the maximum chlorophyll a level. There was also a zone of low fluorescence beneath the primary pycnocline corresponding to the unidentified green pigment within the low–oxygen zone. There was only fair agreement between the *in vivo* chlorophyll a values and the extractable chlorophyll a values. An approximate conversion is 100 fluorescence units to 1 μg of extractable chlorophyll a per litre.

The distribution of biomass (Fig. 2e) is in agreement with that of extractable chlorophyll a. Attention is drawn to the apparent discrepancy between biomass and fluorescence: highest fluorescence values were found above the main phytoplankton population as indicated by both chlorophyll a and biomass. The biomass maximum (> 29 μm^3 ml^{-1} or 29 ppm vol/vol) lying just above the de–oxygenated water at 10 m, was composed largely (85%) of the dinoflagellate *Ceratium furca* (25 – 35 μm diameter, 40 – 60 μm long). These dinoflagellates dominated the protoplankton in close proximity to the low–oxygen water throughout the estuarine period.

4. DISCUSSION

This study clearly demonstrates that precipitation can dictate the state of the aquatic ecosystem in SWA. Small infrequent rains have little impact. However, a heavy rainfall (> 20 mm day^{-1}), initiating the formation and subsequent deepening of a primary pycnocline, followed by a second storm and prolonged precipitation, can lead to complete isolation and de–oxygenation of the bottom water. Judging from the dissolved oxygen profiles, only occasionally did oxygenated marine water penetrate the primary pycnocline and mix with the isolated water. The saltatory and unpredictable nature of precipitation in the Sydney region would seem to preclude any prediction of when the basin will become isolated and when it will be flushed.

The biological effects of these rainfall events are only superficially similar to those reported by Loftus *et al.* (1972) for Chesapeake Bay and ascribed by them to an increase in nutrient from the watershed. Nutrient pulses from the land are unlikely in SWA because the watershed soils are generally impoverished. The total biomass in the aerobic water column did not change greatly, only its diversity and distribution. One of the most interesting effects was the development of a dense monospecific protoplankton population just above or within the 10% oxypleth. This was composed almost entirely of the large, mobile *Ceratium furca* which at other times is usually present but does not dominate the protoplankton biomass.

In vivo chlorophyll *a* fluorescence was highest above the extractable chlorophyll *a* maximum. The chlorophyll *a* maximum coincided with the biomass maximum. It may be argued that fluorescence is the complement of photosynthesis; quanta absorbed and not converted into sensible heat are nearly equal to the number used in photochemistry plus the number fluoresced. In this sense fluorescence represents an inefficiency, and the more fluorescence per unit of absorbing pigment the less the system is capable of utilizing absorbed photosynthetically active radiation. Fluorescence per unit of extractable chlorophyll *a* (Fig. 2f) had a maximum which coincided with the fluorescence maximum. Evidently, phytoplankters occupying the level of maximum fluorescence were photosynthesizing at a very low quantum efficiency. This result may be caused by low nutrient concentrations (Samuelsson & Öquist, 1977; Slovacek & Hannan, 1977; Samuelsson *et al.*, 1978; Cullen & Renger, 1979). The reduction of NO_3^- and excretion of NH_4^+ in the oxygen–limited zone by the heterotrophic bacterial population associated with the accumulating detritus may have created a light–limited rather than a nutrient–limited zone. It appears that this nutrient pool was being exploited by the motile dinoflagellates by cycling downward against increasing nutrient but decreasing light and oxygen gradients and upward into increasing oxygen and light but low nutrient. The gradients were evidently quite sharp. High cell density and low nutrient supply in the euphotic stratum would lead to extreme depletion and hence a high fluorescence : chlorophyll *a* ratio. This explanation needs to be examined with detailed nutrient profiles which are not available at this time.

The development of a monoculture of dinoflagellates in association with these altered water–column conditions suggests that *Ceratium furca* can dip into low–oxygen water for nutrients and then photosynthesize in a nutrient–depleted euphotic zone. This has been postulated previously by Eppley *et al.* (1968) but for quite different conditions. The lack of detailed nutrient data precludes any definite statement on nutrient cycling. Total chlorophyll *a* in the oxygenated water column was reduced from about 50 mg m^{-2} prior to the study period to about 25 mg m^{-2} by Day 18. Presumably this was the first response to the heavy rain and light–limiting conditions which followed. By Day 35 the column total had returned to about 50 mg m^{-2} and fluctuated ($\pm$ 10 mg m^{-2}) around that level for the rest of the period. Presumably, the ecosystem was operating on a fixed total amount of nutrient. Because of a decrease in nutrient availability in the euphotic and oxygenated zone and accumulation in the anaerobic zone the diverse community of marine phytoplankton species was superseded by a monospecific population of *Ceratium furca* . This species was able to adapt and flourish under the altered regime, generally adverse to most other endemic species. The concentration of the protoplankton at the aerobic-anaerobic interface reduced the volume of water in which herbivores and detritivores could feed; consequently the detrital "rain" dropped out of the euphotic zone and was mineralized anaerobically. Little nutrient was recycled within the euphotic zone, and hence a low biomass of autotrophs was observed in the upper water column during the double–pycnocline period.

The unidentified green pigment which accumulated in the reducing environment under the primary pycnocline may be present in other estuarine basins. Checks should be built into any environmental monitoring program to detect such anomalies. The choice between chlorobium chlorophyll and an altered chlorophyll *a* degradation product cannot be made from the evidence available. The two possibilities are not exclusive or even mutually exclusive. The distinction is, however, important to the objectives of the Port Hacking Estuary Project. Czeczuga & Gradski (1973) reported that production of both particulate organic carbon and dissolved organic carbon by *Chlorobium limicola* may be as great as from the phytoplankton biomass in surface waters of several Polish lakes. Berman (1976) reported substantial production of both POC and DOC in a dense layer of *Chlorobium phaeobacteriodes*. If the unidentified green pigment in SWA is from a *Chlorobium* sp. population then a primary production system has been missed in our study. If it is due to an altered chlorophyll *a* it signifies a short-cutting of several steps in the heterotrophic remineralization process, steps which represent a food chain leading to the larger biota in the water column.

These results underline the dynamic status of the ecosystem. The protoplankton responded to changes in the physical environment forced by the pattern of precipitation. These responses were rapid. From the rainfall patterns experienced in the area the concept of a climax community or a steady–state ecosystem does not seem valid for the water column.

REFERENCES

Berman, T.: Release of dissolved organic matter by photosynthesizing algae in Lake Kinneret, Israel. *Freshwater Biology* 6, 13–18 (1976)

Cullen, J.A., Renger, E.H.: Continuous measurement of the DCMU–induced fluorescence response of natural phytoplankton populations. *Marine Biology (Berlin)* 53, 13–20 (1979)

Czeczuga, B., Gradski, F.: Relationship between extracellular and cellular production in the sulfuric green bacterium *Chlorobium limicola* Nads. (Chlorobacteriaceae) as compared to primary production of phytoplankton. *Hydrobiologia* 42, 85–95 (1973)

Eppley, R.W., Holm–Hansen, O., Strickland, J.D.H.: Some observations on the vertical migration of dinoflagellates. *Journal of Phycology* 4, 333–340 (1968)

Früngel, F., Koch, C.: Practical experience with the Variosens equipment in measuring chlorophyll concentrations and fluorescent tracer substances, like rhodamine, fluorescein, and some new substances. *IEEE Journal of Oceanic Engineering* OE–1(1), 21–32 (1976)

Gaffron, H. Energy storage: photosynthesis. In: F.C. Steward, ed. *Plant Physiology. Vol. 1B. Photosynthesis and Chemosynthesis.* New York: Academic Press, 3–277 (1960)

Godfrey, J.S.: Tidal flushing and vertical diffusion in South West Arm, Port Hacking. In: W.R. Cuff and M. Tomczak jr, eds *Synthesis and Modelling of Intermittent Estuaries.* Berlin, Heidelberg, New York: Springer (1983)

Godfrey, J.S., Parslow, J.: Description and preliminary theory of circulation in Port Hacking estuary. *CSIRO Division of Fisheries and Oceanography Report* 67 (1976)

Hasle, G.R.: An analysis of the phytoplankton of the Pacific Southern Ocean: abundance, composition, and distribution during the Brategg Expedition, 1947–1948. *Hvalradets Skrifter* 52 (1969)

Jeffrey, S.W.: Profiles of photosynthetic pigments in the ocean using thin-layer chromatography. *Marine Biology (Berlin)* 26, 101–110 (1974)

Jeffrey, S.W., Humphrey, G.F.: New spectrophotometric equations for determining chlorophylls a, b, c_1 and c_2 in higher plants, algae and natural phytoplankton. *Biochemie und Physiologie der Pflanzen* 167 191–194 (1975)

Kovala, P.E., Larrance, J.D.: Computation of phytoplankton cell numbers, cell volume, cell surface and plasma volume per liter, from microscopical counts. *University of Washington Department of Oceanography Special Report* 38 (1966)

Lockwood, D.R.: Portable temperature–chlorinity bridge (S–T meter) instruction manual. *CSIRO Division of Fisheries and Oceanography Report* 47 (1970)

Loftus, M.E., Subba Rao, D.U., Seliger, H.H.: Growth and dissipation of phytoplankton in Chesapeake Bay: Response to a large pulse of rainfall. *Chesapeake Science* 13, 282–299 (1972)

Lund, J.W.G., Kipling, C., Le Cren, E.D.: The inverted microscope method of estimating algal numbers and the statistical basis of estimations by counting. *Hydrobiologia* 11, 143–170 (1958)

Paerl, H.W.: Effectiveness of various counting methods in detecting viable phytoplankton. *New Zealand Journal of Marine and Freshwater Research* 12, 66–72 (1978)

Parker, R.R., Vaudrey, D.J.: A proposed reference standard for *in vivo* chlorophyll *a* fluorometry. *CSIRO Division of Fisheries and Oceanography Report* 125 (1980)

Rochford, D.J.: Sediment trapping of nutrients in Australian estuaries. *CSIRO Division of Fisheries and Oceanography Report* 61 (1974)

Samuelsson, G., Öquist, G.: A method for studying photosynthetic capacities of unicellular algae based on *in vivo* chlorophyll fluorescence. *Physiologia Plantarum* 40, 315–319 (1977)

Samuelsson, G., Öquist, G., Halldal, P.: The viable chlorophyll *a* fluorescence as a measure of photosynthetic capacity in algae. *Mitteilungen Internationale Vereinigung für Theoretische und Angewandte Limnologie* 21, 207–217 (1978)

SCOR Working Group 33: A review of methods used for quantitative phytoplankton studies. *UNESCO Technical Papers in Marine Science* 18 (1974)

Scott, B.D.: Phytoplankton distribution and light attenuation in Port Hacking Estuary. *Australian Journal of Marine and Freshwater Research* 29, 31–44 (1978)

Slovacek, R.E., Hannan, P.J.: *In vivo* fluorescence determinations of phytoplankton chlorophyll *a*. *Limnology and Oceanography* 22, 919–925 (1977)

Stanier, R.Y.: On the existence of two chlorophylls in green bacteria. In: M.B. Allen, ed. *Comparative Biochemistry of Photoreactive Systems.* New York: Academic Press, 60–72 (1969)

Strickland, J.D.H., Parsons, T.R.: A practical handbook for seawater analysis. *Bulletin of the Fisheries Research Board of Canada* 167 (1972)

Sverdrup, H.U., Johnson, M.W., Fleming, R.H.: *The Oceans, their Physics, Chemistry and General Biology.* New York: Prentice-Hall (1942)

Vaudrey, D.J., Griffiths, F.B., Sinclair, R.E.: Data base for the Port Hacking Estuary Project: Parameters, monitoring procedure, and management system. In: W.R. Cuff and M. Tomczak jr, eds *Synthesis and Modelling of Intermittent Estuaries.* Berlin, Heidelberg, New York: Springer (1983)

Wood, E.J.F.: The relative importance of groups of Protozoa and algae in marine environments of the Southwest Pacific and East Indian Oceans. In: C.H. Oppenheimer, ed. *Symposium on Marine Microbiology.* Springfield: Thomas (1963)

Synthesis and Modelling of Intermittent Estuaries
(W.R. Cuff and M. Tomczak jr. eds) Berlin, Heidelberg,
New York: Springer (1983), pp. 147–166.

Primary Production of Benthic Micro–organisms in South West Arm, Port Hacking, New South Wales

Max S. Giles

Australian Atomic Energy Commission
New Illawarra Road, Lucas Heights, N.S.W. 2234, Australia

Summary. Net photosynthetic incorporation of dissolved inorganic carbon into the micro–organisms of shallow benthic areas of South West Arm, Port Hacking, New South Wales was measured *in situ* and in the laboratory. Methods for calculating productivity from continuously recorded light readings using mathematical models based on laboratory studies were evaluated. Three sediment types were investigated and shown to have incorporation rates which varied between 130 and 310 mg of carbon per square metre per day. High levels of carbon uptake by incubations done in the dark and the concentration of dissolved inorganic carbon in sediment interstitial waters were also investigated.

Key words: benthic micro–organisms, primary production, modelling, Port Hacking, South West Arm

1. INTRODUCTION

The intertidal flats and sublittoral regions of estuaries have been recognized ecologically as highly productive areas and a number of workers have made measurements of carbon fluxes through the sediment surface in an attempt to estimate "energy" flow. Gas respirometry (Pomeroy, 1959; Risnyk & Phinney, 1972) or ^{14}C isotopic dilution techniques (Steele & Baird, 1968; McIntyre & Wulff, 1969; Leach, 1970; Boucher, 1972; Luchini, 1972; Joint, 1978) have been used to measure the photosynthetic productivity of samples such as intact cores, subsamples of sediment surface layers, and growths of organisms on artifical substrates. The experiments were lit either by natural light at various times of the day or by artificial light in specially constructed incubators.

The majority of these workers favoured the ^{14}C tracer technique and generally followed the methods set out by Marshall *et al.* (1973); most of these placed importance on the measurement of net photosynthetic incorporation of carbon into the sediment. Only Pomeroy (1959) and Risnyk & Phinney (1972) measured the loss of carbon by respiration in the dark.

All of the above studies were done under specific conditions on particular days and the calculation of integrated annual carbon flux based on such data must contain some element of uncertainty owing to the continual variation of a number of important conditions such as temperature, photosynthetic biomass, nutrient concentration, and light intensity. Of these conditions, light intensity is the most variable, changing markedly from day to day.

The main purpose of the work reported here was to supply data for a carbon flow model (see Cuff *et al.*, 1983). For this reason the results are in a form which is amenable to mathematical manipulation and also assesses the importance of the sources of variability mentioned above, as far as possible.

2. METHODS

In this study the ¹⁴C technique was used, and to overcome the uncertainty associated with fluctuating light a series of measurements was made over a range of known light intensities to establish a mathematical relationship between production and light intensity, *i.e.* a P *v.* I curve. To monitor variation in photosynthesizing biomass and temperature, chlorophyll *a* concentrations were measured and experiments were conducted at ambient temperature. The problem of nutrient status could not be addressed because of the complexity of the micro–environment and the uncertainty about which layers of water were being utilized by the micro–organisms.

To assess its variability, daily readings of the incident light were taken at the CSIRO Division of Fisheries and Oceanography's laboratory, located about 5 km east of South West Arm (see Fig. 1 of Vaudrey *et al.*, 1983); these were then used in conjunction with the P *v.* I curves to calculate real–time production. As a check on the accuracy of this method, *in situ* productivity measurements were made under actual–light conditions and compared with the calculated values.

Choice of the ¹⁴C tracer method, which is an isotope dilution technique, led to a further problem in establishing the level of stable carbon available to the micro–organism during photosynthesis. For the laboratory measurements, this could be controlled by choosing appropriate experimental conditions, but, for the intact sediments used for *in situ* experiments, the concentration of dissolved inorganic carbon (DIC) in the micro–environment of the photosynthesizing cells was unknown. To get an estimate of the importance of this parameter the levels of DIC in the interstitial water of the upper centimetre of sediment were measured.

Samples were collected on intertidal flats from a cockle bed, an exposed flat and a *Zostera* bed at points marked CB, EF, and ZB, respectively, on Fig. 1 of Vaudrey *et al.* (1983). Collections were made in 33–mm diameter polypropylene pill packs which were inverted and pushed into the sediment by hand at low tide. Each pill pack sampler had a hole drilled into the base to allow water to escape as it was pushed into the sand; this hole was sealed with a rubber bung before the sampler was capped in place and then withdrawn from the sediment. Care was taken not to disturb any flocculent layer that might exist during sampling.

The methods for labelling and incubating the samples were slightly modified versions of those described by Baird & Wetzel (1973) and Marshall *et al.* (1973). For *in situ* experiments, 0.37 MBq (10 µCi) of NaH¹⁴CO₃ was injected through the hole in the pill pack and the cores incubated, between noon and sunset, in racks set out on the sediment at the depth from which they were sampled. Six "light" bottles and four "dark" bottles were incubated for each sampling station. After several experiments "dark" bottle uptake of ^{14}C appeared to be high so four extra "dark" bottles which had been killed by adding 10% formalin were added to the array.

At the end of incubation, the labelled solution was drained from the vials and samples were returned to the laboratory where the top 3 mm of sand was scraped from the sediment core, rinsed with filtered seawater, and placed in an 18 ml aliquot of a liquid scintillant made by mixing 4.5 g of 4,5–diphenyloxazole, 400 cm^3 toluene, 400 cm^3 dioxan, and 200 cm^3 ethyl alcohol. Sediment and scintillant were vibrated in an ultrasonic disintegrator; Cab–O–Sil thixotropic agent (Packard Instruments Inc.) was added before counting the mixture in a Beckman Model LS11 liquid scintillation spectrometer. Samples were not fumed with nitric acid because the presence of large amounts of shell neutralized its effect in removing inorganic ^{14}C. It was assumed that "dark" bottle counts, which were subtracted, would allow for inorganic ^{14}C. Each sample was then checked for quenching effects by adding a carefully weighed amount of the original spiking solution and recounting.

For controlled light studies, sample cores were returned to the laboratory, and drained, the top 3 mm layer of sand was then scraped off and placed into clear, 66 cm^3 polypropylene pill packs. After filling with filtered sea water, 0.37 MBq (10 μCi) of NaH^{14}CO$_3$ was added to the incubators which were then capped and inverted several times before being placed in a temperature–controlled bath. Samples were exposed for 3 h to light ranging in intensity from 44 to 2400 μeinst m^{-2} s^{-1}, (photosynthetically available radiation), delivered by quartz iodide lamps and filtered through Balzer absolute filters. Analysis and counting procedures were the same as for *in situ* samples.

The uptake of DIC was calculated by the formula:

$$\left(\frac{C_s\, W_i}{C_i\, W_e} - \frac{C_D\, W_{iD}}{C_{iD}\, W_{eD}} \right) \frac{DIC}{t\; A} \quad = \quad \text{Productivity}$$

where C_s = counts/min for sample, C_i = counts/min from internal standard, W_i = weight of internal spike (g), W_e = weight of label added to incubator (g), C_D = counts/min for "dark" bottle, C_{iD} = counts/min from internal standard for "dark" bottle, W_{iD} = weight of internal standard for "dark" bottle, W_{eD} = weight of label added to "dark" bottle, DIC = weight of dissolved inorganic carbon in the seawater in the incubator (mg), t = incubation time, and A = area of sediment incubated (m^2). ("Dark" bottle counts were subtracted as the average of four incubations.) No allowance was made for isotopic fractionation.

Chlorophyll a determinations were made by scraping the top 3 mm of a core onto a Millipore HA membrane and sucking it dry. The membrane and sediments were then ground in 100% acetone using a mortar and pestle. Four extractions were made for each sample and each determination was done in triplicate. A number of samples, particularly those from station ZB, had to be rejected because the absorption of the acetone extract at 750 nm was too high. The chlorophyll concentration was calculated by the method of SCOR–UNESCO (1966).

2.1 Determination of DIC in interstitial water

Samples were collected at station EF and CB by pushing a 3–cm diameter corer into the sediment. After sealing the base of the corer, the sampler was removed and the surface water was drained through a side hole by depressing a plunger onto the sand surface. The sampler was then inverted and the sample pushed up the corer by the plunger until only the top cm of the core remained inside. The protruding portion was scraped off and the remainder placed in a capped plastic vial. Four vials each containing seven 1–cm slices were collected at each sampling station. Samples with black anaerobic sediment in the top cm were rejected.

At each sampling point 4.5 dm^3 of free water was collected in clean polythene bottles 0.3 m above the sand. Interstitial water was separated from the sand by centrifuging subsamples with Fluorinert FC78, an inert heavy liquid (3M Company). The subsamples were spun at 2000 rev/min for 3 min to recover about 85% of the water. Details of this method have been published by Batley & Giles (1979). The DIC of interstitial and free water was determined by a modification of the Van Slyke method (Scott, 1974), each experiment being in triplicate.

To measure light attenuation in the sand, carefully weighed layers of sand were spread evenly on the bottom of a vial and moistened; collimated light from a quartz iodide lamp was then transmitted through the sample and was measured with a Lambda Instruments quantum meter (Model LI–185A). The sediment weight–to–depth ratio was calculated during a separate experiment.

Salinity was measured with a platinum electrode salinometer (Autolab Model 602). Sediment characteristics were determined by drying and sieving on standard sieves, boiling in 10 M HCl until all shell fragments had dissolved, and by ignition in a muffle furnace for 16 h at 450°C.

3. RESULTS

Variations in the physical characteristics of sediments from the three sampling sites are shown in Table 1.

Table 1.

Physical properties of sediments at sampling sites CB, EF, and ZB.

Physical Property	CB	EF	ZB
Percentage of grains > 1200 μm	0.5	0.6	0.5
Percentage of grains < 1200 μm > 500 μm	1.1	19.7	17.6
Percentage of grains < 500 μm > 250 μm	17.5	71	58
Percentage of grains < 250 μm > 125 μm	78.7	7.7	18.6
Percentage of grains < 125 μm	2.1	0.5	5.2
% weight loss after 16 h at 450°C	2.9	0.5	4.0
% weight loss after boiling in HCl	35.8	1.1	3.1

The bulk of the substrate in all samples was quartz sand. At Stations EF and ZB, the granules were large and angular whereas those at CB were smaller and rounded having been redeposited from the sea. Sediment from Station CB also contained a large proportion of small shell fragments; sediment from ZB contained a larger proportion of fine organic particulates which made it darker in colour than that from the other two sites.

The depth at which a dark anaerobic layer could be seen in the sediment profile occurred at about 1.5 cm at location CB, 3 cm at EF, and 0.3 cm at ZB.

Results of the measurement of light penetration into the sediments are shown in Fig. 1. Lack of linearity is due to refraction from the walls of the vessel holding the sediment during measurement, but the relative opacity of the different sediments is clearly indicated. Greater light absorption of ZB sediments was due to the fine black organic silt content, whereas CB sediment absorbed more light than that from EF because of the increased shell fraction.

The results of chlorophyll *a* determinations are shown in Fig. 2. Since unknown amounts of pigment degradation products are present, the absolute values are in doubt, but the relative trends can be seen and suggest that there were no marked seasonal changes.

The laboratory experiments on productivity were normalised for chlorophyll *a* content and data for the whole year were combined to produce P *v.* I curves. Because of the scatter of points obtained (see Fig. 3) no obvious relationship with temperature could be established and this effect had to be ignored.

To obtain the best fitting mathematical curve to model the data a computer program (BMDX85) originally compiled by Sampson was used. This method obtains a least squares fit, by Gauss–Newton iterations, for a user-specified function. The iterations are continued until further analysis does not decrease the residuals. Asymptotic relationships are obtained which indicate the goodness–of–fit of the function being tested.

Three mathematical models were tested:

$$P = m\alpha I \,/\, (\alpha^2 + m^2 I^2)^{1/2} \tag{1}$$

$$P = \alpha \tanh (mI/\alpha) \tag{2}$$

$$P = \alpha \,(1 - e^{-mI/\alpha}) \tag{3}$$

where P = productivity, I = illuminance, m and α are constants. These functions saturate as I becomes large (*i.e.* P → α). These models were chosen because Jassby &

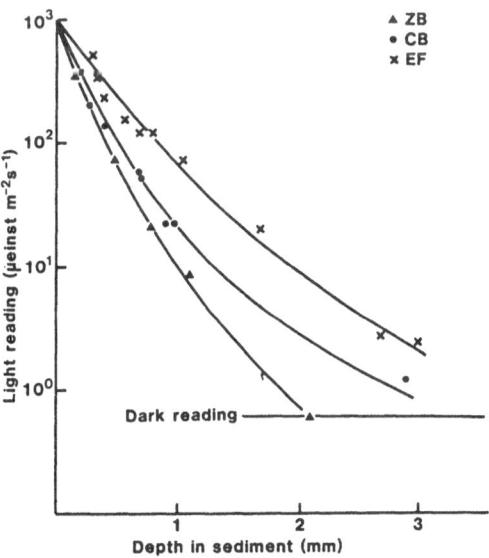

Fig. 1: Light attenuation in sediments.

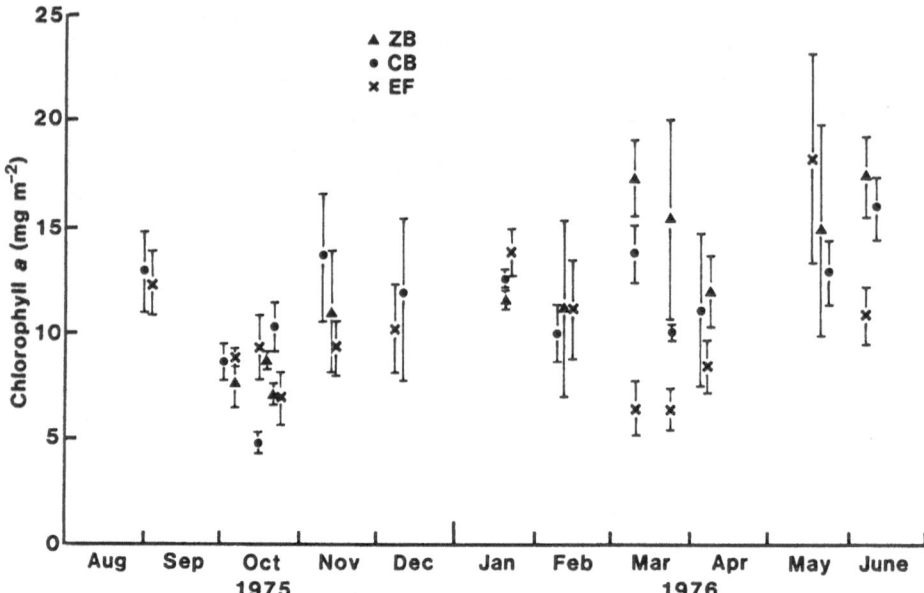

Fig. 2: Chlorophyll *a* content of sediments.

Platt (1976) found these functions most useful to describe data relating productivity and illuminance for phytoplankton studies. The results of this analysis are given in Table 2.

This analysis suggests that Eq. 2 fits the data more closely since the asymptotic standard deviations are the lowest. Eq. 1 may be the better choice however since the parameters are more weakly correlated. Fig. 3 shows the data points plotted against the models represented by Eq. 1 & 2.

Photo–inhibition is not expressed by these functions and was not observed in the experiments, probably owing to the motility of many of the diatoms in the sand, which makes them capable of selecting optimum light conditions at all times.

Table 2.
Results of non–linear regression analyses.

Equation	Station	α	m	Asymptotic Relationships			Error Mean Square
				sd[†] (α)	sd[†] (m)	Correlation of Parameters	
$P = \dfrac{m\alpha l}{(\alpha^2+m^2l^2)^{\frac{1}{2}}}$	CB	1.7	0.01	0.14	0.0017	−0.41	0.33
	ZB	1.9	0.0029	0.2	0.00035	−0.52	0.12
	EF	3.1	0.011	0.19	0.0015	−0.27	0.58
$P = \alpha \tanh (ml/\alpha)$	CB	1.7	0.0056	0.13	0.0011	−0.62	0.33
	ZB	1.7	0.0016	0.17	0.0003	−0.83	0.12
	EF	3.0	0.0035	0.17	0.0005	−0.56	0.6
$P = \alpha(1-e^{-ml/\alpha})$	CB	1.7	0.007	0.14	0.0016	−0.71	0.33
	ZB	1.9	0.0017	0.2	0.00036	−0.88	0.12
	EF	3.1	0.004	0.18	0.00067	−0.6	0.58

[†] Standard deviation.

The results of *in situ* productivity measurements are shown in Fig. 4 with error bars representing one standard deviation. The dotted line joins points calculated from the P *v.* I curves obtained in the laboratory using Eq. 1. One experiment on 25 March 1976 was carried out deliberately under heavily overcast skies to test the method under extreme conditions. The low *in situ* reading suggests that incident light is a major controlling factor since temperature and chlorophyll *a* concentrations were similar on that day to those during the prior and subsequent experiments.

Three methods of calculating daily productivity can be applied to the data: firstly since they represent productivity for a half day (*i.e.* noon to sunset), the *in situ* values can be doubled; secondly, the light readings taken during the experiments can be used to make a second estimate of the productivity during the afternoon from the P *v.* I curves which can then be doubled; and thirdly, the P *v.* I curves can be used to calculate the whole day's productivity from the light charts recorded at the laboratory. The results from these three methods are shown in Table 3.

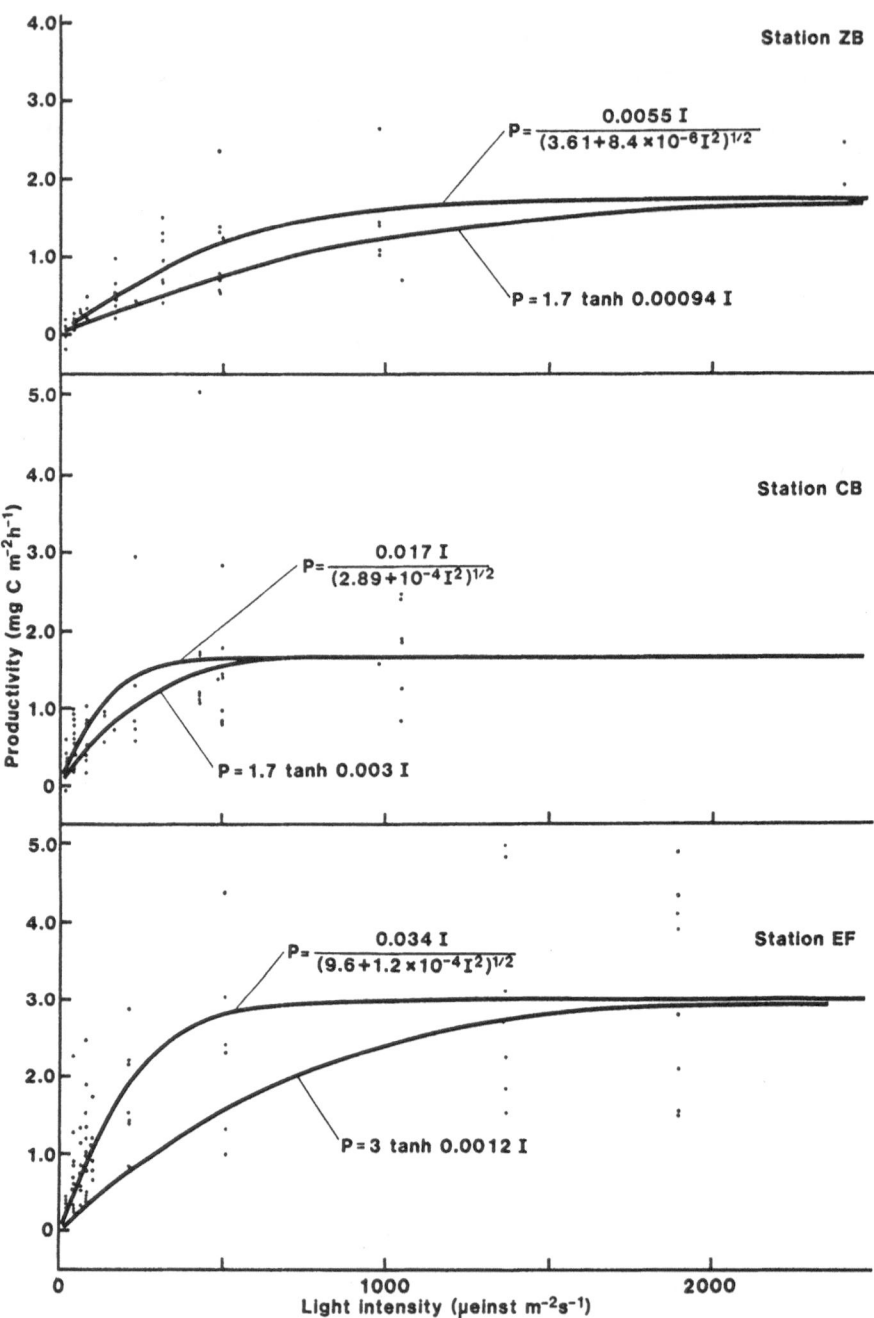

Fig. 3: P v. 1 curves for sediments.

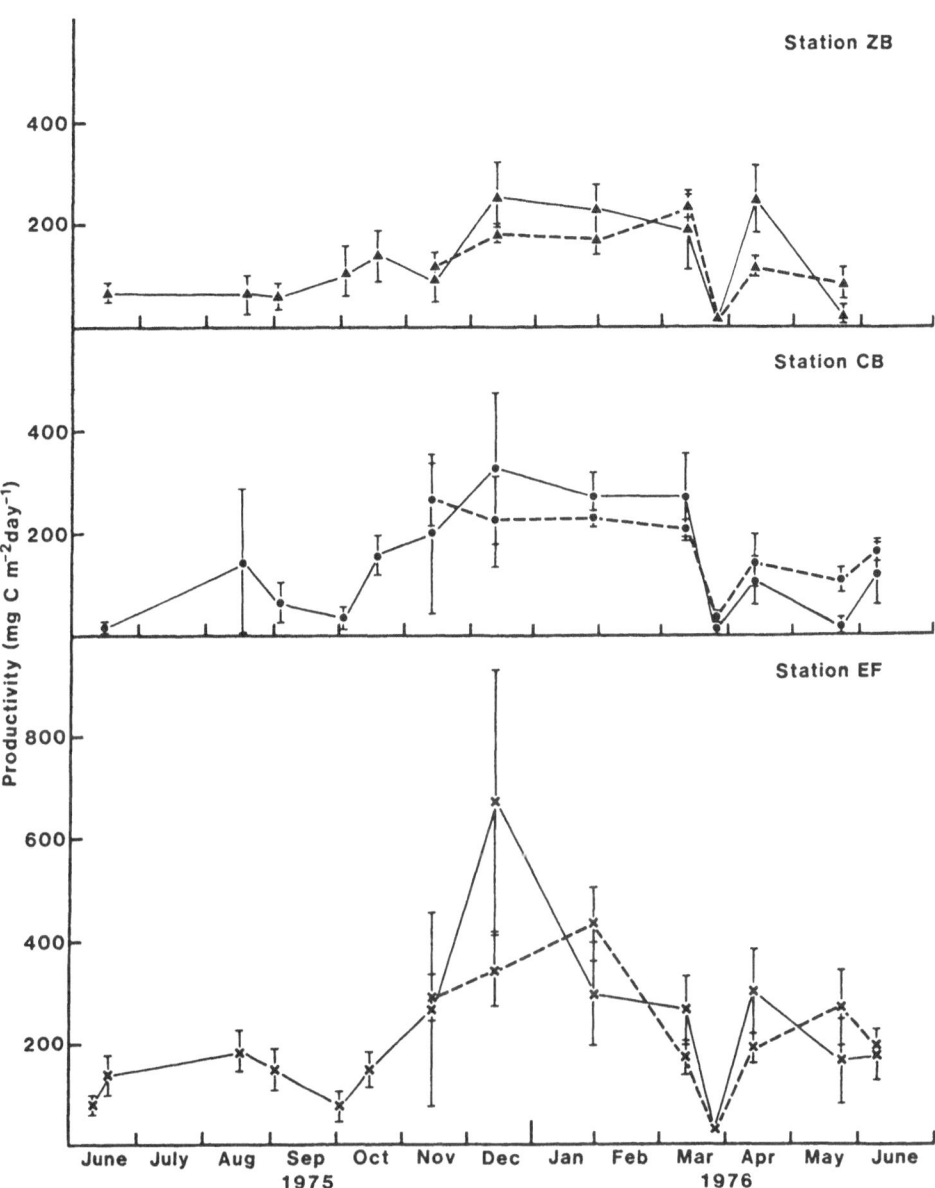

Fig. 4: Results of *in situ* measurements and calculated values of production (Full lines, *in situ* results; broken lines, calculated results).

Table 3.

Productivity on specific days calculated by different methods

(mg C m^{-2} day^{-1}).

Date	In situ (1/2 day doubled)			Calculated [†] (1/2 day doubled[§])			Calculated [‡] (1/2 day doubled[§])			Calculated [†] (whole day[¶])			Calculated [‡] (whole day[¶])		
	CB	ZB	EF	CB	ZB	EF	CB	ZB	EF	CB	ZB	EF	CB	ZB	EF
13.11.75	200	94	268	130	64	142	242	116	260	136	68	154	265	159	364
12.12.75	328	258	670	132	82	172	230	138	286	124	77	161	254	192	402
27. 1.76	268	226	298	190	98	260	233	172	436						
11. 3.76	270	190	264	104	38	188	194	68	346	93	33	168	198	86	457
25. 3.76	14	18	28	26	4	24	48	8	44	32	5	30	92	18	161
12. 4.76	108	252	300	78	34	96	142	60	170	75	30	87	162	86	280
24. 5.76	18	26	162	90	60	114	158	122	202	83	61	103	174	179	330
10. 6.76	120		176	100	28	112	172	54	182	88	28	93	198	84	335

[†] Calculated using Eq. 2.
[‡] Calculated using Eq. 1.
[§] Results from noon to sunset doubled.
[¶] Results from full day's light readings.

Comparison of the results in column 2 of Table 3 with those in columns 3 and 4 allows a test of the efficacy of the two mathematical models. Column 2 (*in situ*) results regressed against column 3 (Eq. 2: $P = \alpha \tanh(mI/\alpha)$) gives an intercept of 48 and a slope of 1.5 ($r^2 = 0.62$, 23 *df*, $p = 0.001$). Column 2 results regressed against column 4 (Eq. 1: $P = m\alpha I/(\alpha^2 + m^2 I^2)^{\frac{1}{2}}$) gives an intercept of 44 and a slope of 0.9 ($r^2 = 0.6$, 23 *df*, $p = 0.001$). This analysis suggests, on the basis of slope estimates, that Eq. 1 most closely fits the experimental findings. Columns 5 and 6 of Table 3 cannot be directly correlated with column 2 since the light readings used to calculate those values include the morning light conditions which cannot be taken into account in the *in situ* studies. In general Eq. 2 tends to underestimate productivity although on 25 March 1976 when low light intensities prevailed, it gave a closer estimation than Eq. 1. Calculations of integrated monthly and annual productivities, using Eq. 1 are given in Table 4.

Table 4.

Monthly and annual integrated productivities calculated from the function

$$P = \frac{m\alpha I}{(\alpha^2 + m^2 I^2)^{\frac{1}{2}}} \quad (g\ C\ m^{-2}).$$

Month	CB	EF	ZB
July 75	4.88	9.68	5.25
August 75	4.84	8.49	2.69
September 75	5.88	5.71	4.30
October 75	4.18	6.81	3.38
November 75	7.65	10.49	4.49
December 75	6.96	10.53	4.69
January 76	6.85	10.53	5.37
February 76	4.99	9.24	4.28
March 76	5.79	13.21	2.37
April 76	4.88	8.67	2.81
May 76	5.36	9.97	5.28
June 76	5.88	9.89	2.49
Annual total	68.1	113.2	47.4

These values are corrected for chlorophyll *a* concentration for individual months but do not reveal any marked seasonal changes. Since chlorophyll *a* concentrations did not vary markedly during the year, the major factor affecting the calculations is the incident light. Although there was a generally higher productivity during the months of longer day length (*i.e.* November, December and January), individual months with shorter day lengths gave high results (*e.g.* May and September).

Of the three stations, EF represents the largest area of intertidal flat in the South West Arm system, having an area of approximately 215×10^3 m², whereas the area of CB is about 130×10^3 m². The substrate at ZB occupies only a few small areas and may be ignored.

"Dark" bottle counts were relatively high for all experiments and equivalent to about 10% of the "light" bottle uptake of DIC. Table 5 shows the average results obtained during laboratory and *in situ* measurements.

Table 5.
Average uptake during dark incubations
(mg C m⁻² h⁻¹).

Station	*In situ*	Laboratory
CB	2.1	3.5
EF	0.6	0.9
ZB	1.1	2.8

In each case, there was a statistically significant difference (28 *df*, $p = 0.001$) between the *in situ* and laboratory experiments, probably due to the different exposure times (3 h for laboratory and 5–8 h for *in situ* experiments) and the core slices used (3 mm for laboratory and 25 mm for *in situ* experiments). The results of experiments in which "killed–dark" bottles were used are given in Table 6.

Table 6.
Average results of "dark" bottle experiments comparing
normal treatments to "killed" treatments for *in situ* experiments
(mg C m⁻² h⁻¹).

Station	Normal	"Killed"	Significance
CB	1.6	1.4	$p = 0.8$, 8 *df*,
EF	0.6	0.1	$p = 0.01$–0.001, 8 *df*.
ZB	0.9	0.3	$p = 0.01$–0.001, 8 *df*,

The significant differences between "killed" and normal treatments at EF and ZB suggests that chemoautotrophic micro–organisms may be active at these more "estuarine" sites where there are larger deposits of organic matter (Table 1). The greater uptake at CB could be explained by increased physical exchange of ^{14}C onto the $CaCO_3$ component of the large, finely–ground shell fraction of the sediment, not inhibited by the presence of formalin, at that sampling site.

Measurements of DIC in interstitial water were made on 13 sets of samples collected between March 1976 and November 1977 at CB and EF; the results are shown in Table 7.

Table 7.
Dissolved inorganic carbon in interstitial water and
free water 0.3 m above the sediment.

Date	Sampling Point	Dissolved Inorganic Carbon (mg dm^{-3})		Percentage Difference $\frac{100\ (IW-FW)}{FW}$
		Interstitial Water (IW)	Free Water (FW)	
16.3.76		27.3	21.6	26
29.3.76		24.0	15.7	53
8.4.76		27.1	21.0	29
22.4.76		26.4	22.3	18
20.5.76		30.0	22.8	32
7.6.76	CB	29.7	23.0	29
6.7.76		26.1	18.1	44
4.8.76		27.7	23.7	17
22.11.76		26.6	20.7	27
30.11.76		28.7	21.6	33
21.9.77		32.6	25.3	29
22.9.77		27.4	23.5	17
26.9.77		32.8	22.7	44
			Mean	31
16.3.76		21.1	16.9	25
29.3.76		15.1	11.4	33
8.4.76		22.5	21.6	4
22.4.76		21.4	24.0	-10
20.5.76		25.5	21.8	17
7.6.76	EF	27.6	24.3	14
6.7.76		21.3	21.0	1
4.8.76		27.1	22.5	20
22.11.76		22.0	17.0	29
30.11.76		25.2	18.9	33
21.9.77		26.2	26.1	4
22.9.77		32.4	21.8	49
26.9.77		33.5	26.5	27
			Mean	19

A consistent difference is obvious in samples from CB whereas in the EF samples there is a greater degree of variability.

Changes in free water DIC can be related to changes in salinity over the range of salinities ($10 - 35 \cdot 10^{-3}$) encountered during the measurements. From 18 measurements taken during this study and 30 by Scott (pers. comm.) a linear relationship DIC (g m^{-3}) = 0.43 x salinity (10^{-3}) + 8.5 was found (r = 0.9; std. dev. slope = 0.03, std. dev. intercept 1.1). At first, it was thought that the differences in DIC could be explained by the fact that sediment samples and free water were always collected at low tide, so interstitial water was being compared to the least–saline water of the column profile. This would have been more important at EF which is closer to the freshwater source. However, given the above relationship, salinities in the region of $40 - 60 \cdot 10^{-3}$ would be necessary in the interstitial water to maintain these differences. Evaporative increase of salinity of interstitial water at low tide at some of the rates seen by Capstick (1957), *i.e.* between 1.5 and $4.5 \cdot 10^{-3}$ h^{-1}, could explain the differences observed at ZB and EF, but this is unlikely at CB since that station is rarely exposed to the air.

It is possible that being denser, water of higher salinity could be moved within the sediment at low tide under a hydraulic gradient in a similar way to ground water movement on dry land, producing the difference at CB. Another influencing factor at CB could be the increased content of finely–ground $CaCO_3$ due to the shell fraction.

Although the actual micro–layer utilized by the photosynthesizing micro–organisms is not known, the effect of higher DIC in the interstitial water would be to raise the value of the productivity figures quoted here, as free water values were used when calculating carbon incorporation.

4. DISCUSSION

The results of this work lie within the range of the findings of most of the other authors, as shown in Table 8, despite the different methods used and the range of substrates examined.

Table 8.
Comparison of results of various studies.

Author	Sediment Type	Production	Comments
Luchini (1972)	Artificial Pebbles	27.5–130 [†]	[14]C
Boucher (1972)	Sand	0.5–100 [‡]	[14]C *in situ*
Risnyk & Phinney (1972)	Sand	275–325 [§]	Respirometry in laboratory
	Mud	0–125 [§]	
Pomeroy (1959)	Mud	200 [§]	Respirometry *in situ*
Leach (1970)	Mud	31 [‡]	[14]C *in situ* (5–h incubation)
Joint (1978)	Mud	143 [§]	[14]C *in situ* (4–h incubation)
This work	Sand (EF+CB)	91 [§]	Calculated real time from P v. I
	Mud (ZB)	47 [§]	
	Sand(EF+CB)	65 [§]	Average of *in situ* measurements
	Mud (ZB)	46 [§]	

[†] mg C m^{-2} h^{-1}
[‡] mg C m^{-2} day^{-1}
[§] g C m^{-2} yr^{-1}

Substrate type is important, however, from another aspect because light penetration into fine grained sediments is poor. This in turn controls the types of photosynthesizing micro–organisms which can inhabit such substrates. Most authors working with these substrates noted that "slicks" of large diatoms formed on the mud surface when the intertidal flats became exposed and began active photosynthesis; Williams (1965) described this motility in detail. Joint (1978) could find no ^{14}C uptake if incubations were done when the sediments would normally have been covered by the tide.

In this study incubations were done during periods of in–flowing tide and no similar effect was noticed. In the case of sandy sediments, deeper light penetration allows a greater assemblage of micro–algae (largely diatoms) and photosynthetic bacteria to operate. No tidal effect was noticed during this study.

Large seasonal changes of productivity were not apparent in the results calculated from the P v. I curves, but the results of the *in situ* experiments suggest that there was a peak during the summer period. Some bias may be present in the *in situ* results since most of the days selected for experimentation had only a light cloud cover. Results for the one day where heavy overcast conditions were chosen (25 March 1976) substantially reduced the average production rate when calculating from the *in situ* results. There may also be bias in the P v. I curves because the samples were generally collected at 1000 hr but not incubated until the afternoon which may have been a time of reduced production capacity if the periodicity of production capacity seen in phytoplankton studies (Doty & Oguri, 1957; Shimada, 1958), applies to benthic micro–algae. Most of the workers reported seasonal changes, each based on *in situ* studies, which may have had the same bias as the *in situ* results reported here.

Seasonal trends, in this study, were smoothed out by the calculations done on a day–to–day basis because:
(a) mornings were cloudy more often than afternoons (*i.e.* when *in situ* studies were done),
(b) the value of α (where the P v. I curve saturates) cuts out a large part of the P v. I curve on cloudless days during summer, and
(c) the winter days during the study period tended to be cloudless while many of the summer days were overcast.

The models revealed the importance of m, which roughly represents the initial slope of the P v. I curves, when calculating total production.

When Eq. 2 was applied to the data, it gave a total figure for the year which was about half that obtained when Eq. 1 was used as the model. α was similar in both models while m was lower in Eq. 2. This indicates that light values which fall below that which saturates the photosynthetic capacity of the micro–organisms are very important. Early morning and late afternoon light, together with the light intensities prevailing on overcast days, must have comprised a large proportion of the total hours of light. The common practice of conducting *in situ* studies between 1000 and 1500 hr often overlooks this effect.

Productivity of benthic micro–organisms in South West Arm was about half the average phytoplankton production per unit area seen by Scott (1979).

One important aspect of benthic production is that it takes place within a very small volume, *i.e.* within the upper 3 mm of sediment. For each gram of carbon incorporated during photosynthesis, about 2.7 g of oxygen is released, this means that approximately 0.74 g O_2 m^{-2} would be released into the pores of the sediment and the micro–layer of water at the boundary of the sediment at EF each day using the average results obtained in this study.

Since diffusion rates for oxygen in water are relatively low, there would be a tendency for saturated conditions to build up. Using micro–electrodes, Revsbech *et al.* (1980) measured oxygen tensions up to five times air saturation levels in sandy sediments after 6 h of illumination.

Such fluxes must play a large part in supporting bacterial oxidation, and supplying oxygen to tube dwelling and burrowing fauna. Since there are highly anoxic conditions in the sediment within 2–3 cm of the surface layer, oxygen from this source must also play an important role in maintaining the equilibria which exist in this chemically–active region.

ACKNOWLEDGEMENTS

I would like to acknowledge the help of A. Dudaitis who measured the levels of dissolved inorganic carbon and assisted in many other aspects of this study. S. Wong and G. Clarke carried out much of the computer analysis.

REFERENCES

Baird, I.E., Wetzel, R.G.: Primary production on a sandy beach. In: *A Guide to the Measurement of Marine Primary Production under some Special Conditions. Oceanographic Methodology Monograph No. 3.* Paris: UNESCO/SCOR (1973)

Batley, G.E., Giles, M.S.: Solvent displacement of sediment interstitial waters before trace metal analysis. *Water Research* 13, 879–886 (1979)

Boucher, D.: Evaluation de la production primaire benthique en Baie de Concarneau. *Comptes Rendus Hebdomadaires des Séances de l'Academie des Sciences, Série D Sciences Naturelles* 275, 1911–1914 (1972)

Capstick, C.K.: The salinity characteristics of the middle and upper reaches of the River Blyth Estuary. *Journal of Animal Ecology* 26, 295–315 (1957)

Cuff, W.R., Sinclair, R.E., Parker, R.R., Tranter, D.J., Bulleid, N.C., Giles, M.S., Godfrey, J.S., Griffiths, F.B., Higgins, H.W., Kirkman, H., Rainer, S.F., Scott, B.D.: A carbon budget for South West Arm, Port Hacking. In: W.R. Cuff and M. Tomczak jr, eds *Synthesis and Modelling of Intermittent Estuaries*. Berlin, Heidelberg, New York: Springer (1983)

Doty, M.S., Oguri, M.: Evidence for a photosynthetic daily periodicity. *Limnology and Oceanography* **2**, 37–40 (1957)

Jassby, A.D., Platt, T.: Mathematical formulation of the relationship between photosynthesis and light for phytoplankton. *Limnology and Oceanography* **21**, 540–547 (1976)

Joint, I.R.: Microbial production of an estuarine mudflat. *Estuarine and Coastal Marine Science* **7**, 185–195 (1978)

Leach, J.H.: Epibenthic algal production in an intertidal mudflat. *Limnology and Oceanography* **15**, 514–521 (1970)

Luchini, L.: Etude qualitative et quantitative d'une population de diatomées du microphytobenthos épilithe (Anse des Cuivres, Marseille). *Téthys* **3**, 459–505 (1972)

Marshall, N., Skauen, D.M., Lampe, H.C., Oviatt, C.A.: Primary production of benthic microflora. In: *A Guide to the Measurement of Marine Primary Production under some Special Conditions. Oceanographic Methodology Monograph No. 3*. Paris: UNESCO/SCOR (1973)

McIntyre, C.B., Wulff, B.L.: A laboratory method for the study of marine benthic diatoms. *Limnology and Oceanography* **14**, 667–678 (1969)

Pomeroy, L.R.: Algal productivity in salt marshes of Georgia. *Limnology and Oceanography* **4**, 386–397 (1959)

Revsbech, N.P., Sorensen, J., Blackburn, T.H., Lomholt, J.P.: Distribution of oxygen in marine sediments measured with microelectrodes. *Limnology and Oceanography* **25**, 403–411 (1980)

Risnyk, R.Z., Phinney, H.K.: Manometric assessment of interstitial micro–algae production in two estuarine sediments. *Oecologia* **10**, 193–203 (1972)

SCOR–UNESCO Working Group 17: *UNESCO Monographs on Oceanographical Methodology Monograph No. 1*. Paris: UNESCO (1966)

Scott, B.D.: A total dissolved inorganic carbon analyser. *CSIRO Division of Fisheries and Oceanography Report* **60** (1974)

Scott, B.D.: Seasonal variations of phytoplankton production in an estuary in relation to coastal water movements. *Australian Journal of Marine and Freshwater Research* **30**, 449–461 (1979)

Shimada, B.M.: Diurnal fluctuation in photosynthetic rate and chlorophyll *a* content of phytoplankton from Eastern Pacific waters. *Limnology and Oceanography* **3**, 336–339 (1958)

Steele, J.H., Baird, I.E.: Production ecology of a sandy beach. *Limnology and Oceanography* **13**, 14–25 (1968)

Vaudrey, D.J., Griffiths, F.B., Sinclair, R.E.: Data base for the Port Hacking Estuary Project: Parameters, monitoring procedure, and management system. In: W.R. Cuff and M. Tomczak jr, eds *Synthesis and Modelling of Intermittent Estuaries*. Berlin, Heidelberg, New York: Springer (1983)

Williams, R.B.: Unusual motility of tube dwelling pennate diatoms. *Journal of Phycology* **1**, 145–146 (1965)

Synthesis and Modelling of Intermittent Estuaries
(W.R. Cuff and M. Tomczak jr. eds) Berlin, Heidelberg,
New York: Springer (1983), pp. 167–176.

Size–Specific Respiration Rate of Port Hacking Zooplankton

David J. Tranter[†], Gillian Kennedy[‡]

[†] Division of Fisheries Research
CSIRO Marine Laboratories
P.O. Box 21, Cronulla, N.S.W. 2230, Australia

[‡] Capricornia Institute of Advanced Education
Yaamba Road, Parkhurst, Rockhampton, Qld 4702, Australia

Summary. The respiration rates of natural zooplankton assemblages from Port Hacking, measured 4 – 5 h after capture, at temperatures of 18 – 22°C, are described by the equation

$$R' = 0.857 \ W^{-0.306} \qquad \mu\text{g-atom } O_2/\text{mg dry weight/h}$$

where W is dry body weight and R' is the respiration rate per unit body weight. These weight–specific respiration rates are higher than those recorded by other workers for larger zooplankton, 24 h after capture. The difference may be due to the small size of our experimental animals (0.8 – 29 μg) or to the fact that we were measuring active, rather than basal, metabolic rates.

Key words: zooplankton, respiration, Port Hacking, South West Arm

1. INTRODUCTION

There is a large body of evidence, summarized by Zeuthen (1947), which shows that, in the animal kingdom as a whole, respiration rate per unit mass is higher in small organisms than in large. More than 94% of the total respiration in seawater samples, for example, is due to organisms small enough to pass through a screen of 366 μm mesh width (Pomeroy & Johannes, 1966). To be comparable, therefore, respiration estimates need to be size–specific. This study deals with the zooplankton.

The aim of the study was to provide size–specific respiration data for a carbon flow model of South West Arm, Port Hacking (CSIRO, 1976; Cuff *et al.*, 1983; Cuff, 1983). Surface temperatures in the study area range seasonally from 14 to 20°C (Godfrey & Parslow, 1976).

Ideally, respiration should be measured *in situ* under normal physical and biological conditions; laboratory measurements impose constraints. Marshall *et al.* (1935), for example, showed that respiration rate is depressed when oxygen tension falls to 30 – 40% of the saturated value. Most workers have used single species for respiration experiments. Being more concerned with size–specific than with species–specific differences, we used species mixtures from the wild, and measured their respiration rates as soon as possible after their collection.

2. MATERIALS AND METHODS

Plankton was collected from South West Arm with nets of 50, 100, 200, or 500 μm mesh according to the size of animals required. Each haul was placed in a 20-l polythene container and transported to the laboratory within 10 min of collection. After allowing debris to settle for about 30 min, the plankton was siphoned off into a separate container and passed through a size fractionating column primed with clean, unused, filtered seawater. The final separation was carried out by flushing the column with another 20 l of clean seawater. The screens ranged from 10 to 800 μm (mesh width), graded logarithmically (see Fig. 3). Each size category of plankton was transferred to a 1–litre jar and aerated in dim light in an air–conditioned room (20 – 22°C). The time taken for collection and fractionation of the plankton was 3 – 4 h. Only active and apparently undamaged plankton organisms were used in the experiments.

The smaller size fractions were dominated by dinoflagellates and nauplii, and the larger by cladocerans and copepodite developmental stages (*Oithona ?oculata, Penilia avirostris, Paracalanus* sp., *Acrocalanus* sp., *Temora turbinata, Acartia bispinosa, Evadne nordmanni,* and *Oithona similis*). The three largest size categories (10 – 20 μg dry weight) contained adult stages.

During the course of the experiments, which lasted several months, the plankton changed from dominance by calanoid copepods to dominance by cladocerans. The latter have a large carapace relative to their body size. Also at times, chains of diatoms were retained by screens that would have been too coarse to retain individual cells; their inclusion in the experiments was either avoided or kept to a minimum.

The methods available for measuring oxygen concentration fall into two categories, chemical and physical. In the former (for example, the Winkler method), oxidation–reduction reactions are involved and these can be quantified. Their main disadvantage is the time needed to effect a measurable reduction in oxygen content, during which bacterial growth and respiration could be taking place and the plankton could be starving. However, such methods do have the advantage that low plankton densities can be used.

Alternative methods involve polarographic detection of changes in oxygen tension or partial pressure. There are basically two types of oxygen electrodes: the probe for use with large volumes of water and the electrode cell for small quantities of cellular or sub–cellular particles. The design by Clark (1956) of an efficient oxygen electrode initiated adoption of polarography for biological work. Pomeroy & Johannes (1966), Nival *et al.* (1971), Razouls (1972), and Smith *et al.* (1973) have all used polarographic methods for measuring plankton respiration.

In this study, polarographic methods were used (Rank Bros, Bottisham, Cambridge, U.K.). The electrode cells were connected in parallel with a 12–channel Speedomax recorder (Leeds and Northrup 1120). To minimize stress due to stirring, the apparatus was modified by inserting a gauze separator between the flea and the 5–ml respiration chamber (Fig. 1). A Haake control unit maintained the temperature of the circulating water to within 0.05°C of constant. A water–cooled plankton storage container was used when the air temperature rose above 20°C. Initially, the experimental temperature was 20°C; subsequent work was done at 18°C, the ambient seawater temperature. Reduction of the working temperature decreased the rate of diffusion of oxygen through the teflon membranes (as shown by the slope of the recorder trace), but did not reduce its sensitivity to oxygen uptake.

The electrodes were allowed to equilibrate for about 1 h before use and they were then used for 1 h in a calibration run. Calibration was done with air–saturated seawater of known salinity and temperature. Oxygen content was derived from standard International Oceanographic Tables. The calibration was maintained at the working temperature of the water bath to avoid de–saturation due to temperature

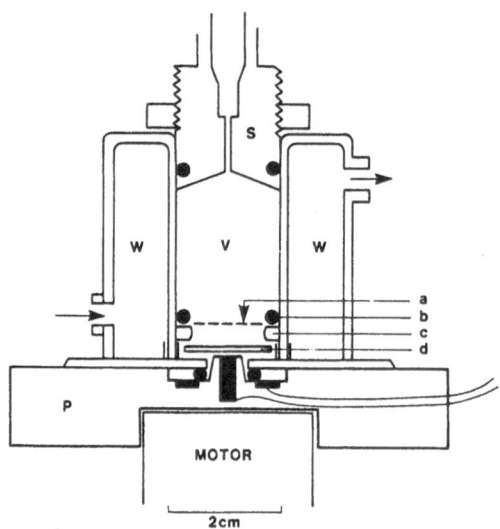

Fig. 1: Modification to Rank Bros respiration cell to protect zooplankton from damage by the stirring flea. (Scale approximate). S – stopper; V – plankton chamber; W – water jacket; P – perspex electrode unit; a – nylon or metal gauze (25 – 75 μm) mesh; b – O–rings; c – perspex split–ring; d – flea.

change. Cells with an electrode consumption of more than 0.1 ml O_2 h^{-1} were not used. Electrode consumption was subtracted from total oxygen uptake to give plankton oxygen demand. Included in the electrode consumption value is oxygen uptake due to phytoplankton and bacteria in the water. Experiments were done in the dark.

The experimental water was filtered, either through a Millipore filter (pore size 0.22 μm) (MFW) or through the 65–μm screen of the fractionation column (FSW), some difficulty being experienced with re–saturating water from Millipore filtration. No appreciable difference was observed over a period of several hours between the oxygen uptake of MFW and FSW. Experiments were run for about 45 min (or less, if the O_2 saturation fell to 80%). All measurements were taken for a 30–min period of the experimental run, or from the initial 10% drop in O_2 level.

Dry weights were determined as follows: the total numbers of individuals of each size were counted, in four aliquots, after cooling for 1 h at 0°C to immobilize them. Each aliquot was then poured through a pre–dried, pre–weighed screen rinsed with distilled water. They were then blotted dry from below, dried overnight at 75°C, cooled in a desiccator, and weighed on a Mettler Microbalance. In later work, an oven temperature of 65°C was used (Lovegrove, 1966), and fine Monel (75 μm) or bronze screens (45 μm) were used instead of nylon. From these results, a function was

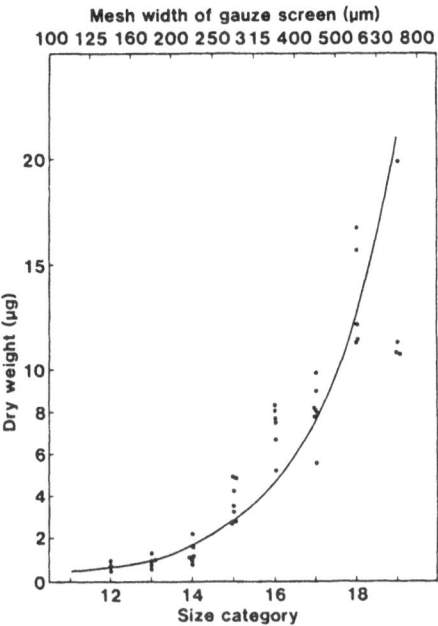

Fig. 2: Size – dry weight relationship for size categories of zooplankton separated by the multiple–screen separator.

derived of dry weight (DW) per unit size (Fig. 2). This function was used for experimental batches as a check on the accuracy of weight determinations which were frequently difficult to determine.

Tests for a range of plankton concentrations (0.003 – 0.50 mg ml^{-1}) showed that specific respiration rate (R') (315 – 500 μm) was a function of the degree of crowding in the experimental chamber. Consequently, in the following experiments, concentrations were standardized.

3. RESULTS

Table 1 gives weight equivalents for size categories of plankton separated by screens ranging in mesh width from 160 μm to > 800 μm. The relationship is exponential (Fig. 2), the regression equation being

$$W_i = 0.370 \text{ x } 1.688^{(i-1)} \tag{1}$$

where W_i is the mean dry weight of plankton organisms in size category i.

The carbon : dry weight ratio was 0.27 (Table 1). This is low compared with other published values (e.g. Ikeda, 1974: 0.35–0.55), but replicate analyses on a later occasion confirm that our ratio is correct.

Table 1.
Dry weight and carbon equivalents for zooplankton of various sizes. The separation of zooplankton into categories of uniform size was carried out with a multiple–screen size separator.

Size Category (μm)	DW [1]/Animal (μg)	Carbon Content/Animal (μg)	Carbon/DW
> 800	29.0	8.4	0.29
630 – 800	15.0	4.9	0.33
500 – 630	8.7	3.1	0.36
400 – 500	7.9	1.6	0.20
315 – 400	4.8	0.71	0.15
250 – 315	1.7	0.42	0.25
200 – 250	0.9	0.36	0.40
160 – 200	0.8	0.125	0.16
Mean			0.27

[1] DW = dry weight

The relationship between weight–specific respiration rate R' and size (μm) is shown in Fig. 3.

Fig. 4 shows the logarithmic relationship between specific respiration rate (R') and dry weight (W). The equation is

$$\log R' = -0.306 \log W + \log (0.857) \tag{2}$$

$$R' = 0.857 \, W^{-0.306} \quad \mu g\text{–atom } O_2/mg \text{ DW}/h \qquad (r = -0.79) \tag{3}$$

where r is the product–moment correlation coefficient.

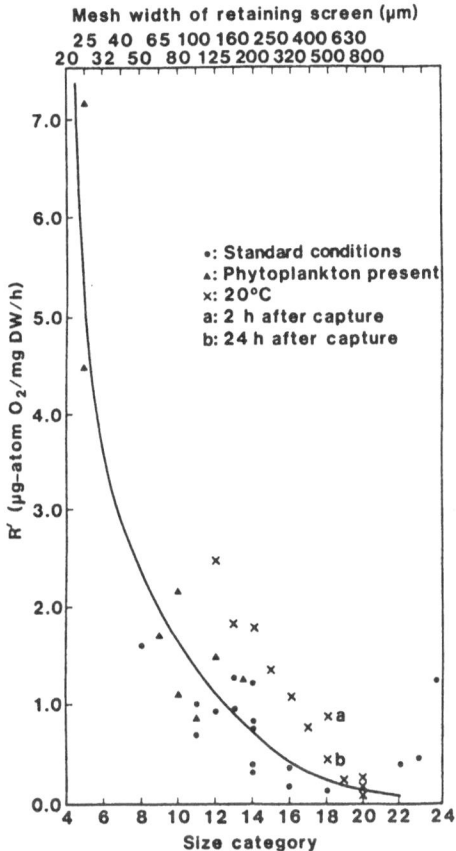

Fig. 3: Relation between weight-specific respiration rate (R') and zooplankton size. The different symbols denote different experiments.

4. DISCUSSION

Table 2 lists some specific respiration rates previously recorded for marine copepods and other zooplankton by other workers.

Ikeda (1970, 1974) has made a study of zooplankton respiration in relation to body size and environmental temperature. He has shown that temperature affects the size of the exponential index (b) in the equation $R' = aW^b$, where R' is the specific respiration rate. His equations (Ikeda, 1974) are as follows:

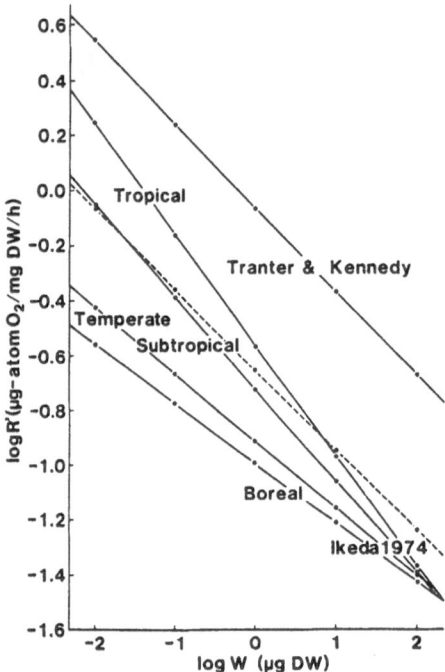

Fig. 4: Comparison of the present respiration data for zooplankton in South West Arm study area with those obtained by Ikeda (1974). Log (specific respiration rate) is plotted as a regression on log (dry weight). The broken line is the (recomputed) regression for those observations of Ikeda (1974) at 18 – 20°C, the range of temperatures prevailing in our experiments.

(1) Boreal: $R' = 1.14 \; W^{-0.217} \; \mu l \; O_2 \; mg^{-1} \; h^{-1} = 0.102 \; W^{-0.217} \; \mu g\text{–atom} \; mg^{-1} \; h^{-1}$
(2) Temperate: $R' = 1.34 \; W^{-0.244} \; \mu l \; O_2 \; mg^{-1} \; h^{-1} = 0.120 \; W^{-0.244} \; \mu g\text{–atom} \; mg^{-1} \; h^{-1}$
(3) Subtropical: $R' = 2.09 \; W^{-0.336} \; \mu l \; O_2 \; mg^{-1} \; h^{-1} = 0.187 \; W^{-0.336} \; \mu g\text{–atom} \; mg^{-1} \; h^{-1}$
(4) Tropical: $R' = 3.03 \; W^{-0.405} \; \mu l \; O_2 \; mg^{-1} \; h^{-1} = 0.270 \; W^{-0.405} \; \mu g\text{–atom} \; mg^{-1} \; h^{-1}$

Re–computation of his data for the temperature range 18 – 20°C (excluding salps) yields the equation

$$R' = 0.224 \ W^{-0.295} \quad \mu g\text{--atom}/mg \ DW/h \qquad (r = -0.927)$$

which may be compared with Eq. 3 derived from the present study.

The exponents are much the same, but the constant 'a' is significantly different (p <0.001). Our value of 'a' is higher by a factor of 3.8.

    There were differences between our experimental conditions and those of Ikeda. Ikeda worked with single species, we used species mixtures; Ikeda used larger (oceanic) animals; we used smaller estuarine animals.

    For the purpose of the Port Hacking model we recommend Eq. 3.

Table 2.
Specific respiration values for planktonic organisms obtained by various authors. (Temperatures 10 – 20°C).

Author	Organism	$R'(\mu g\text{--atom} \ O_2 \ mg^{-1} \ h^{-1})$
Marshall *et al.* (1935)	*Calanus finmarchicus*	0.06
		0.19
Marshall *et al.* (1966)	*Acartia*	1.25
Razouls (1972)	*Calanus finmarchicus*	0.87
Lance (1965)	*Acartia clausi*	0.89
Ikeda (1970)	*Tortanus*	0.41
Ikeda (1974)	*Undinula* (28°C)	0.34
Zeiss (1963)	*Calanus* sp.	0.23
Zeiss (1963)	*Daphnia* sp.	0.76
Satomi & Pomeroy (1965)	"Total" plankton	0.59

ACKNOWLEDGEMENTS

We are grateful to D.D. Reid, CSIRO Division of Mathematics and Statistics, for advice and   assistance in analysing the results arising from this study.

REFERENCES

Clark, L.C.: Monitor and control of blood and tissue O$_2$ tension. *Transactions of the American Society for Artificial Internal Organs* **2**, 41 (1956)

CSIRO: *Estuarine Project Progress Report 1974–1976*. Sydney: CSIRO Division of Fisheries and Oceanography (1976)

Cuff, W.R., Sinclair, R.E., Parker, R.R., Tranter, D.J., Bulleid, N.C., Giles, M.S., Godfrey, J.S., Griffiths, F.B., Higgins, H.W., Kirkman, H., Rainer, S.F., Scott, B.D.: A carbon budget for South West Arm, Port Hacking. In: W.R. Cuff and M. Tomczak jr, eds *Synthesis and Modelling of Intermittent Estuaries*. Berlin, Heidelberg, New York: Springer (1983)

Cuff, W.R.: An evaluation of the Port Hacking Estuary Project from the viewpoint of applied science. In: W.R. Cuff and M. Tomczak jr, eds *Synthesis and Modelling of Intermittent Estuaries*. Berlin, Heidelberg, New York: Springer (1983)

Godfrey, J.S., Parslow, J.: Description and preliminary theory of circulation in Port Hacking estuary. *CSIRO Division of Fisheries and Oceanography Report* **67** (1976)

Ikeda, T.: Relationship between respiration rate and body size in marine plankton animals as a function of the temperature of habitat. *Bulletin of the Faculty of Fisheries Hokkaido University* **21**, 91–112 (1970)

Ikeda, T.: Nutritional ecology of marine zooplankton. *Memoirs of the Faculty of Fisheries Hokkaido University* **22**, 1–97 (1974)

Lance, J.: Respiration and osmotic behaviour of the copepod *Acartia tonsa* in diluted seawater. *Comparative Biochemistry and Physiology* **14**, 155–165 (1965)

Lovegrove, T.: The determination of the DW of plankton and the effect of various factors on the values obtained. In: H. Barnes, ed. *Some Contemporary Studies in Marine Sciences*. London: George Allen and Unwin (1966)

Marshall, S.M., Nicholls, A.G., Orr, A.P.: On the biology of *Calanus finmarchicus* VI. Oxygen consumption in relation to environmental conditions. *Journal of the Marine Biological Association of the United Kingdom* **20**, 1–28 (1935)

Marshall, S.M., Nicholls, A.G., Orr, A.P.: Respiration and feeding in some small copepods. *Journal of the Marine Biological Association of the United Kingdom* **46**, 513–530 (1966)

Nival, P., Nival, S., Leroy, C.: Essai d'utilisation de la méthode polarographique de desage d'oxygène pour l'estimation de la respiration d'*Acartia clausi*. *Rapports et Procès-Verbaux des Réunions Conseil International pour l'Exploration de la Mer* **20**, 301–303 (1971)

Pomeroy, L.R., Johannes, R.E.: Total plankton respiration. *Deep-Sea Research* **13**, 971–973 (1966)

Razouls, S.: Influence des conditions expérimentales sur le taux respiratoire des copépodes planctoniques. *Journal of Experimental Marine Biology and Ecology* **9**, 145–153 (1972)

Satomi, M., Pomeroy, L.R.: Respiration and phosphorus excretion in some marine populations. *Ecology* **46**, 877–881 (1965)

Smith, K.L., Burns, K.A., Carpenter, E.J.: Respiration of the pelagic *Sargassum* community. *Deep-Sea Research* **20**, 213–217 (1973)

Zeiss, F.R.: Effects of population densities on zooplankton respiration rates. *Limnology and Oceanography* **8**, 110–115 (1963)

Zeuthen, E.: Body size and metabolic rate in the animal kingdom with special regard to the marine micro-fauna. *Comptes Rendus des Travaux Laboratorie Carlsberg Série Chimique* **26**, 1–161 (1947)

Synthesis and Modelling of Intermittent Estuaries
(W.R. Cuff and M. Tomczak jr. eds) Berlin, Heidelberg,
New York: Springer (1983), pp. 177–192.

Data Base for the Port Hacking Estuary Project:
Parameters, Monitoring Procedure, and Management System

David J. Vaudrey[§], F. Brian Griffiths[†], Richard E. Sinclair[‡]

[§] Division of Oceanography
CSIRO Marine Laboratories
P.O. Box 21, Cronulla, N.S.W. 2230, Australia

[†] Division of Fisheries Research
CSIRO Marine Laboratories
P.O. Box 21, Cronulla, N.S.W. 2230, Australia

[‡] CSIRO Division of Computing Research
P.O. Box 1800, Canberra, A.C.T. 2601
Australia

Summary. The data base which resulted from two of the major monitoring activities of the Port Hacking Estuary Project is described and discussed: monitoring stations, variables, sampling schedules, field/laboratory analytical procedures, and management system.

Key words: data base, monitoring, management system, estuary, Port Hacking, South West Arm

1. INTRODUCTION

In order to provide an over–view of the dynamics of Port Hacking, monitoring of various variables (parameters) was carried out over 21 months. This paper describes the collection, analysis, and management of the data from two time series.

The first series was carried out by J.S. Godfrey from 6 September 1974 to 30 December 1975 (Series 1, stations a–i). The parameters measured were salinity (10^{-3}) and temperature (°C). The second series was carried out from 2 June 1975 to 10 March 1977 (Series 2, stations A–I). The overall logistics of this series was co–ordinated by R.R. Parker; he was also responsible for the salinity, temperature, dissolved oxygen, nitrate, nitrite, phosphate, and fluorescence measurements. H.W. Higgins was responsible for the chlorophyll measurements; N.C. Bulleid for the adenosine triphosphate, dissolved organic carbon, and total particulate organic carbon measurements; and F.B. Griffiths for the zooplankton–biomass measurements.

2. COLLECTION AND ANALYSIS OF SAMPLES

2.1 Sampling positions

The sampling positions are illustrated in Fig. 1. These positions can generally, if somewhat inelegantly, be referred to using grid map co–ordinates, based on the Australian Map Grid (Universal Transverse Mercator) using a km² grid. Other, shorter, methods of referring to the stations have also been introduced (Fig. 1).

The majority of sampling was carried out at Stn C, Grid Reference (G.R.) 27.1,25.2, in South West Arm (SWA), which is over the deepest water in the Arm. Samples taken at this and the other stations are shown in Figs 2 – 14. Each dot in these illustrations represents a sample of the indicated variable taken on the associated date, for the station indicated.

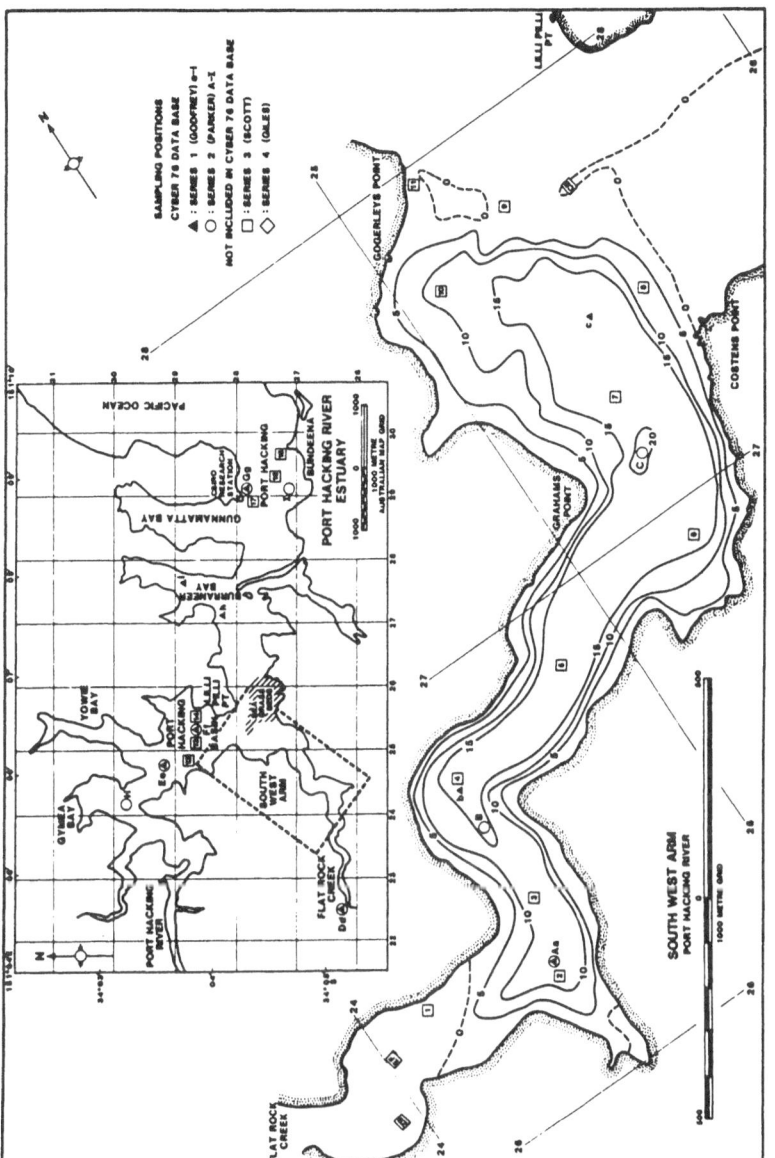

Fig. 1: Bathymetry of South West Arm, station positions, and reference grid. The inset shows the location of additional sampling positions within Port Hacking. G.R. co-ordinates define the corner of a square that contains the station and give the north–south reference first. (In the main part of the figure the north–south reference line slopes up to the left.)

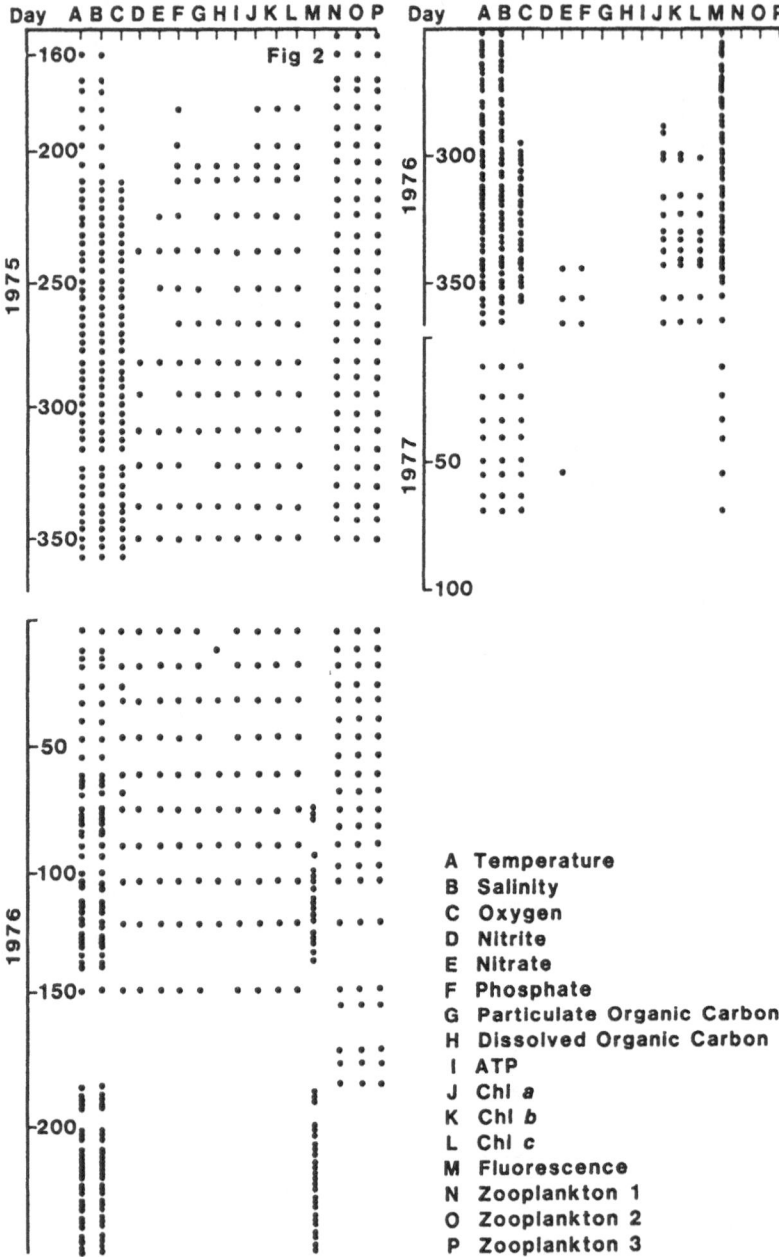

A Temperature
B Salinity
C Oxygen
D Nitrite
E Nitrate
F Phosphate
G Particulate Organic Carbon
H Dissolved Organic Carbon
I ATP
J Chl *a*
K Chl *b*
L Chl *c*
M Fluorescence
N Zooplankton 1
O Zooplankton 2
P Zooplankton 3

Fig. 2: Data available from Stn C, G.R. (27.1,25.2) from 2 June 1975 to 10 March 1977.

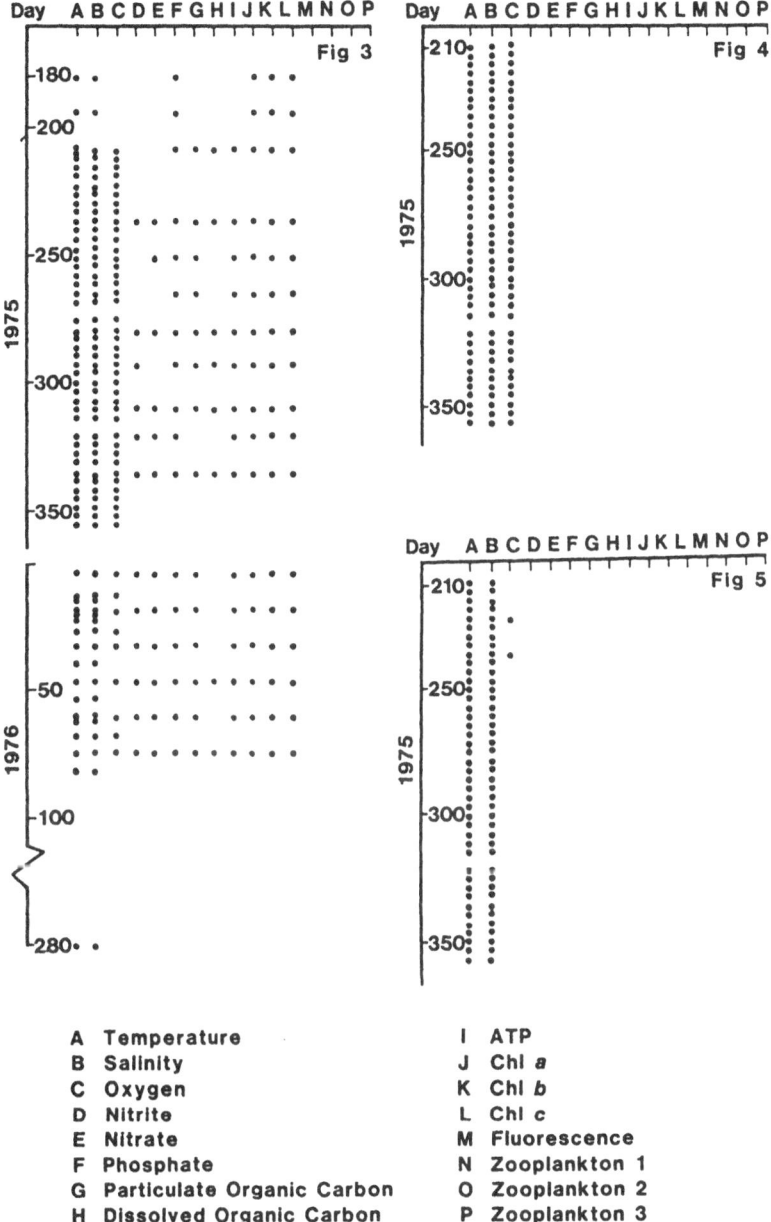

A   Temperature
B   Salinity
C   Oxygen
D   Nitrite
E   Nitrate
F   Phosphate
G   Particulate Organic Carbon
H   Dissolved Organic Carbon

I   ATP
J   Chl *a*
K   Chl *b*
L   Chl *c*
M   Fluorescence
N   Zooplankton 1
O   Zooplankton 2
P   Zooplankton 3

Fig. 3: Data available from Stn B, G.R. (26.6,24.5) from 30 June 1975 to 6 Oct 1976.

Fig. 4: Data available from Stn H, G.R. (29.8,24.2) from 29 July to 22 Dec 1975.

Fig. 5: Data available from Stn I, G.R. (27.3,29.0) from 29 July to 22 Dec 1975.

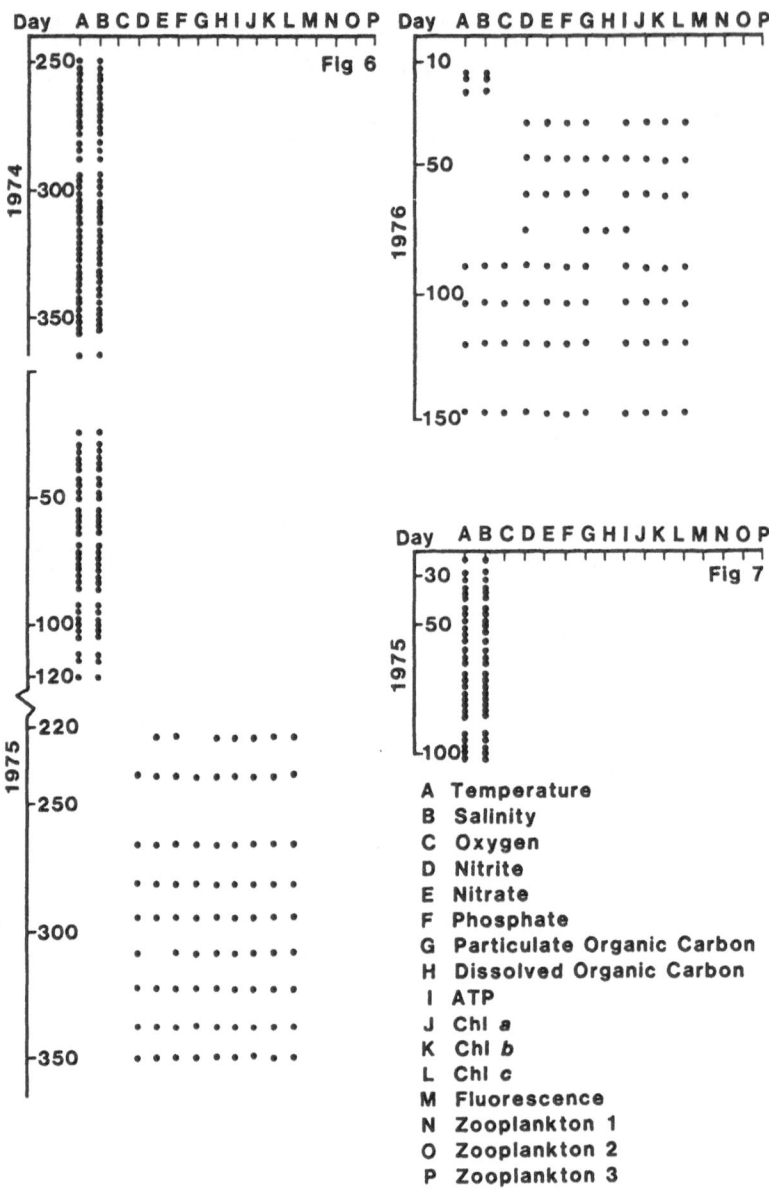

Fig. 6: Data available from Stn Aa, G.R. (26.3,24.4) from 6 Sept 1974 to 25 May 1976.

Fig. 7: Data available from Stn b, G.R. (26.7,24.4) from 24 Jan to 11 April 1975.

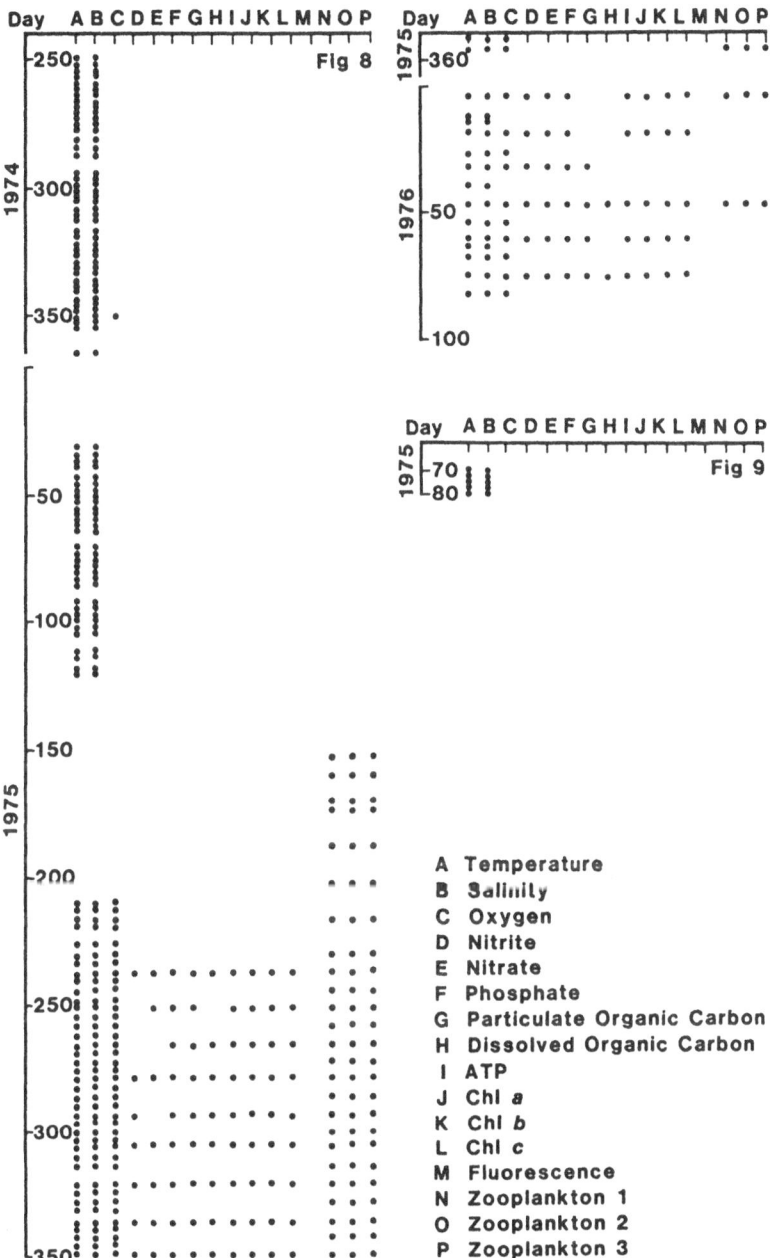

Fig. 8: Data available from Stn Ff, G.R. (28.6,25.4) from 6 August 1974 to 22 March 1976.

Fig. 9: Data available from Stn i, G.R. (28.7,27.5) from 12 to 21 March 1975.

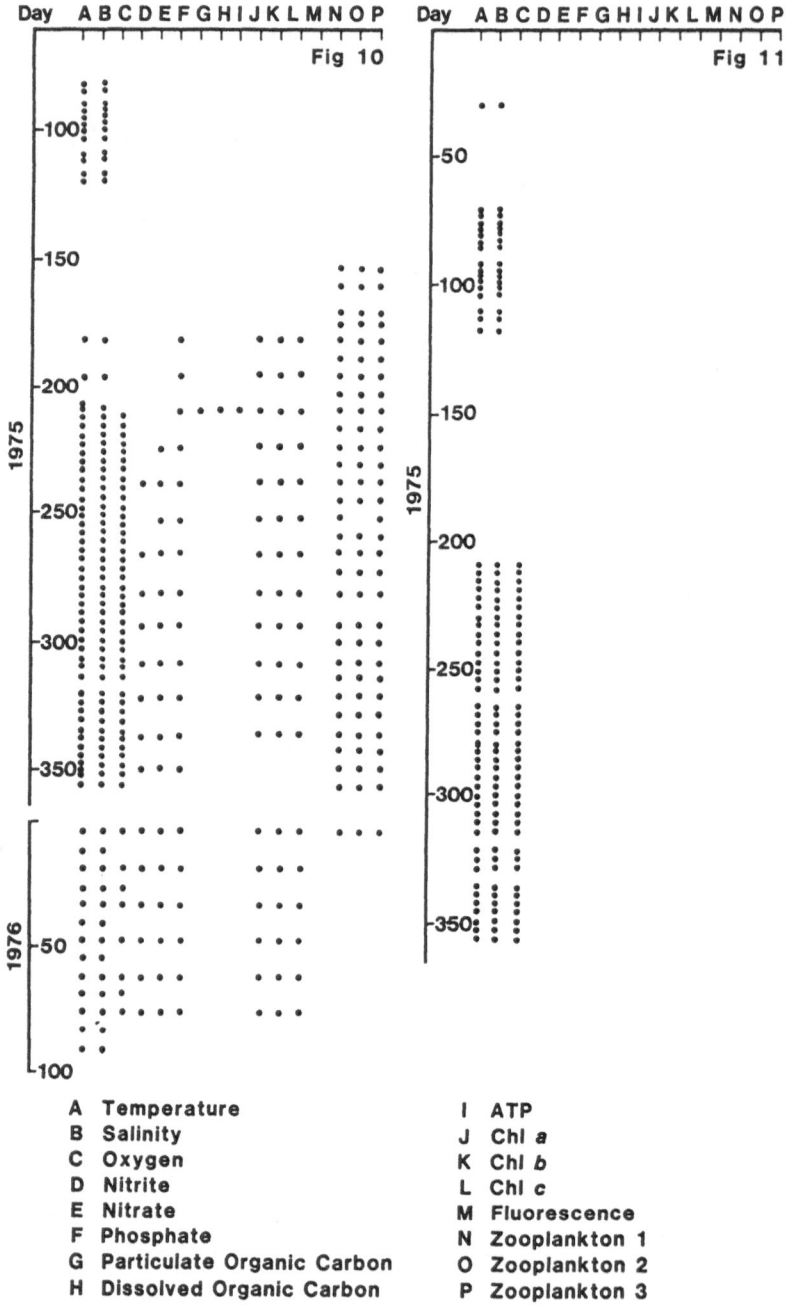

A   Temperature   I   ATP
B   Salinity   J   Chl *a*
C   Oxygen   K   Chl *b*
D   Nitrite   L   Chl *c*
E   Nitrate   M   Fluorescence
F   Phosphate   N   Zooplankton 1
G   Particulate Organic Carbon   O   Zooplankton 2
H   Dissolved Organic Carbon   P   Zooplankton 3

Fig. 10: Data available from Stn Gg, G.R. (27.9,29.0) from 24 March 1975 to 30 March 1976.

Fig. 11: Data available from Stn Ee, G.R. (29.2,24.7) from 29 Jan to 22 Dec 1975.

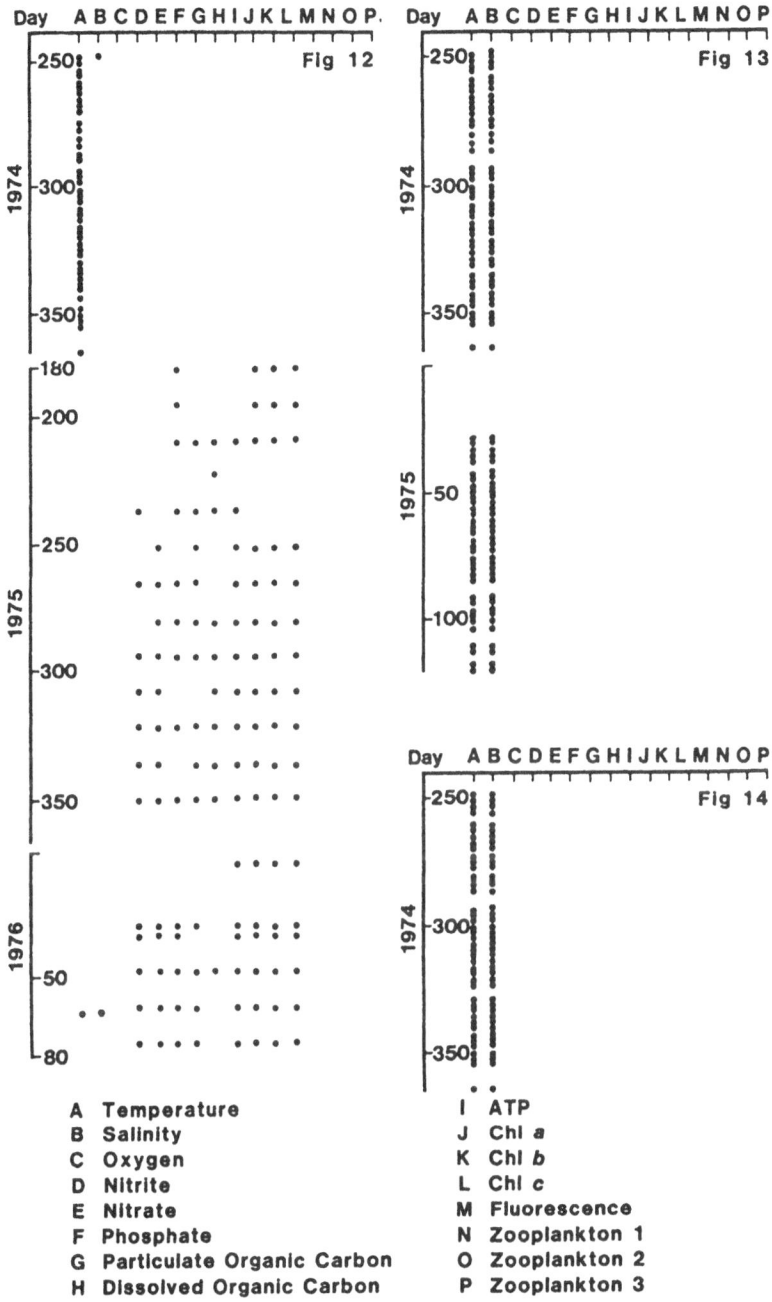

A   Temperature
B   Salinity
C   Oxygen
D   Nitrite
E   Nitrate
F   Phosphate
G   Particulate Organic Carbon
H   Dissolved Organic Carbon

I   ATP
J   Chl *a*
K   Chl *b*
L   Chl *c*
M   Fluorescence
N   Zooplankton 1
O   Zooplankton 2
P   Zooplankton 3

Fig. 12: Data available from Stn Dd, G.R. (26.1,22.6) from 6 Sept 1974 to 15 March 1976.

Fig. 13: Data available from Stn c, G.R. (27.5,25.3) from 6 Sept 1974 to 30 April 1975.

Fig. 14: Data available from Stn h, G.R. (28.2,27.1) from 6 Sept to 30 Dec 1974.

2.2 Field collections and measurements

2.2.1 Water samples – Water was usually raised from depth using a Flygt Model B2050.131 submersible pump. At the surface the pumped water was split into two streams, one for filling sample bottles; for Series 2 collections the second stream ran continuously through a 50–ml chamber in which temperature, salinity, and oxygen probes were mounted. Alternatively, water samples were taken with a modified Van Dorn Sampler (Jitts, 1964).

Water samples were stored in 2.5–l brown glass carboys which were kept in closed padded boxes during transport to the laboratory.

Samples for the measurement of dissolved oxygen (for Winkler titration analysis – see Section 2.2.3 and 2.3) were fixed immediately with 0.5 ml $MnSO_4$ followed by 1.0 ml NaOH–KI and stored in 250–ml B.O.D. bottles in the dark before return to the laboratory.

2.2.2 Temperature and salinity – Temperature and salinity measurements were taken with a Hamon temperature–salinity bridge (Hamon & Brown, 1958) Model 602, manufactured by Auto Lab Industries Pty Ltd, Chatswood, N.S.W., Australia. Temperature (°C) and salinity (10^{-3}) (measured as conductivity) were read directly from the instrument, with a precision of $\pm$ 0.25°C for temperature, $\pm$ $0.05 \cdot 10^{-3}$ in the $32 - 36 \cdot 10^{-3}$ salinity range and $\pm 0.25 \cdot 10^{-3}$ in the $20 - 32 \cdot 10^{-3}$ range. The calibration and reading procedures are given by Lockwood (1970). Daily, before and after use, calibrations were made using a seawater sub–standard. Occasionally, large discrepancies were observed and these were usually traceable either to a poisoned electrode due to H_2S in anaerobic bottom water or to oil on the electrode.

Series 1 measurements were taken by lowering the sensor to depth, while Series 2 measurements were taken in conjunction with a submersible pump (see Section 2.2.1). Temperature and salinity data are available from all stations (Figs 2 – 14).

2.2.3 Oxygen saturation – A Kent E.I.L. Portable Oxygen Meter Model 1520 manufactured by Electronic Instruments Ltd, Chertsey, Surrey, England was used to profile relative oxygen saturation in the field. Two samples for Winkler titration were taken in each profile, one from relatively high O_2 water, the other from relatively low O_2 water. These were used to scale the oxygen meter data. Full profiles of Winkler oxygen samples were done on some occasions. Oxygen data are available from all Series 2 stations except Stn D, *i.e.* from Stns B, C, H, I, Aa, Ee, Ff, and Gg.

2.2.4 Fluorescence – Measurements of *in vivo* chlorophyll *a* were made with a Variosens 1 *in situ* fluorometer (Impulsphysik GmbH, Hamburg, West Germany) (Früngel & Koch, 1976) attached to the submersible pump. The instrument was suspended for 2 – 3 min at each 1–m depth interval from surface to bottom to obtain an average fluorescence with depth. The mV output of the instrument was then modified to a linear scale of relative intensity of fluorescence (Parker & Vaudrey, 1980). Fluorescence data is available from Stn C only.

2.2.5 Zooplankton biomass (1975–1976 only) – Zooplankton biomass samples were obtained with a 0.25 m², 100–μm mesh aperture net, hauled from near the bottom to the surface. Medusae and ctenophores were separated from the zooplankton by screening the sample through a 2–cm mesh. Live zooplankton and the medusae/ctenophore fraction were concentrated on a 100–μm mesh aperture metal screen, rinsed briefly with fresh water, and dried at 60°C to a constant weight (Lovegrove, 1966). The results are presented as milligrams dry weight per cubic metre (mg DW m⁻³) with the live zooplankton fraction represented in Figs 2–14 by Zooplankton 1, the ctenophore fraction by Zooplankton 2: Zooplankton 3 denoting the sum of the two fractions. The zooplankton biomass from Stn C is an estimate of the average water column biomass produced by combining samples from five stations within South West Arm (Stns B,C, G.R. 27.4,25.2, G.R. 26.9,25.0, G.R. 26.7,24.6). Similarly the zooplankton biomass from Stn Ff is a combination of two hauls, and from Stn Gg only a single haul was used.

2.3 Laboratory analysis of oxygen samples

Dissolved oxygen of field–collected samples was determined by a modification of the Winkler titration method (see Major *et al.*, 1972). Where complete profiles were not available, percent oxygen saturation data, corrected for salinity and temperature with titrated samples were used to interpolate the dissolved oxygen concentration linearly. The error limit on the dissolved oxygen data is $\pm\ 0.5\ \mu M\ kg^{-1}$ of seawater.

2.4 Laboratory analysis of nutrient, DOC, chlorophyll, POC, and ATP samples

Aliquots of the 2.5–l seawater samples (see Section 2.2.1) were filtered through Watman GF/C glass–fibre filters using not more than 12.5 mm Hg vacuum. Filtrates were used for determination of nitrate, nitrite, phosphate, and dissolved organic carbon (DOC). The particulate material was used for determination of the chlorophylls, particulate organic carbon (POC), and adenosine triphosphate (ATP). Data is available from Stns Aa, B, C, Dd, Ff, and Gg.

2.4.1 Nitrate and nitrite – Nitrate and nitrite concentration was determined using the cadmium reduction method of Wood *et al.* (1967) as described in Strickland & Parsons (1972) and Major *et al.* (1972).

2.4.2 Phosphate – Dissolved orthophosphate in the samples of filtered seawater was determined by the $SnCl_2$ reduction method (Major *et al.*, 1972).

2.4.3 Dissolved organic carbon – DOC was determined by a modified Menzel–Vaccaro method (Menzel & Vaccaro, 1964) using the total carbon system (Oceanographic International Corp., College Station, Texas, USA). The inorganic carbon was removed by bubbling nitrogen through the sample. The carbon was oxidized to CO_2 by heating in a sealed ampoule. The concentration of CO_2 was estimated by non–dispensive infra–red spectroscopy (Strickland & Parsons, 1972).

Contamination of DOC samples with oil occurred because of leaking seals on the submersible pump used for sampling. Spillage and exhaust products from power boats may have also caused high readings. Any interpretation or further use of these data should bear these facts in mind.

2.4.4 Chlorophylls *a*, *b*, and *c* – Chlorophylls were determined by a modified SCOR–UNESCO procedure (SCOR–UNESCO, 1966). Absorption values of extracted pigment in 90% acetone were employed in the trichromatic equations of Jeffrey & Humphrey (1975). The presence of chlorophyll degradation products and/or other interfering substances was not taken into account.

2.4.5 Particulate organic carbon – POC was determined by the dry combustion of the particulate material at 650°C in a heated furnace under an oxygen stream. The CO_2 produced was absorbed in a pre–weighed "ascarite" (soda asbestos) packed tube. The tube was re–weighed and the amount of CO_2 produced determined by difference (Airey & Hogan, 1980).

2.4.6 Adenosine triphosphate – ATP was extracted from the particulate matter by Bulleid's (1977, 1978) modification of the Holm–Hansen method (Holm–Hansen & Booth, 1966).

3. STORAGE AND RETRIEVAL OF THE DATA

3.1 Description

As they were collected, data were entered into the CSIRO CDC Cyber 76 computer. The FORDATA data base management system (DCR, 1980) was used for data storage and retrieval. FORDATA is an implementation of the Codasyl proposals, using extensions to the Fortran language system; for further details see Mackenzie & Smith (1977).

Data are stored in "records", each of which contain a number of "data items". Basic data items include station codes, serial number (record identifier), date (year : month : day), time, zooplankton biomass, depth to bottom, cloud cover, rainfall, wind speed and direction. These were normally followed by depth–specific data items. The "depth records" include the following items: depth, temperature, salinity, oxygen, nitrate, nitrite, phosphate, chloryphylls a, b and c, particulate organic carbon (POC), dissolved organic carbon (DOC), adenosine triphosphate (ATP), and fluorescence. The data base design (or "schema") allowed for extra items but this facility was not used.

Data entry was not automated but was carried out manually from data sheets. The data were punched onto cards before being transferred to the data base. Verifying was done by returning data listings to the originators for scrutiny.

FORDATA provided data retrieval and a standard report form, being organized tables with headings, on the line printer. Other tables and data analysis were provided by computing staff, using specially written software.

3.2 Development and use

The data base was designed 6 months after the start of data collection. The experimentalists worked independently and designed their own data sheets. The data base designer specified sampling sites and items for inclusion following discussion with experimentalists, but the data base design had to be altered three times to cater for changed requirements (e.g. new variables), and to improve computing efficiency.

At the time of implementation the only suitable and available data base management system was FORDATA. In fact, this Project was one of the first major users of this package. Many of the currently available updating and reporting (and schema) features of FORDATA and similar packages were not then available. The computing staff wrote programs to produce special output reports.

Little use was made of the data base during the Project (1973–1978). The procedure of experimentalists verifying their own data, necessitated by inadequate numbers of computing staff, resulted in delays of up to 18 months in getting reliable data. A more basic problem was ignorance on the part of experimental staff of what was on the computer data base and how to access it. Excessive requests were put to the computing staff; but even reasonable requests for data retrieval and collation and analysis that were made could not be met quickly because of inadequate staff. Only one–fifth of the programmer's time was available and one–half of the time of a data editor. In a very few cases experimentalists were able (with help) to retrieve data and do their own analyses. A graphical listing of data available by the end of the Project is given in Figs 2 – 14.

In two instances effective use was made of the data base, but only after the Project was finished. In preparing the carbon budget synthesis (Cuff et al., 1983) the data base was used extensively to compute average conditions and to look for correlations among the parameters. In modelling activities (Sinclair et al., 1983) the data base was used for model validation.

4. COMMENTS

There were several problems in the development and use of the data base. Perhaps the most basic problem was that there was not a careful consideration of the role that such a data base could play, especially given the available computing staff. It is estimated that the construction and use of an effective data base for a project of this size requires a staff of two: a full–time programmer and a full–time data editor. In our Project, the lack of these basic requirements resulted in improperly designed data sheets, poor verification of data (sometimes brought about by long delays from data collection to submission for entry), lack of complete data sets (due to inadequate explanation of the usefulness of putting data on the data base), inadequate referencing of sampling sites, and an *ad hoc* approach to data entry and data base use. The expectations of some experimentalists were excessive relative to the efforts required to produce very rough, but adequate, outputs; scientists wanted unavailable levels of data security (on the computer data base) until publication.

The *potential* usefulness of such a data base seems to be restricted largely to the later stages of the program, especially for synthesis and modelling work. It is probably only at this late stage that the data base would have reached a stable form. It may also be of value to other researchers, who may, or may not, be directly concerned with the specific ecosystem to which the data base relates.

In short: a data base management system seemed to be able to meet some – but only a small fraction – of the expectations associated with it. In particular, it would seem necessary to have a substantial number of computing staff to provide a quick and efficient level of data retrieval and analysis at the request of experimentalists. As a long term storage and retrieval system it seems adequate.

ACKNOWLEDGEMENTS

We acknowledge the work of A.D. Crooks and R.A. Irrgang in the computer storage and retrieval of data. R.R. Parker and W.R. Cuff provided help with the manuscript.

REFERENCES

Airey, D., Hogan, M.: Method for the determination of dissolved and particulate organic carbon in marine samples. *CSIRO Division of Fisheries and Oceanography Report* 127 (1980)

Bulleid, N.C.: Adenosine triphosphate analysis in marine ecology: a review and manual. *CSIRO Division of Fisheries and Oceanography Report* 75 (1977)

Bulleid, N.C.: An improved method for the extraction of adenosine triphosphate from marine sediment and seawater. *Limnology and Oceanography* 23, 174–178 (1978)

Cuff, W.R., Sinclair, R.E., Parker, R.R., Tranter, D.J., Bulleid, N.C., Giles, M.S., Godfrey, J.S., Griffiths, F.B., Higgins, H.W., Kirkman, H., Rainer, S.F., Scott, B.D.: A carbon budget for South West Arm, Port Hacking. In: W.R. Cuff and M. Tomczak jr, eds *Synthesis and Modelling of Intermittent Estuaries*. Berlin, Heidelberg, New York: Springer (1983)

DCR: *FORDATA User's Guide, Edition 1.1*. Canberra: CSIRO Division of Computing Research (1980)

Früngel, F., Koch, C.: Practical experience with the Variosens equipment in measuring chlorophyll concentrations and fluorescent tracer substances, like rhodamine, fluorescene and some new substances. *IEEE Journal of Oceanic Engineering* 1, 21–32 (1976)

Hamon, B.V., Brown, N.L.: A temperature–chlorinity–depth recorder for use at sea. *Journal of Scientific Instruments* 35, 452–458 (1958)

Holm–Hansen, O., Booth, C.: The measurement of adenosine triphosphate in the ocean and its ecological significance. *Limnology and Oceanography* 11, 510–519 (1966)

Jeffrey, S.W., Humphrey, G.F.: New spectrophotometric equations for determining chlorophylls *a*, *b* and *c* in higher plants, algae and natural phytoplankton. *Biochemie und Physiologie der Pflazen* 167, 191–194 (1975)

Jitts, H.R.: A twin six–litre plastic water sample. *Limnology and Oceanography* 4, 452 (1964)

Lockwood, D.J.: Portable temperature–chlorinity bridge (S–T meter) instruction manual. *CSIRO Division of Fisheries and Oceanography Report* 47 (1970)

Lovegrove, T.: The determination of the dry weight and the effect of various factors on the values obtained. In: H. Barnes, ed. *Some Contemporary Studies of Marine Science*. London: Allen & Unwin (1966)

Mackenzie, H.G., Smith J.L.: The implementation of a data base management system. *The Australian Computer Journal* 9, 138–144 (1977)

Major, G., Dal Pont, G., Klye, J., Newell, B.: Laboratory techniques in marine chemistry. *CSIRO Division of Fisheries and Oceanography Report* 51 (1972)

Menzel, D.W., Vaccaro R.F.: The measurement of dissolved and particulate carbon in seawater. *Limnology and Oceanography* 9, 138–142 (1964)

Parker, R.R., Vaudrey, D.J.: A proposed reference standard for *in vivo* chlorophyll *a* fluorometry. *CSIRO Division of Fisheries and Oceanography Report* 125 (1980)

SCOR–UNESCO: Determination of photosynthetic pigments in seawater. In: *Monographs on Oceanography 1*. Paris: UNESCO (1966)

Sinclair, R.E., Cuff, W.R., Parker, R.R.: Ecosystem modelling of South West Arm, Port Hacking. In: W.R. Cuff and M. Tomczak jr, eds *Synthesis and Modelling of Intermittent Estuaries*. Berlin, Heidelberg, New York: Springer (1983)

Strickland, J.D.R., Parsons, T.R.: A practical handbook of seawater analysis. *Bulletin of the Fisheries Research Board of Canada* 167 (1972)

Wood E.D., Armstrong, F.A., Richards, F.A.: Determination of nitrate in sea water by cadmium–copper reduction to nitrite. *Journal of the Marine Biological Association of the United Kingdom* 47, 23–31 (1967)

Synthesis and Modelling of Intermittent Estuaries
(W.R. Cuff and M. Tomczak jr. eds) Berlin, Heidelberg,
New York: Springer (1983), pp. 193–232.

A Carbon Budget for South West Arm,
Port Hacking

Wilfred R. Cuff[§], Richard E. Sinclair[ɪ], Robert R. Parker[ɪ], David J. Tranter[ɪ],
Nicholas C. Bulleid[ɪ], Max S. Giles[⊕], J. Stuart Godfrey[a], F. Brian Griffiths[ɪ],
Harry W. Higgins[a], Hugh Kirkman[ɪ], Sebastian F. Rainer[ɪ], Barry D. Scott[a]

[§] CSIRO Division of Computing Research
P.O. Box 1800, Canberra, A.C.T. 2601, Australia

[ɪ] Division of Fisheries Research
CSIRO Marine Laboratories
P.O. Box 21, Cronulla, N.S.W. 2230, Australia

[⊕] Australian Atomic Energy Commission
New Illawarra Road, Lucas Heights, N.S.W. 2234, Australia

[a] Division of Oceanography
CSIRO Marine Laboratories
P.O. Box 21, Cronulla, N.S.W. 2230, Australia

[ɪ] Division of Fisheries Research
CSIRO Marine Laboratories
P.O. Box 20, North Beach, W.A. 6020, Australia

Summary. A multidisciplinary study of the structure and dynamics of a small (~78 ha) Australian marine embayment (South West Arm of Port Hacking, New South Wales) was conducted during 1973–1978. Compatible data were obtained by studying processes in terms of the flow of carbon. The carbon budget developed in this paper represents an attempt at a synthesis of that information. The chemical and biological species contained in each of 10 compartments are described; as data allows, the average carbon mass within each compartment and the average flow rates between the compartments, with variances, are estimated. This information is used to piece together the distribution of carbon among the compartments and to ascertain the major flow paths of carbon into, within, and out of South West Arm.

Key words: carbon, budget, synthesis, Port Hacking, South West Arm

1. INTRODUCTION

From August 1973 until 1978 scientists from CSIRO's Division of Fisheries and Oceanography and from several other organizations conducted a study aimed at understanding the principles underlying the structure and dynamics of Australian estuarine systems, as explained by Allen (1983). The participating scientists chose to focus their studies on a small marine embayment (~78 ha, low water) of Port Hacking called South West Arm (SWA); it is located about 30 km south of the centre of Sydney, New South Wales, at 34°05'S and 151°06'E. Detailed maps can be found in Albani *et al.* (1983) and Vaudrey *et al.* (1983).

Much of the work conducted under the auspices of the Port Hacking Estuary Project was directed towards understanding the chemical and biological processes measured in terms of the distribution and flow of carbon. In this paper we make a serious attempt at synthesizing this information by using it to construct as complete a picture as possible of the distribution of carbon and of its flow into, within, and out of SWA. (The way in which carbon was initially visualized to flow through SWA is described by Parker *et al.*, 1983.)

The available information relates to a set of "compartments", being chemical and biological groupings. The most comprehensive data set, residing on a computer data base (Vaudrey *et al.*, 1983), contains time series of water column compartment masses for various sites and depths within SWA. Information about the masses of sediment compartments also exists, but not on the data base. Information relating to the rates of flow of carbon between compartments exists mainly in manuscript form. These data take a variety of forms. In some cases only average flow rates have been estimated: these estimates range from those based on small sample size through to those adequately replicated to provide good estimates of variation as well. In other cases flow–rate information is suitable to the construction of a dynamic model of carbon flow, with experiments having been conducted for purposes of functional identification and parameter estimation.

There exists inadequate information about SWA to construct a dynamic model of carbon flow on the basis of local data, even though this was the initial aim of the study. This being the case, we feel that it is most useful to synthesize the available information in terms of a static model – a carbon budget. Firstly, when there is inadequate data, one is obliged to choose the synthesis/modelling tool which admits the greatest proportion of available data. Budgets require, in addition to estimates of average compartment masses, estimates of average flow rates. Functional information obtained for purposes of dynamic modelling can be used to estimate these flow rates but average flow rates are generally not sufficient for functional identification, although they can be useful for parameter estimation for a limited range of simple functional types (Sinclair *et al.*, 1983). Secondly, the greatest value of dynamic models lies in their ability to clarify the logical consequences of the component functions and parameters. But this power virtually requires a complete specification of all of the components: features of dynamic models such as feedback loops can enable a large variety of types of model behaviour dependent on the structure of the missing functions. The flow rates of budgets are not, however, influenced by the missing flow rates and hence it is possible to make some reliable inferences regarding ecosystem behaviour in the face of incomplete information.

2. IDEALIZATION OF SOUTH WEST ARM

SWA shows considerable variation in both time and space and this variation needs to be assessed to enable the idealization required for meaningful budget parameters to be made.

2.1 Morphological aspects

SWA has a central basin (see Fig. 1 of Vaudrey *et al.*, 1983) which rises steeply (within about 100 m) from its maximum depth of greater than 20 m to a wide, shallow (<2 m) sill of sand, separating SWA from the tidal channel of Port Hacking. The central basin also rises steeply along its sides. But at the Flat Rock Creek end the basin shallows slightly to a maximum depth of 10 – 15 m, before rising to exposed sand banks of about 9 ha. The central basin occupies about 53% of the total area of SWA.

The sediments of the central basin consist of unconsolidated silt, which generally do not occur above 10 m below the low–water surface (Rainer, 1980). On the other hand, the sediments of the intertidal region vary from rock walls and ledges to gently sloping sand or muddy sand beaches. Subtidally the sides slope fairly steeply (10° or more) to the central basin, with sediments of sand (~34% of the area of SWA), shell (10%), and bedrock (3%).

Due to the predominance of the central basin, it is convenient to define SWA as that area extending from a line between Gogerleys Point and Costens Point to the sand banks of Flat Rock Creek. This definition, based on the low–water mark, excludes a few exposed sand banks within SWA (generally the intertidal region is small) and at the mouth of the Creek (in total about 22 ha; Giles, 1983) which are known to contain a community of benthic micro–organisms (Giles, 1983) and some seagrasses (see Fig. 3 of Rainer, 1980) and hence contribute some primary production to SWA. Further, a large seagrass bed (about 30 ha and 70% *Posidonia australis*) exists outside the mouth of SWA (roughly contained within the dashed region (0 depth) of Fig. 1 of Vaudrey *et al.*, 1983; see also Kirkman & Reid, 1979). It is bounded on one side by land and on two sides by sand banks, one of which separates the seagrass bed from the tidal channel of Port Hacking. At low tide the water level falls below the sand banks; this leaves the seagrass in a shallow lagoon which, because it shelves off into the central basin, drains slowly towards SWA. Both of these regions will be considered in terms of their effects on SWA.

2.2 Chemical and hydrodynamic aspects

Time series of various chemical parameters were collected by some of the participants and stored on the data base; Vaudrey *et al.* (1983) give a detailed account of the temporal and spatial sampling schedules for these parameters. Average values, along with variability estimates, for each parameter are given in Table 1.

Table 1.
Chemical properties of South West Arm.

Property	Mean	Std. Dev.	Observations
Temperature, °C	19	3	6789
Salinity, 10^{-3}	34	2	6766
O_2 saturation, %	80	32	1879
Phosphate, mg–atom m^{-3}	0.4	0.7	1018
Nitrate, mg–atom m^{-3}	1.2	3.5	964
Nitrite, mg–atom m^{-3}	0.2	1.3	279
Silicate, mg–atom m^{-3}	5.8	6.8	590

Temperature varied by both season and depth. Average monthly temperatures (°C), from January, were 22.3, 22.7, 22.6, 20.5, 18.6, 15.6, 15.6, 14.7, 15.6, 17.3, 19.2, and 21.5. During the months of September through February temperature decreased consistently with depth but during the remainder of the year a variety of profiles occurred including consistent increases in temperature with depth.

A statistically significant difference in salinity among months was observed ($F_{11,6754} = 36$) but no interesting temporal trend was apparent in the means. Mean salinity generally increased slightly with depth but variation about the means was generally highest in the 0 – 2 m region. This pattern is a reflection of the behaviour which follows the influx of fresh water onto the surface following rainstorms, as detailed below.

Mean oxygen values tended to be low and variable at depths below 10 m. This pattern is known to be associated with freshwater input onto the surface following rainstorms, as detailed below. A statistically significant difference in percent oxygen saturation among months was observed ($F_{11,1867} = 17$), the period November to April having a mean value of $72 \pm 5\%$ (std. dev.) with the other months having a value of $89 \pm 5\%$.

The overall mean phosphate value is higher than the average $0 - 2$ m value (0.16 mg–atom m^{-3}) because of large and variable values in the $15 - 20$ m region, associated with rainfall, as detailed below. The same pattern occurs in nitrates and silicates with a mean $0 - 2$ m nitrate value of 0.9 mg-atom m^{-3} and silicate value of 3.4 mg-atom m^{-3}. Except for average February values ($n=8$ for $10 - 14$ m values) nitrites showed neither depth structure nor seasonal pattern.

Tidal exchange is generally simple as long as surface salinities in SWA range from 33.0 to $35.5 \cdot 10^{-3}$, during dry weather. Such conditions occur 75% of the time (Godfrey & Parslow, 1976). (Seventy percent of the surface–salinity records on the data base are greater than $33.0 \cdot 10^{-3}$, and 75% are greater than $32.5 \cdot 10^{-3}$.) On a rising tide (typical range of 2 m spring tide; 1.3 m neap tide) surface waters from the channel of Port Hacking flood over the largely exposed sill of sand located at the mouth of SWA and spread over the water column. On the falling tide virtually the same waters are exited to Port Hacking.

During the rest of the year (overall about 25%), SWA exhibits a series of transient behaviours following rainstorms, which do not have any strong annual pattern (Figs 1,2 of Bulleid, 1983). The influx of fresh water from the catchment ($\sim$ 20 km^2) and from Flat Rock Creek spreads as a homogeneous sheet ($50 - 100$ cm thick) over SWA within 2 or 3 h. Following strong rainstorms the surface salinity in parts of SWA can fall to nearly zero. This results in a strong pycnocline between the fresh and saline layers.

On the rising tide dense and saline water enters SWA and an intense front develops between it and the less saline water inside. Salinity at the front can change by $20 \cdot 10^{-3}$ within a metre horizontally. The tidal inflow immediately drops several metres down the steep inner sandbanks of the central basin in a turbulent plume; it entrains some water from inside the basin and the mixture then spreads throughout the Arm at an intermediate depth (Godfrey & Parslow, 1976). As high tide approaches, the rising–tide front moves downstream and is broken up by turbulence. During the falling tide, less saline surface waters flow into and mix with the tidal–channel waters of Port Hacking. On the following rising tide much of this mixed water returns, sinking at the rising-tide front. The net effect is to indirectly mix a portion of the freshwater layer down to 4 m or deeper. The remaining fresh water is lost seaward. Over a number of tide cycles this process results in the progressive deepening and lessening of the pycnocline until a more–or–less uniform salinity structure is attained. Occasionally strong winds can assist in the vertical mixing. Turbulent mixing is, however, generally not important; tidal flow is laminar and slow, about 5 – 10 cm s^{-1} (Godfrey & Parslow, 1976).

In an intense rainstorm which began on 10 March 1975 and lasted for about 4 days (Fig. 1c of Scott, 1978b), the fresh water was removed at an exponential rate with a half–life of about 5.1 days. More recently, Godfrey (1983) developed a mathematical model to describe tidal flushing and vertical diffusion in SWA and successfully applied it to predicting the effects of this rainstorm on salinity.

Following the influx of stormwater and the detritus it brings with it from the catchment, the bottom part of the water column (> 10 m; Fig. 2b of Scott, 1978b) can become de–oxygenated (Rochford, 1974). These effects were noted above, from analysis of data–base oxygen concentrations. In extreme cases the water column to as high as the 13–m level (measured as depth from the surface) and sediment of SWA can go anoxic. Scott (1978b) observed anoxic conditions only once in 1975; his Fig. 2 suggests that the anoxic state may have lasted up to about 10 days. The data base shows that anoxic conditions occurred in the water column on 13 days in 1976.

Scott (1978b, Fig. 3) pointed out that after a rainstorm, the relatively dense oxygenated water that is brought into SWA on a rising tide and enters the column at intermediate depths isolates the waters below from oxygen input and allows de–oxygenation to proceed. The pycnocline associated with this input of mixed saline and fresh water at intermediate depth discourages turbulence in the bottom waters; however, a novel mechanism involving a double pycnocline (Parker, 1983) may be necessary for the complete anoxia of the central basin of SWA. Scott (1978b) attributed the consumption of available oxygen to the activities of the decomposition micro–organisms which break down the continually settling detritus at, or slightly above, the sediment–water interface. A recent experiment (Bulleid, 1983) has confirmed that intense microbial activity does occur at the sediment–water interface. Scott (1978b) suggested that the rates of de–oxygenation will be greater after heavy rainfalls than at other times, and he has evidence to suggest that the introduced suspended or colloidal material entrains suspended phytoplankton and detritus as it sinks. He estimated that about 35% of the de–oxygenation results from oxidation of the older accumulated detritus and 65% from recent detritus.

During periods of de–oxygenation, nutrient concentrations in the bottom waters can increase dramatically, as noted above. In addition to these values, examples may be found in Scott (1978b), Bulleid (1983), and Godfrey (1983). Based on linear correlations between the rates of apparent oxygen utilization and the rates of nutrient regeneration, Scott (1978b) concluded that nutrients were being released as a by–product of the activities of the decomposition micro–organisms on the detritus. A fuller study of this phenomenon is given by Bulleid (1983). Parker (1983) describes the effects of the freshwater influx on the protoplankton of SWA. Scott (1978b) also noticed that as the density stratification broke down nutrient (phosphorus) concentrations decreased, at roughly the same rate as they built up – about 3.7% per tidal cycle. (The rate can, at times, be faster with some PO_4 re–precipitated to the bottom as ferric compounds.) The loss was attributed to removal to the ocean, to some of the nutrients being biologically fixed, and to long–term sedimentation.

The experimental results summarized above examine in some depth a number of interesting and important processes that occur after the influx of fresh water. But a budget is a static model best representing average steady–state conditions and these fresh water–induced processes are transient conditions of relatively short duration. As the inclusion of these transient periods would only make the budget parameters more variable, it seems preferable to postpone further discussion of them to a paper describing a dynamic ecosystem model of SWA (Sinclair *et al.*, 1983). Hence, the budget presented below will assume marine salinities, with associated oxygen and nutrient concentrations.

2.3 Biological aspects

Not all biological groupings were uniformly distributed throughout SWA. In some sediment–related taxa this variation is extreme. For instance, macrobenthos is abundant and diverse between the intertidal zone and 5 m; is less diverse on slopes deeper than 5 m (to about 10 m); and is virtually absent in the silt of the central basin at depths greater than 10 m (Rainer, 1980). Rainer also found that seagrasses cover about 14% of the total area of SWA, although they occur only to depths of 6 m. Macroalgae are present only between the intertidal zone and about 10 m. This suggests that the shallow sediments should be considered separately from the deep sediments. Ideally, distinctions should be made for macrobenthos from 0 – 5 m, 5 – 10 m, and > 10 m; for seagrasses above and below 6 m; and for macroalgae above and below 10 m.

Spatial variation in the density of biological taxa living in the water column is not as extreme. The only known case of some biological grouping being excluded from part of the water column is phytoplankton below the 1% light level, as net photosynthesis is near zero or negative below this level. (This statement is not strictly true as phytoplankton can obviously survive limited periods in the dark, and in fact one SWA plankton species is known (see Parker, 1983) to move for short periods below the 1% light level to obtain nutrients and then move back into the lighted region to photosynthesize.)

Scott (1978a) found that the distribution of SWA phytoplankton is almost uniform horizontally, but non–uniform vertically with maxima often occurring at a depth of 2 – 6 m.

2.4 Overall idealization of South West Arm

Combining the above idealizations it is clear that the carbon budget of SWA is best based on data collected during that 75% of the time when approximately uniform salinities are present. The water column compartments are assumed, probably correctly, to exhibit no significant horizontal variability. Variation with depth will be considered, where known, but attempts will be made to obtain average per cubic metre values. Sediment values will be given as per square metre but the values will be restricted to those depths to which they apply. A carbon budget for the seagrasses outside of SWA will be reproduced from Kirkman & Reid (1979) and interpreted in terms of the potential effects of this seagrass bed on the Arm. The effects of the benthic micro–organisms at the sand flats of Flat Rock Creek will also be considered.

In Sections 3 – 5 we provide full details of the components (compartments, flow rates, inflows and outflows) of the budget. These details, although tedious, allow one to judge the adequacy of our synthesis of the data and of our conclusions about carbon flow within SWA. On a first, or casual, reading it might be appropriate to move directly to Section 6, and return as detailed interest dictates.

3. COMPARTMENTS

In this Section we give a brief description of the chemical and biological species contained within each compartment and present estimates of compartment mass for SWA.

The water column and the sediment are treated as separate subsystems, with each having the following compartments: DIC (dissolved inorganic carbon), DOC (dissolved organic carbon), detritus, autotrophs, heterotrophs.

3.1 Water column

3.1.1 DIC – Carbon dioxide is the only inorganic chemical species of interest within the present context.

Seventy–nine times during 1975, Scott (1983) measured total DIC concentrations at each of depths 1, 2, 3, 4, 6, 10, and 14 m at a central basin station (Stn 7 of Series 3: Vaudrey *et al.*, 1983). The overall mean concentration was 23.9 ± 1.2 g C m^{-3}. (Here and throughout, unless otherwise indicated, the $\pm$ figure represents standard deviation.) However, DIC concentration tended to increase linearly with depth ($r^2 =$ 0.99), with an increase of about 1.34 g C m^{-3} over the 10–m depth of SWA. (There was an average value of 23.17 g C m^{-3} at the 1–m depth and 25.04 g C m^{-3} at the 14–m depth.) Scott found the DIC concentration to be proportional to salinity, with only slight deviations (not more than 1 g C m^{-3}) at the bottom of the water column during periods of de–oxygenation.

3.1.2 DOC – Various types of soluble organic carbon compounds are released into the environment by plants and animals (Parsons & Takahashi, 1973, p. 99). One pathway is via excretion resulting from the digestive processes of animals; another is via decomposition of dead organisms; and a final pathway is via "exudation" during photosynthesis. A variety of soluble compounds are released including amino acids, peptides, fatty acids, glycerol, carbohydrates, polysaccharides and more refractory compounds. Obviously this compartment is chemically heterogeneous, but DOC is conveniently measured as a unit (see Vaudrey *et al.*, 1983).

DOC concentration was measured in SWA at irregular intervals at Stns Aa, B, C (Fig. 1 of Vaudrey *et al.*, 1983). Samples were taken over a period from July 1975 to March 1976 (see Vaudrey *et al.* (1983) for more detail). A mean value of 1.5 ± 0.8 g C m^{-3} (n=97) was calculated.

Wiebe & Smith (1977) estimated the concentration of photosynthetically derived DOC (PDOC) in SWA at 1.2 mg C m^{-3}. (They estimated a PDOC concentration of 1.23 mg C m^{-3} on 18 July 1975 and of 1.20 mg C m^{-3} on 8 August 1975.)

3.1.3 Detritus – Estimates of detrital mass were obtained from the data base as the differences between corresponding particulate organic carbon (POC) and ATP (carbon equivalent) measurements. POC (not shown in Fig. 13 of Vaudrey *et al.* (1983)) was monitored at approximately fortnightly intervals at 4 stations (Stn c and as for DOC) over a period from February 1974 to May 1976; ATP was also measured at approximately fortnightly intervals but only at 3 stations (as for DOC) from July 1975 to May 1976.

POC occurred at an average concentration of 0.52 ± 0.31 g C m^{-3} (n=334). There was a small, but significant, tendency for the concentration to be higher from October to January (0.68 ± 0.5 g C m^{-3}) than during the rest of the year (0.41 ± 0.14 g C m^{-3}, n=7). ATP occurred at an average concentration of 1.42 ± 0.86 mg m^{-3} (n=380). ATP tended to be low during the months of August and September (0.94 mg m^{-3}) and high in January (2.26 mg m^{-3}).

In order to estimate the detrital mass from POC and ATP measurements it is necessary to have an estimate of the C/ATP ratio. There exist, on the data base, measurements of both POC and ATP for various stations, times and depths. With the remarks of Banse (1977) in mind, we manipulated these data in various ways to get a sensible and reliable estimate, but to no avail. (It is known (Wiebe & Smith, 1977) that bacteria attach themselves to detrital particles; hence detrital carbon will not be independent of ATP, as required for a good estimate.) Station–specific and time–specific POC data were regressed against the corresponding ATP data for each depth at which both variables were measured. There are only four stations with more than 10 depth records in these variables and these stations give estimates of 227 (n=19), 327 (n=18), 347 (n=19), and –66 (n=19)! On the other hand, a regression containing all the data (n=329) gave an estimate of 179 ± 20 (std. error). It is felt that a value of 200 – 250 would be a realistic estimate, but given the magnitudes of the computed values, we use a value of 270 in this paper.

Given this ratio estimate we calculate mean detrital mass at 0.25 ± 0.26 g C m^{-3} (n=217). The use of a 270 C/ATP ratio resulted in 30% of the calculated detrital masses being negative. These negative values were eliminated for purposes of calculating the mean detrital mass. Had these values been retained a mass of 0.13 ± 0.17 g C m^{-3} would have been calculated. The largest detritus estimate was 1.6 g C m^{3}.

Since the C/ATP estimate is not very reliable, an alternative estimate seems desirable. Such an alternative can be derived from Scott's (1978a) data relating the light attenuation coefficient to the chlorophyll a concentration. Mean chlorophyll a concentration was 1.84 mg m^{-3}, equivalent to 92 mg C m^{-3} (assuming a C/Chl a ratio of 50 – appropriate for healthy phytoplankton). Scott estimated the relative proportions of phytoplankton : detritus recently derived from phytoplankton : other absorbents not directly related to the phytoplankton as 13 : 25 : 76. With these data a detrital mass can be calculated if it is assumed that the attenuation : carbon ratio is the same for all components. But it is clear that "other absorbents" will be less carboniferous (say by ~50%) since it includes particles of non–organic origin. Incorporating this modification yields a detrital estimate of 446 mg C m^{-3} (177 mg C m^{-3} of phytoplankton derivative; 269 mg C m^{-3} other absorbents) and a POC estimate of 538 mg C m^{-3}. Since the POC estimate is similar to that estimated elsewhere in this paper (520 mg C m^{-3}), we have some confidence in the detrital estimate. The detrital estimate derived in this paragraph is a factor of 1.8 larger than that derived above.

3.1.4 Autotrophs – There exists no published list of the phytoplankton species which inhabit SWA. But for Port Hacking, Wood (1964, p.482) noted that one "may expect dinoflagellate maxima from September through February, and in August – September, with a *Ceratium maximum* (sic) in May–June, but one cannot predict that these will, in fact, occur. In the case of the diatoms, we may expect blooms of *Chaetoceros secundus* and *Asterionella japonica* in March, of *Coscinodiscus granii* and *Rhizosolenia robusta* in June or July and mixed blooms of estuarine and neritic species in September, especially in the lower parts of the estuary". Scott (1978a) noted during two periods of high chlorophyll *a* biomass, a March maximum dominated by the dinoflagellate *Ceratium* sp. and a July maximum dominated by the diatom *Asterionella japonica*. In additon, Scott (1978b) found a seasonal change in the ratio of silicate concentration to the apparent oxygen utilization which he attributed to shifts in community structure. Scott (1979) noted that the existence of dinoflagellate maxima in the summer months is consistent with hydrological events – dinoflagellates are common bloom organisms in coastal waters and intrusions of such waters into Port Hacking are most frequent in summer. During periods of extended estuarine stratification dinoflagellates can reach very high densities.

Scott (1978a) found that the distribution of phytoplankton is almost uniform horizontally, but non–uniform vertically with maxima often occurring at depths of 2 – 6 m. Those few cases of vertical homogeneity observed by Scott (1979) occurred in winter as a result of the mixing effect of persistent strong cold winds.

Scott (1979) measured *in vivo* chlorophyll *a* fluorescence during 1975 and estimated a mean annual concentration of about 2 mg Chl *a* m^{-3}. Monitoring of chlorophyll *a* was also conducted at 3 stations (as for DOC) at fortnightly intervals over a period from June 1975 to about May 1976. These observations suggest a concentration of 2.9 ± 2.5 mg Chl *a* m^{-3} (n=535).

A statistically significant difference in chlorophyll *a* among months was observed ($F_{11,484}$ = 1.91, p=0.036) with a minimum of 1.25 ± 1.25 mg Chl *a* m^{-3} (n=11) occurring in June and a maximum of 3.7 ± 4.1 mg Chl *a* m^{-3} (n=63) occurring in November. Assuming a C/Chl *a* ratio of 60, we calculate a mean carbon biomass of 0.12 g C m^{-3} (Scott's data) or 0.17 ± 0.15 g C m^{-3} (monitoring data), with a winter minimum of about 0.08 g C m^{-3} and a summer maximum of about 0.22 g C m^{-3}.

The C/Chl *a* ratio was estimated from a set of depth–replicated data, in the same manner as the C/ATP ratio described above. The 4 stations gave estimates of 108 (n=19), 171 (n=11), 126 (n=12), and –97 (n=17). The regression using all data (n=304) yielded an estimate of 72 ± 9 (std. error). A value of 50 was used above, as representative of a healthy phytoplankton population, but given the magnitudes of the computed values, we use a value of 60 in this paper.

Scott's (1979) Fig. 2 presents a comprehensive time–depth profile of chlorophyll *a* concentration for 1975. This figure shows that phytoplankton biomass can exhibit substantial variation, consistent through most depths, over a period of a few days. This is also reflected in the high standard deviations, as given above. Chl *a* concentrations ranged up to 12 mg m^{-3} in 1975.

3.1.5 Heterotrophs – These animals tend to fall into distinct size classes and the data collected from SWA reflect this. The largest are, of course, the fishes. In a 1975 survey of benthic biotopes, Rainer (1980) identified 33 species of fish. In the seagrass bed outside of SWA, a total of 40 fish species was sampled (Bell *et al.*, 1978a, 1978b). Leatherjackets, especially *Monocanthus chinensis*, dominated the community (see Table 1 of Bell *et al.*, 1978a), occurring at a density of about 1 fish per 7 m^2. Leatherjackets, notably *M. chinensis*, also occur within SWA (Rainer, 1980). No estimates of the carbon mass of the fishes are available for SWA.

Among the invertebrate heterotrophs three subcategories seem appropriate, related to mesh size for sampling them. Ctenophores and large medusae are individually much larger than most of the zooplankters; they constitute 14% of total zooplankton dry–weight (DW) biomass (n=51), occurring at an average mass of 14 ± 12 mg DW m^{-3}. Assuming the C/DW ratio to be the same for the ctenophores and large medusae as for the relatively smaller zooplankton, we may use the estimate for the intermediate sized zooplankton (see below) to calculate a carbon mass for the "large zooplankton" (ctenophores and large medusae) of 3.6 ± 3.1 mg C m^{-3}.

There also exist the very small heterotrophs. At the beginning of the Project, the participants chose to aggregate all heterotrophs smaller than 124 μm and refer to them as "microheterotrophs". Microheterotrophs include bacteria, yeasts, ciliates and small zooplankton. One feature of these organisms is that they have a large surface–to–volume ratio, facilitating the use of dissolved nutrients (*e.g.* DOC) in the formation of particulate cellular material.

Microheterotroph mass is estimated as the difference between ATP (<200 μm material) and chlorophyll *a* measurements. The mean microheterotroph mass is 0.23 ± 0.18 g C m^{-3} (n=319). Microheterotrophs were significantly more abundant ($F_{10,318} = 18$) in December and January (0.4 g C m^{-3}) than during the rest of the year (0.2 g C m^{-3}). Using the 270 C/ATP and the 60 C/Chl *a* ratios, 8% of values were negative.

Wiebe & Smith (1977) distinguished between autotrophs and microheterotrophs on the basis of size: they found that the production of photosynthetically produced DOC (PDOC by autotrophs) is due mainly to organisms of 20 – 63 μm while its consumption (by microheterotrophs) is due mainly to aggregates of 106 – 124 μm, made up of small organisms of a few micrometres attached to detritus or clumped together. These results were confirmed by Smith & Higgins (1978).

Finally, there are the intermediate–sized zooplankton studied by Griffiths (1983) Griffiths & Caperon (1979) and Tafe & Griffiths (1983). Two communities were identified: marine and estuarine. Diel migrations among the crustacean zooplankton are not well developed (CSIRO, 1976).

Intermediate sized zooplankton (> 124 µm excluding ctenophores and large medusae) occured at a concentration, averaged through the water column, of 89 ± 73 mg DW m⁻³ (n=50). (These values were calculated from the data base. Obviously Griffiths' (1983) marine community estimate of 55.7 is more appropriate but the oversight was realized too late for use herein.) Monitoring was conducted at one station (Stn C) at weekly intervals from June 1975 to July 1976. No significant difference among months was found in the zooplankton mass but sample sizes per month were small (n=2 – 8) and variation high. Tranter & Kennedy (1983) estimated (and re–checked) C/DW ratios for various size classes, in terms of mesh width, of plankton and found no trend with plankton size. The mean (± std. dev.) value for size classes of 160 – 200, 200 – 250, 250 – 315, 315 – 400, 400 – 500, 500 – 630, 630 – 800, and > 800 µm is 0.27 ± 0.09. Thus the data suggest a mean C biomass of 24 ± 20 mg C m⁻³ as an average over both communities. The yearly mean biomass in SWA is about five times higher than that in adjacent coastal waters.

3.2 Sediment

3.2.1 DIC – The concentration of DIC in the interstitial water of that region of SWA where benthic micro–organisms exist (0 – 10 m) has been measured by Giles (1983). He found a DIC concentration of 26.4 ± 4.2 g C m⁻³. However, he also found a DIC concentration of 21.4 ± 3.4 g C m⁻³ in the water column 0.3 m above the sediment.

3.2.2 DOC – The concentration of DOC in the sediments of SWA has not been measured.

3.2.3 Detritus – Bulleid (unpublished) measured the amount of organic detritus in the top 1 cm of sediments with more than 12 m of water above them as 137 g C m⁻², as the difference between sediment POC and ATP measurements. (This value is particularly accurate for deep sediments which have an aerobic layer of about 2 cm deep.) These sediments consist almost exclusively of silt (Fig. 1 of Rainer, 1980).

In shallow (0 – 10 m) regions of SWA, the sediment consists of shell fragments and coarse sand. We have only a rough estimate of detrital mass: POC in the shallow sandy sediments is 0.8 – 1.2% DW, or about 110 g C m⁻². Of this, about 2 g C m⁻² is living carbon or 108 g C m^{-2} detrital. In really clean bare sand these values are reduced by a factor of 2 or 3.

3.2.4 Autotrophs – The dominant type of sediment autotroph on shallow deposit sediments in SWA is seagrasses (Rainer, 1980). The dominant species is *Posidonia australis* f Hook although *Zostera capricorni* Aschers, *Halophila ovalis* (R. Br.) f Hook, and *Halophila decipiens* Ostenfeld also occur.

The total biomass of *P. australis* was estimated for the seagrass beds outside the mouth of SWA by harvesting four randomly located 0.25 m square quadrats in each of 5 locations (Kirkman & Reid, 1979). Samples were obtained on 7 occasions over an 18–month period from April 1974 to September 1975. The samples were separated into leaves and rhizomes, oven dried at ~105°C for 24 h and weighed; total carbon content of the leaves was estimated to be 36% of leaf dry weight.

Kirkman & Reid (1979) present their results as a plot of mean standing crop biomass per square metre over the sampling period. This plot shows an annual mean total biomass of 264 ± 21 g C m^{-2} (grand mean of 7 time dependent means ± the std. dev. of the mean), and of mean leaf biomass of 59 ± 14 g C m^{-2}. Each of the 7 data points used in these calculations is itself a mean of 20 measurements and mean variation around the data points should give some idea of spatial variation; on average the variation (% C.V.) around each mean was 27% for total biomass and 39% for leaf biomass. An estimate of total variance about the grand–mean seagrass mass (44% C.V.) was derived from the average within–time C.V. (39%) and the among–times C.V. (24%) by using the theory of partitioning of sums of squares within a one–way AOV (Sokal & Rohlf, 1969).

Rainer (1980) notes that *P. australis* occupies 58% of the area of SWA covered by seagrasses (10.6 ha), *Z. capricorni* occupies 34%, and *H. ovalis* occupies 8%. He also suggests that *Z. capricorni* is distributed down to 2 m; *P. australis* below *Z. capricorni* to 2 – 3 m and occasionally to 5 m; and *H. ovalis* between 2 and 5.8 m. Using the uniform–coverage density estimates for the seagrass beds outside of SWA (given above); using our knowledge that the three species do not overlap much (see Fig. 3 of Rainer, 1980); and estimating that, where it occurs, *Z. capricorni* has a biomass of 50% that of *P. australis* and *H. ovalis* a biomass of 1%, we calculate an average density of seagrasses above the 6–m depth of 69 ± 2 g C m^{-2} (total), or 15 ± 7 g C m^{-2} (leaves). (The standard deviation given here was calculated from the data of the previous paragraph by assuming no change in C.V. under multiplicative coding.)

A second form of sediment autotroph in SWA is benthic micro–organisms, consisting of micro–algae (mostly diatoms), and chemotrophic and autotrophic bacteria. These benthic micro–organisms live within the interstices of the bottom sediments, probably within the top few millimetres. They occur both around the edge of the central basin of SWA (0 – 10 m) and in the intertidal region. Using the trichromatic method, Giles (CSIRO, 1976) estimated an average chlorophyll *a* concentration of 9.9 mg Chl *a* m^{-2} and gives a time series in his 1983 paper. No obvious seasonal change was found (Giles, 1983). This concentration of chlorophyll *a* includes both living and dead matter; we guess that about 3 – 4 mg Chl *a* m^{-2} is living matter. And further assuming a C/Chl *a* ratio of 60, we estimate a benthic micro–organism biomass of about 0.2 g C m^{-2}.

3.2.5 Heterotrophs – These animals are customarily recognized in terms of living on (epifauna) or in (infauna) the sediments. The infauna have been subdivided (Fenchel, 1978) in terms of microbenthos (customarily defined as benthic organisms that pass through a 100 µm seive), meiobenthos (organisms that are retained by a 100 µm seive but pass through a 1.0 mm seive), and macrobenthos (organisms retained by a 1.0 mm seive).

The epifauna of the deposit substrates are different from those of the hard substrates (Rainer, 1980). Intertidal rock substrates are dominated by the Sydney rock oyster, *Saccostrea commercialis*. Between low water and 2 m, rock surfaces are often completely covered by the hairy mussel, *Trichomya hirsuta*. A wide variety of species is present on the vertical rock faces and under overhangs. The diversity is, however, limited below 10 m; on rock faces at about 15–m depth only one or two species are present, usually including *Balanus* sp. On silty rocks the fauna is also restricted, the dominant species being massive or branching forms of sponge and the sessile ascidian *Pyura stolonifera*.

Characteristic epifauna of deposit substrates are asteroid starfish (*Patiriella* sp., *Astropecten polyacanthus*, and *Coscinasterias calamaria*) and the occasional large sponge, mud oyster (*Ostrea* sp.), scallop (*Pecten* sp.), and sea tulip (*Pyura pachydermatina*).

Infaunal macrobenthic species included the bivalve *Theora fragilis*; the polychaetes *Australonereis ehlersi*, *Chaetopterus variopedatus*, *Spiochaetopterus* sp., and *Arenicola* sp.; the ghost shrimp *Callianassa arenosa*; and the heart urchin *Echinocardium cordatum*. *T. fragilis* and *E. cordatum* are most common in the deeper silty sediments around the central basin.

Estimates of carbon mass of epifauna are available only for the Sydney rock oyster *Saccostrea commercialis* and the hairy mussel *Trichomya hirsuta* (Colquhoun–Kerr, 1977). The bay–wide DW biomass of the oyster species, measured on February 1976, was 2 ± 0.8 t and of the mussel species was 3.7 t. Assuming a C/DW ratio of 0.4 (even though there is no evidence from SWA to support this value), we get carbon mass estimates of 0.8 ± 0.3 t C for the oyster species and 1.5 t C for the mussel species. Diving experience in SWA suggests that the total epifaunal mass may be twice that estimated for these two species.

Very little information is available about the species composition of the infaunal micro– and meiobenthos; about all that is known is that the sill of sand at the mouth of SWA contains very large amounts of foraminiferans, even though the central basin is largely devoid of them.

Using ATP measurements, Bulleid (1983) estimated the mass of micro– and meiobenthos in the central basin of SWA at ~2.8 g C m^{-2}.

No direct estimates of macrobenthos are available for SWA but a mean value of 0.35 g C m^{-2} is available for nearby Gunnamatta Bay (see Fig. 1 of Vaudrey *et al.*, 1983); applying this value to the non–silty regions of SWA suggests a macrobenthic mass of 0.12 t C for SWA. Estimates for Cabbage Tree Basin (unnamed basin immediately south of Burraneer Bay in above mentioned Figure) are larger, suggesting this figure to be a minimum.

Rainer collected data on epifaunal and macrobenthos abundance in the central basin of Cabbage Tree Basin. His estimates are relevant to the central basin of SWA because both basins go anoxic periodically. Sampling 6 sites over 18 months at 2 month intervals yielded a total value of 0.06 g C m^{-2}.

4. FLOWS BETWEEN COMPARTMENTS

Those flows between compartments for which data from SWA are available (14 of them) are shown in Table 2. There are, no doubt, other flows that exist within SWA, but only those shown could be estimated from the available SWA data. It is important that the reader understand our approach, as a number of obviously desirable additions to our information are apparent. We recognize the potential desirability of such inclusions; our aim, however, is simply to synthesize what is known at this time about carbon flow in SWA.

4.1 Water column

4.1.1 DIC – The flow to water column autotrophs has been extensively studied in SWA (Scott, 1978a, 1978b, 1979, 1980). Primary production was measured by ^{14}C uptake at 2 – 4 day intervals from February to December 1975, at depths of 1, 2, 3, 4, 6, 10, and 14 m. All samples were from one station in SWA (Stn 7 of Series 3: Vaudrey *et al.*, 1983) whose mid–tide depth was 20 m (Scott, 1978b). The samples were incubated *in situ* at the depth from which they were collected, from local noon to sunset. The depth–integrated, yearly–averaged, net primary production rate during the sampling period was 50 ± 31 (mean ± std. dev., n=75) mmol C m^{-2} day^{-1} (data from Fig. 4 of Scott, 1978b), or 0.60 ± 0.4 g C m^{-2} day^{-1}. (This rate is actually somewhere between the net and gross rates but here we assume it to be the net rate – incubations lasted about 2 h.) Water column maxima were observed to vary up to 1.7 g C m^{-2} day^{-1} (Scott, 1979).

The influence of season on this production rate was determined by subdividing the data: December, January and February (summer); March, April and May (autumn); June, July and August (winter); September, October and November (spring). Production means (mmol C m^{-2} day^{-1}, $\pm$ std. error) were: summer 82.39 ± 15.00 (n=8), autumn 51.40 ± 4.72 (n=23), winter 27.70 ± 5.00 (n=21), and spring 57.61 ± 5.81 (n=23). As expected, production is high in the summer, low in the winter, and intermediate in spring and autumn. A one–way AOV declared significant differences among the four seasons, p<0.01. Perhaps the most interesting *a posteriori* comparison is between spring and autumn; no significant difference was found by an appropriately modified test of comparison of treatment means (Snedecor & Cochran, 1967; p. 271). The variation in the rate of primary production per m^2 could be attributed to both solar irradiance and temperature variation but Scott (1979) found it to be more strongly correlated with solar irradiance.

The largest rates of primary production per m^3 were found at the surface of SWA (Scott, 1979), with values up to 0.4 g C m^{-3} day^{-1}. The rates decrease with depth throughout the euphotic zone; the 1% light level is at about 15 m, with no seasonal trend, but for short durations following runoff from rainstorms the 1% light level is shallower. The seasonal trend, mentioned above for the water column production data, is evidenced in the per m^3 data by the more numerous lower production rates which occur in the winter and by the shallow depths to which the productive zone extends (Scott, 1979).

An annual mean rate of primary production per m^3 will obviously be subject to considerable variation. This said, the mean rate can be calculated by dividing the water–column rate by the depth of the euphotic zone during dry–weather conditions (15 m), to get a value of 0.040 ± 0.03 g C m^{-3} day^{-1}. (The standard deviation given here was calculated from the per m^2 estimate.)

Table 2.
Interaction matrix of flows estimated for SWA.
Flow is from the horizontal to the vertical compartment.

WATER COLUMN

(from → to)	WATER COLUMN: DIC	DOC	Detritus	Autotrophs	Heterotrophs	SEDIMENT: DIC	DOC	Detritus	Autotrophs	Heterotrophs
DIC				X					X	
DOC					X					
Detritus								X		
Autotrophs		X			X					
Heterotrophs	X									

SEDIMENT

(from → to)	WATER COLUMN: DIC	DOC	Detritus	Autotrophs	Heterotrophs	SEDIMENT: DIC	DOC	Detritus	Autotrophs	Heterotrophs
DIC	X								X	
DOC										
Detritus										X
Autotrophs		X						X		
Heterotrophs						X		X		

A second flow of water–column DIC is that to sediment autotrophs. Kirkman & Reid (1979) have estimated the yearly–averaged net growth rate of the leaves of *Posidonia australis* in the seagrass beds outside of SWA to be 388.7 mg C m^{-2} day^{-1}. Variation of mean growth rates (each mean based on n=70) over 9 time samples was ± 90 mg C m^{-2} day^{-1}. Assuming that this estimate applies equally to the seagrass beds as they occur in SWA (and with appropriate modifications for a different species mix), we calculate a flow rate of 101.6 ± 24 mg C m^{-2} day^{-1} for the region in which seagrasses occur in SWA – *i.e.* above the 6–m depth. (See Section 3.2.4 for details on the method.)

4.1.2 DOC – The flow to water column heterotrophs has been studied in SWA (Wiebe & Smith, 1977; Smith & Higgins, 1978) in terms of the incorporation of photosynthetically–produced DOC (PDOC) into microheterotrophs. A compartmental model with linear donor–dependent flow rates (see Fig. 1; from Wiebe & Smith, 1977) was hypothesized and tracer kinetic experiments conducted to estimate the "fractional turnover rate" constants, λ_{ij} (flow from compartment j to i). Table 2 of Wiebe & Smith (1977) gives parameter estimates from light and dark experiments, 18 July 1975 and 8 August 1975. Since the parameter estimates do not differ with the presence or absence of light, average estimates (± std. error) are given in Fig. 1.

Since there is 1.2 mg C m^{-3} of PDOC in SWA and $\lambda_{21} = 0.028$ ks^{-1}, the flow rate of PDOC to microheterotroph POC is 2.9 mg C m^{-3} day^{-1}.

4.1.3 Detritus – The settling rate of detritus has been measured for a 16–m water column station in SWA. The sediment tube experiment gave a mean annual settling rate of 0.56 g C m^{-2} day^{-1} (see also CSIRO, 1976; Bulleid, 1983).

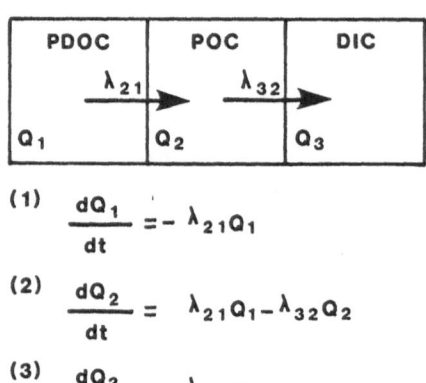

(1) $$\frac{dQ_1}{dt} = -\lambda_{21}Q_1$$

(2) $$\frac{dQ_2}{dt} = \lambda_{21}Q_1 - \lambda_{32}Q_2$$

(3) $$\frac{dQ_3}{dt} = \lambda_{32}Q_2$$

$$\overline{\lambda}_{21} = 0.028 \pm 0.001 \ ks^{-1}$$

$$\overline{\lambda}_{32} = 0.0009 \pm 0.0002 \ ks^{-1}$$

Fig. 1: Model devised by Wiebe & Smith (1977). See text for more details.

4.1.4 Autotrophs – The production of PDOC by water column autotrophs was estimated by Wiebe & Smith (1977). They found that PDOC was produced at a rate of 0.104 mg C m^{-3} h^{-1} on 18 July 1975 and 0.116 mg C m^{-3} h^{-1} on 8 August 1975. It is not clear whether PDOC is produced in the dark. Firstly, assume it is not: there is about 10 h of light per day in late July, early August (Spencer, 1975); thus daily PDOC production rates are 1.04 and 1.16 mg C m^{-3} day^{-1}. This rate is 38% of the flow rate from PDOC to microheterotrophs; in an experiment by Smith & Higgins (1978) untreated water samples from SWA produced DOC at the same rate at which it was incorporated into microheterotrophs. Now assume that PDOC is produced at the same rate in the light and in the dark: then daily PDOC production rates are 2.50 and 2.78 mg C m^{-3} day^{-1}. These rates are close enough to the 2.9 mg C m^{-3} day^{-1} rate of microheterotroph incorporation of PDOC to suggest that the second assumption is the correct one.

During the second half of 1975, the *in situ* production of total DOC by phytoplankton was measured and found to be about 10% of the rate of particulate production by phytoplankton (CSIRO, 1976). Since the net production rate of phytoplankton autotrophs is about 0.04 g C m^{-3} day^{-1} (see above), this suggests a total DOC production rate of 4 mg C m^{-3} day^{-1}.

The flow to water column heterotrophs has been studied by Griffiths & Caperon (1979). In the third of three experiments they measured the grazing rate of a natural population of zooplankton on a natural population of phytoplankton, to which a tracer amount (10% ambient concentration) of labelled phytoplankton had been added. Both the phytoplankton and the zooplankton were collected in a 6–l water bottle (Jitts, 1964) from the 6–m depth of SWA, on 25 November 1975.

In their first experiment, Griffiths & Caperon (1979) showed that, for constant zooplankton density, the grazing rate of zooplankton varied as a linear function of phytoplankton density:

$$-dD/dt = kD$$

where D is phytoplankton density, t is time, and k is the grazing rate constant. Assuming this relationship to hold in their third experiment, they measured dD/dt and D and used these estimates to calculate k. They reported that, for the first subsample, $k = 0.28$ day^{-1} and, for the second subsample, $k = 0.42$ day^{-1}, assuming that zooplankton graze at the same rate both day and night. They did not report values of D, but for the second experiment, conducted on 6 November 1975, a phytoplankton density of 80 μg C l^{-1} (mg C m^{-3}) was given. This is in agreement with the chlorophyll measurement for that date at a 6–m depth given by Scott (1979) of $1 - 2$ mg Chl a m^{-3}. But Scott's data also suggest a density of $2 - 4$ mg Chl a m^{-3} on 25 November 1975. Hence we assume a value of 2 mg Chl a m^{-3} on 25 November 1975 at 6–m depth, or 120 mg C m^{-3}. This implies a flow rate due to the zooplankton grazing rate of $34 - 50$ mg C m^{-3} on that day. Estimates for other dates are not available and the model does not treat zooplankton mass explicitly, allowing one to calculate the average grazing rate from average phytoplankton and zooplankton masses.

Apparently there also exists in SWA a group of "epibenthic zooplankton" that migrate diurnally between the sediments and lower water column (Smith et al., 1979). Only 5 of 22 kinds of these organisms were shown by these authors to feed on phytoplankton. Grazing rates in terms of carbon are not available.

4.1.5 Heterotrophs – The flow from microheterotrophs to DIC was estimated by Wiebe & Smith (1977). Using Eq. 3 of Fig. 1 and knowing that there is about 0.23 g C m^{-3} as microheterotrophs (see above), we estimate a flow rate of 17.9 ± 14.8 mg C m^{-3} day^{-1}.

4.2 Sediment

4.2.1 DIC – The flow to sediment autotrophs has been studied in SWA by Giles (1983). He measured *in situ* the net primary production of the benthic micro–organisms from July 1975 through to June 1976 for three stations (Series 4 in Fig. 1 of Vaudrey *et al.*, 1983). Monthly and annual productivities are listed in Table 4 of Giles (1983). For two of the larger areas of benthic micro–organism abundance, annual estimates of production were 68.1 (in seagrass beds outside the mouth of SWA) and 113.2 g C m^{-2} (in sand banks in Flat Rock Creek). There were generally higher productivities during the months of longer day length (*i.e.* November, December and January).

Since there is no long term build up of DIC in the sediments the rate of diffusive flow into the water column from the central basin sediments is assumed to be equal to the rate of DIC production in these sediments.

4.2.2 DOC – No information is available on flows from DOC.

4.2.3 Detritus – The manner in which the sediment, or, at least, the detritus near to the sediment, is recycled has been studied for two epifaunal suspension feeding species, the Sydney rock oyster and the hairy mussel, by Colquhoun–Kerr (1977). He found, in experiments conducted in February and March 1976, that the oysters consumed 5600 g C day^{-1} and the mussels consumed 7000 g C day^{-1} in SWA. No estimates of variation were obtained by Colquhoun–Kerr (1977).

Kirkman (unpublished data) found that detrital seagrass leaves decompose fairly rapidly. He found that about 59% of carbon is lost from the decaying leaves after one month.

4.2.4 Autotrophs – Kirkman & Reid (1979) note that *Posidonia australis* contributes DOC to the water column both by excretion from living leaves and by decay of the non–growing or senescent portions. From samples taken between August and October 1977, they estimated that *P. australis* produces, on average, 211.3 mg C m^{-2} day^{-1} of (water column) DOC. The 95% confidence limits of DOC release are 0.744 – 1.752 mg C/g DW/day; and since there is a mean of 163.9 g DW m^{-2} of *P. australis* leaves in the seagrass beds outside of SWA, this suggests limits of 121.9 – 287.2 mg C m^{-2} day^{-1}. Assuming that these estimates apply equally to the seagrass beds as they occur within SWA (and with appropriate modifications for a different species mix), we estimate a DOC production rate of 55.2 (32 – 75) mg C m^{-2} day^{-1}.

A novel flow from the seagrasses to the detrital compartment occurs through the grazing activities of leatherjacket fishes (*Monocanthus chinensis, Meuschenia freycineti,* and *Meuschenia trachylepis*). Seagrasses and algae form major dietary components in these fishes (Bell *et al.*, 1978a) but the seagrasses are taken only incidentally with the attached organisms. These encrusting organisms are digested and the seagrass pieces are apparently expelled undigested, although the labile carbon in the seagrasses may be incorporated in the tissues of the fishes.

Kirkman & Reid (1979) harvested leaves from ten 0.5 m² plots on 19 May 1975 and from six plots on 10 January 1978 to estimate the dry weight of missing pieces of the leaves, which they assumed to be eaten by herbivores. This estimate was then divided by the replacement time of the leaves (being the reciprocal of the average relative growth rate) to provide a grazing rate estimate of 13.5 mg C m^{-2} day^{-1} and hence of 3.5 mg C m^{-2} day^{-1} for SWA seagrasses.

This flow is obviously from the seagrasses to the water column detrital compartment. But since *Monocanthus chinensis*, the most abundant leatherjacket in the seagrass beds, is totally estuarine dependent (Bell *et al.*, 1978a) and relies on seagrasses and algae for about 60% of its food by volume, the expelled seagrass detritus is probably in the water column for only a short period of time. On the other hand, leatherjackets stay in weed beds and in areas where there is moving water. Hence it is reasonable to assume only half of the flow to go from seagrasses to sediment detritus directly, the rest being lost to the sea from the tidal channel.

A second flow to sediment detritus occurs as old leaves of *P. australis*, as a result of cellular decomposition, leaching of DOC, and growth of encrusting epibiota, sink and contribute to the detrital mat accumulating on the substratum. Kirkman & Reid (1979) covered 10 squares each of area 0.25 m² with a layer of 1 cm diameter white pebbles and after collecting the detrital leaves (> 9 mm) at two–weekly intervals, determined the dry weights of the leaves (minus the epibiota). Data were collected for 9 months from 17 February 1978. They found an annual average estimate of 162 mg C m^{-2} day^{-1} and hence of 42 mg C m^{-2} day^{-1} for SWA seagrasses, using the method described in Section 3.2.4. There was an average variation (% C.V.) over time (of means) of about 45% and over space of about 53% (n=10). Using these estimates and the theory of partitioning of sums of squares within a one–way analysis of variance (Sokal & Rohlf, 1969), we calculated an estimate of the total variation of the observations around the grand mean at 67%.

Rainer (1980) shows flora and sediment type for 80 transects through SWA. There is no obvious correlation between seagrass presence and a silt component in the sediment.

These two sources provide a flow of seagrasses to sediment detritus of 43.8 mg C m^{-2} day^{-1}.

4.2.5 Heterotrophs – It has been shown experimentally (Bulleid, 1983) that sediment (plus biota) from the central–basin region of SWA consumes oxygen at a rate of 567 ± 50 mg O$_2$ m^{-2} day^{-1} (n = 12). Since most biota would be heterotrophs this suggests that micro/meiobenthos respire 0.27 g C m^{-2} day^{-1}. The chemical oxygen demand has not been subtracted because it is itself a consequence of past biological activity, in particular sulphate reduction. The sulphate reducers are breaking down organic (detrital) matter. Thus the estimate given above includes both the aerobic respiration in the surface of the sediments and the anaerobic respiration that is going on beneath it.

Colquhoun–Kerr (1977) estimated the biodeposition rate of the Sydney rock oyster and the hairy mussel in SWA; that is, the flow rate from two epifaunal species to sediment detritus. He found for experiments conducted in February and March, 1976, that the oysters deposit 1200 g C day^{-1} and the mussels deposit 1780 g C day^{-1}. Probably shell deposition following death of these organisms represents an important flow of carbon but this rate was not measured.

5. INFLOWS AND OUTFLOWS

During normal conditions there seems to be no net change of carbon mass due to tidal movements. Scott (1978a) presents data on the rate of photosynthesis at constant irradiance for 1–m stations at the entrance of, and within, SWA. Finding mainly similar rates over the period from May to October 1975, he concluded that there is probably no significant net change in the amount of SWA phytoplankton due to tidal exchange.

A few divergent rates were, however, observed and Scott (1978a) attributed the higher biomass at the mouth of SWA to the effects of coastal upwelling introducing nutrient–rich waters on the rising tide and removing the surface waters containing less phytoplankton and nutrients on the falling tide. The qualitative correlation between nitrate (commonly assumed to be the limiting nutrient in marine waters) 6 km east of Port Hacking and the phytoplankton biomass of SWA was found to be good, but the quantitative correlation quite low, about 0.35 – 0.40 for those biomass maxima thought to be due to a coastal enrichment process (Scott, 1979). Influencing factors include wind strength, tidal amplitude and season.

To further investigate the possibility of net changes in SWA compartments due to tidal exchange, we used the information on the data base from stations throughout Port Hacking to do a numerical classification of the surface salinity, oxygen, nitrate, nitrite, phosphate, silicate, POC and DOC records. Temperature was excluded because a preliminary analysis indicated its dominance, but a dominance reflecting only sampling dates. A divisive polythetic algorithm (POLYDIV; see p. 118 of Williams, 1976) was used in the analysis.

The first split of the data separated the records on the basis of (in order of importance) salinity, silicate, and nitrate. The low ·salinity, high silicate and nitrate records tended to come from SWA and nearby stations. A further split of these data, more–or–less equally on the basis of these variables and in the same directions, showed about equal percentages of records from SWA and Port Hacking stations being assigned to the low salinity subcategory. This indicates that the characteristics of SWA are typical of much of Port Hacking, although near the mouth of Port Hacking more marine conditions – high salinity and low nutrients – prevail. A net change in the compartment masses of SWA due to tidal exchange with Port Hacking seems unlikely.

A weekly estimate of floating *Posidonia australis* debris (> 9 mm) was obtained by Kirkman & Reid (1979) over the period from September 1975 to September 1976. They give a mean floating rate of 50.6 mg C m^{-2} day^{-1} but their data suggest a lognormal distribution of values over time, with mean value of 31 mg C m^{-2} day^{-1} (8 for SWA seagrasses) with 95% confidence limits of 22 – 42 mg C m^{-2} day^{-1} (6 – 11 for SWA seagrasses). Leaves of *P. australis* were not found in large drifts on the beaches near the beds, nor was there much seasonal change in biomass; this behaviour is in contrast with other seagrass beds but the seagrass bed outside of SWA is well sheltered from storms (Kirkman & Reid, 1979), as are the beds within SWA. We assume that all floating leaves are exported to the sea.

This flow combines with that mentioned in the previous Section to suggest that 9.8 mg C m^{-2} day^{-1} of seagrass detritus is exported to Port Hacking.

Of that detritus which sinks to the sediment, probably some is lost to the deeper sediments. Scott (1978b) implies that the accumulation rate found for Californian coastal basins, *viz.* 5 mmol C m^{-2} day^{-1} (0.06 g C m^{-2} day^{-1}), is of the same order as the accumulation rate for SWA. But these basins are open–sea basins, so we will assume the value for SWA to be about an order of magnitude larger.

6. THE CARBON BUDGET

The required components of the carbon budget have been given above: information relating to the nature of the compartments and our best estimates of average compartment masses (Section 3), of average flow rates between compartments (Section 4), and of flow rates into and out of SWA (Section 5). In these Sections, we attempted to provide sufficient detail on the derivation of the estimates, and on their variability, to allow the reader to judge the adequacy of our synthesis of the data, and of our conclusions about carbon flow within SWA. But this detail needs to be

presented in a form which helps one to see the overall picture and this is the purpose
of this Section.

In the following calculations, the low–water volume of SWA is taken to be
8.33×10^6 m^3. Average compartment sizes and flow rates within the sediment
subsystem (except for heterotroph respiration) are given in terms of the aerobic
sediment only. In the shallow (0 – 10 m) regions of SWA the substrate consists of
sand and shell fragments (Rainer, 1980) and here the top 5 cm is assumed to be
aerobic; in deeper regions (> 10 m) the sediment consists of silt and only the top 1 cm
is taken to be aerobic. (We do not consider, given the absence of suitable quantitative
data, the effects of the infaunal macrobenthos: local areas of oxygenation will be
present around the tubes and burrows of such animals.) On the other hand, in fine silt
only the top 2 – 3 mm is oxygenated. Seagrasses are considered only in terms of their
leaves.

A summary of the average distribution of carbon, ± C.V., into compartments is
given in Table 3.

Mean estimates represent bay wide values and hence account for the restricted
distributions of certain compartments. Variance estimates represent variation of the
observations around the grand mean. These data will be discussed in Section 7.2.

To improve understanding of the variation about the mean estimates, we did a
breakdown of this Total MS into its variance components (Sokal & Rohlf, 1969) for
the water column compartments.

The information on the data base is sufficient to allow the analysis of a Depths x
Stations x Times Factorial AOV design for each of water column DOC, detritus,
chlorophyll *a*, and microheterotrophs. Depths are considered in terms of 0 – 2 m,
3 – 5 m, 6 – 9 m, and 10 – 14 m regions; stations are grouped by location and water

Table 3.
Compartment sizes, with C.V.s, in South West Arm.

	10^6 g C ± C.V.			Percent of Total	
Total C	325.3			100.	
Inorganic C	199.6			61.	
Organic C	125.7			39.	
Water Column					
DIC	199.	±	5%	61.2	
DOC	12.5	±	53%	3.8	
Detritus	2.1	±	104%	0.6	
Autotrophs	1.4	±	88%	0.4	
Heterotrophs	2.2			0.7	
−microheterotrophs		1.92 ±	78%		89
−intermediate sized zooplankton		0.20 ±	83%		9
−large zooplankton		0.03 ±	83%		2
−fishes		−			−
Total Water Column	217.2			66.7	
Sediment					
DIC	0.6			0.2	
DOC	−			−	
Detritus	97.			29.8	
Autotrophs	3.6			1.1	
−seagrasses (leaves)		3.5 ±	47%		98
−benthic micro−organisms		0.07			2
Heterotrophs	6.9			2.1	
−epifauna		4.6			66
−macrobenthos		0.12			2
−micro/meiobenthos		2.2			32
Total Sediment	108.1			33.2	

column depth. Time is considered by months, irrespective of the year. The variance components for each of these variables, assuming a pure random–effects model, are given in Table 4. Although only unbalanced designs were used, the degrees of freedom for the Total MS were assumed equal to abcn–1, where a is the number of depth regions, b is the number of station groups, c is the number of months for which data exists, and n is an estimate of the "average" within–cell replication. This n was then used in the calculation of the variance components from the expected MSS for the appropriate balanced design.

Table 4.
Variance components for the water column compartments.

Variable	Percentage variation due to				
	Depth	Station	Time	Interaction	Error
Three–way AOV's					
DOC	0	4	14	– [†]	82
Detritus	0	8	32	–	60
Chlorophyll *a*	5	<1	0	18 [‡]	77
Microheterotrophs	0	0	31	6 [¶]	63
Two–way AOV					
DIC	31	–	17	13	39
One–way AOV's					
Intermediate sized zooplankton	–	–	5	–	95
Large zooplankton	–	–	26	–	74

[†] Not estimated
[‡] Depth x Time
[¶] Sum of Depth x Time (4%) and Depth x Station (2%)

For the DIC compartment sufficient information is available only to investigate a Depths x Times design. DIC information is available at 1, 2, 3, 4, 6, 10, and 14 m depths. Time is, as above, considered in terms of months.

For the zooplankton, only water column data are available; a one–way AOV over various months is the best analysis.

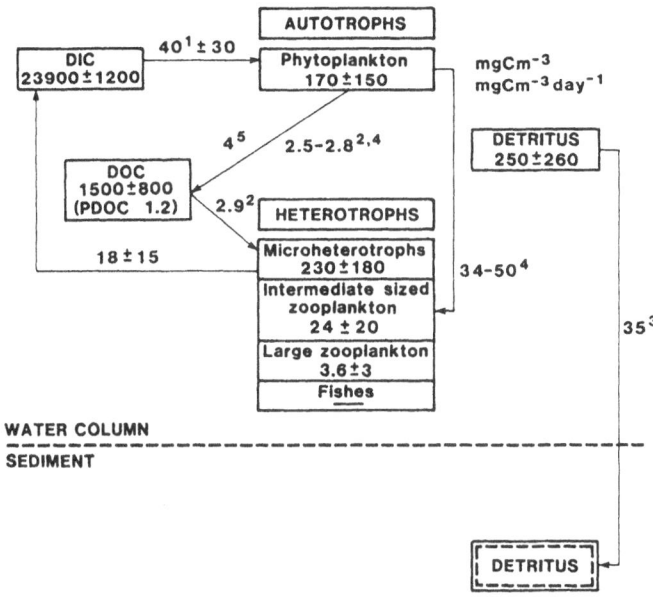

Fig. 2: Carbon budget for the water column of South West Arm. The autotroph and heterotroph
compartments are subdivided; each subdivision has been treated as a separate compartment in terms
of flows. The sediment detritus compartment is indicated by both a dotted and a solid box. Units for
compartment size (mg C m^{-3}) and flow rate (mg C m^{-3}day^{-1}) are given. Superscripts designate the
following: [1]net growth rate; [2]PDOC only; [3]column value from 16-m station, divided by 16; [4]range;
[5]total DOC.

The most interesting result from these analyses (Table 4) is the high proportion
of total variance due to error, or within–cell replication. For the compartments
analysed by a three–way AOV, such "local effects" relate to variation within a 2 – 4 m
region of a particular station of a particular month. Obviously there are relevant
phenomena occurring within very small spatial regions of SWA. (The DIC
compartment appears to be an exception to this rule, and although the interaction is
statistically significant a study of the mean DIC concentrations suggests that the high
interaction variance component is localized and spurious.)

Mean compartment masses and flows among them are detailed in Figs 2 – 4. It
is, unfortunately, not possible to present all data in one illustration. Variation through
space was extreme for some compartments, as mentioned in Section 2.3 above.

The water column was, however, sufficiently homogeneous to allow one
illustration (Fig. 2) to suffice. The flow rates in this figure represent average, daily
rates per cubic metre, plus measures of variation where known. Our best estimates of
variation exist for the compartment masses, given as total variation around the grand
mean. The most reliable estimate for variation around the flow rates is that flow

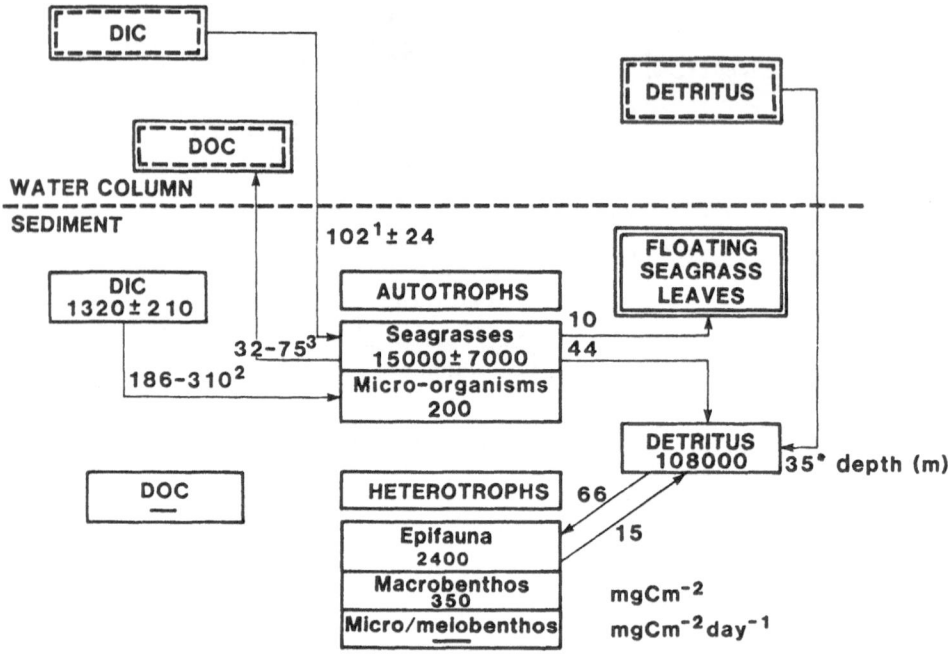

Fig. 3: Carbon budget for sediment regions to a depth of 5 m below low water. The compartments of the water–column subsystem are surrounded by both dotted and solid boxes. Carbon exported to Port Hacking is indicated by a double–sided solid box. Superscripts designate the following: [1]net growth rate; [2]net growth rate, range; [3]95% confidence limits. The "35°depth(m)" indicates that the column flow rate is 35 mg C m⁻³ day⁻¹ times the column height in metres. See also the legend to Fig. 2.

corresponding to net primary production. This standard deviation estimates the variation between times (3–day sampling intervals) at one station over about one year. In two cases, only two estimates are published and these are included as ranges. Also, estimates of variation are available for the parameters of the dynamic model describing the flows from PDOC to microheterotroph POC to DIC (Wiebe & Smith, 1977), but these estimates describe only experimental errors which are relatively small. To obtain an estimate of the variation about the mean flow rate from microheterotrophs to DIC, we used the estimate of variation about the microheterotroph compartment mass.

The sediment subsystem is represented in terms of Figs 3 & 4. In Section 2.3 we noted that ideally the macrobenthos should be described in spatial regions 0 – 5 m, 5 – 10 m, and below 10 m; and seagrasses in regions above and below 6 m. Fig. 3 of Rainer (1980) suggests that the lower limit of seagrasses may be safely taken as 5 m. Thus the sediment subsystem is presented in terms of 0 – 5 m (Fig. 3) and > 10 m

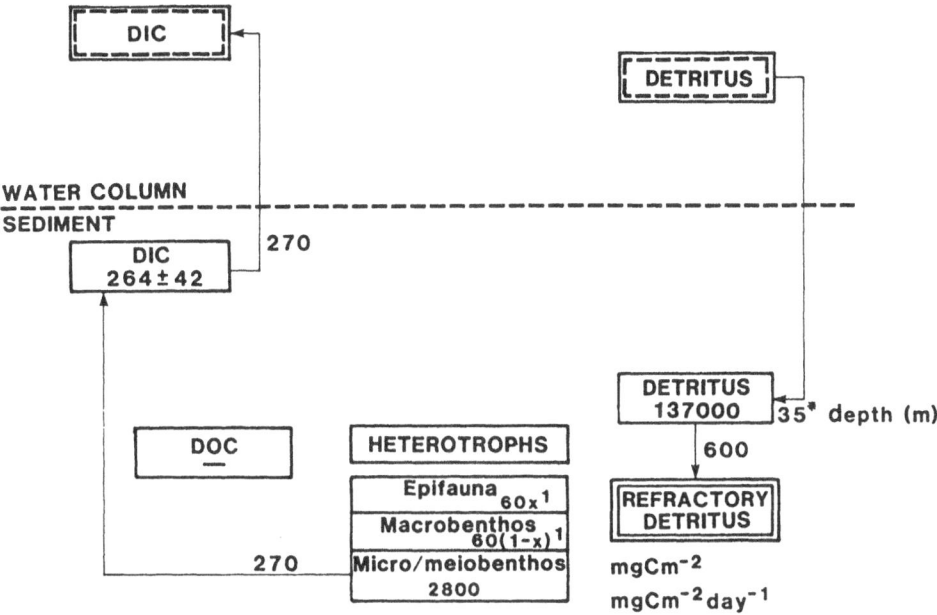

Fig. 4: Carbon budget for the sediment regions at depths 10 m below low water. Superscript [1]designates data for combined epifauna and macrobenthos; x is the proportion that is epifauna. See also the legends to Figs 2 and 3.

(Fig. 4) regions. The 5 – 10 m region is intermediate between the shallow and deep regions in terms of macrobenthos mass, and identical to the deep region in terms of seagrasses. Other than for compartment masses (where variance estimates are as for Fig. 2), variance estimates are available only for seagrass–related flows. The estimate for the mean flow rate from water column DIC to seagrasses is a standard error of 9 mean values (n=70) over time. The confidence limits for the flow rate from seagrasses to water column DOC represent total variation.

The average compartment sizes and average flow rates relating to the seagrasses of the sea flats outside of SWA are given in Fig. 5. All except one estimate of variation around the flow rates are as explained in the previous paragraph. The confidence limits for the flow rate from seagrasses to floating seagrass leaves represent total variation.

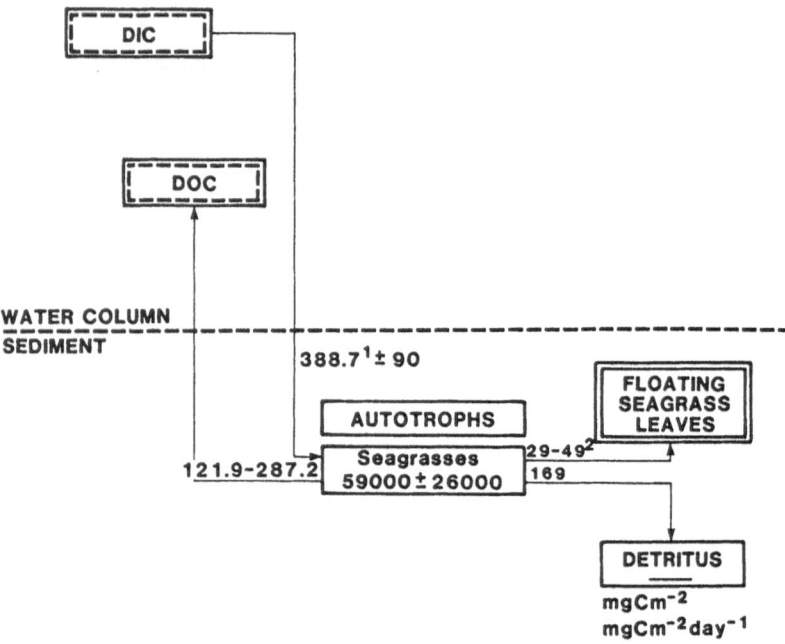

Fig. 5: Carbon budget for the seagrass beds outside of South West Arm. Superscripts designate the following: [1]net growth rate; [2]95% confidence limits. See also the legends to Figs 2 and 3.

7. DISCUSSION

7.1 General aspects of the budget

As is readily seen from the above Sections, our knowledge about carbon flow within SWA is incomplete. The motivation for constructing the carbon budget at this stage of knowledge relates more to the high costs of such research and to organizational constraints than to any misconception regarding the completeness of our knowledge. Expensive and long–term ecosystem studies require, for logical development, a sequence of attempts at synthesizing the available information. Our budget thus represents a first step in an attempt to see how far we have come in terms of understanding the distribution of carbon and its flow into, within, and out of SWA. As such, it should have been constructed early in the Project and revised at set intervals throughout; however, established modelling practice at the time was in a different direction (Sinclair *et al.*, 1983) and we took some time to see how one might effectively synthesize incomplete data sets.

In drawing conclusions from the available data, we consider the possible effects of missing data and of known variation on the conclusions. We do not, however, apologize fór lack of information and will not automatically reject conclusions which may appear to be a function of limited data. Although it would have been possible to incorporate knowledge from other estuaries, or to have speculated on missing data, this would have been in conflict with our goal of synthesizing what we know about SWA. In a sense, we have attempted to redress the balance of much current ecosystem modelling with their combination of relatively small amounts of data from the ecosystem of specific interest and of much information of unknown reliability.

Finally, we recognize that the mass of detail incorporated in this paper makes difficult reading. It is our belief that a full description of the strengths and weaknesses of our present budget information is the only way to a complete and reliable carbon budget of SWA. A less detailed and shorter description of what we think is happening in SWA would certainly be easier to read but would tend to discourage open investigation. We have tried to synthesize what we know in as simple a way as possible without sacrificing realism, so as to encourage changes where needed. The synthesis is thus meant as a tool to help those who work on SWA in the future to build on past experience.

7.2 Specific aspects of carbon flow through South West Arm

In this Section we use the budget to see how carbon is distributed and to ascertain the pathways through which carbon flows, and at what rates.

Table 3 shows that the water column subsystem contains 217 t C and the sediment subsystem contains 108 t C of a total of 325 t C. Thus the water column contains about twice as much carbon as the sediment subsystem. One compartment mass is missing in Table 3. The main body of carbon in the water column subsystem is DIC and in the sediment subsystem is detritus. With these two compartments subtracted the water column contains 1.5 times more carbon than the sediment subsystem. If the seagrass root mass (12 t C) is included in the calculations, then the ratio falls to 1.8 : 1 (0.7 : 1 for DIC and detritus subtracted). The variance estimates, as shown for the water column compartments, range from ~50 to 100% C.V. (except for DIC at 5%); the ratios given above will obviously vary considerably through space and time.

Living organisms contain 14 t C (4% of the total carbon mass) or 26 t C (8%) if seagrass roots are included. The majority of the carbon retained in living organisms (as known) is within the epifauna (5 t C) but this estimate is approximate. The second largest pool of living carbon is the leaves of the seagrasses (3.5 t C), with 95% confidence limits being approximately 2 – 5 t C. Thus seagrasses will, at times, be as large a pool of carbon as the epifauna. If the seagrass roots are included, then the seagrasses represent the largest pool (15.5 t C) of carbon retained in living organisms.

Figs 2 – 4 show a total flow rate of 107 mg C m^{-3} day^{-1} in the water column subsystem, 373 mg C m^{-2} day^{-1} in the shallow (0 – 5 m) sediments, and 270 mg C m^{-2} day^{-1} in the deep (> 10 m) sediments. Thus, in shallow regions of SWA the water column and sediments cycle about the same amount of carbon; in deep regions the water column cycles substantially more carbon. Throughout the whole of SWA the bay–wide water column flows amount to about 0.9 t C day^{-1} while the sediment flows amount only to 0.25 t C day^{-1}. This picture suggests dominance by the water column.

Carbon flow from the water column to the sediment subsystem occurs with the sinking of detritus (Fig. 2), at a rate of 35 mg C m^{-3} day^{-1} (column rate varies with depth), and with the net photosynthetic use of water column DIC by the seagrasses (Fig. 3), at a rate of 102 mg C m^{-2} day^{-1}. Carbon flow in the reverse direction occurs with the production of DOC by seagrasses at a rate of 32 – 75 mg C m^{-2} day^{-1}; and with a diffusion of DIC from the sediment into the water column subsystem, at a rate of 270 mg C day^{-1} in the deep sediments. On a bay–wide basis, these rates suggest a total flow from the water column to the sediment subsystem of 0.3 t C day^{-1} and in the reverse direction of only 0.1 t C day^{-1}. Thus a net flow of 0.2 t C day^{-1} flows to the sediment subsystem; we estimated that roughly 0.5 t C day^{-1} of this sediment undergoes long–term sedimentation (*i.e.* gets buried in the sediments to a depth such that its constituents will not be recycled back into the water column). This value thus overestimates the true value.

Primary producers in SWA include phytoplankton, seagrasses, and benthic micro–organisms. On a per m^2 basis phytoplankton produce 400 mg C day^{-1} (for 10–m average depth) with seagrasses producing 102 mg C day^{-1} (389 mg C day^{-1} for seagrass beds outside the mouth of SWA) and benthic micro–organisms 186 – 310 mg C day^{-1}. On a bay–wide basis phytoplankton produce about 0.15 ± 0.11 t C day^{-1} (60%) with seagrasses producing about 0.01 ± 0.003 t C day^{-1} (4%) and benthic micro–organisms 0.07 – 0.11 t C day^{-1} (36%).

The percentage contribution of benthic micro–organisms is surprisingly high; in calculating the values for benthic micro–organisms, we assume that they occur throughout the 0 – 10 m region of SWA. Had we taken 0 – 5 m as the range of distribution of the benthic micro–organisms (close to that suggested by Giles (1983) for his station type EF), the percentage contributions would have changed to phytoplankton 72%, seagrasses 5%, and benthic micro–organisms 23%. Seagrass growth was measured directly, by clipping, and thus the resulting estimates represent net daily growth; the benthic micro–organism and phytoplankton estimates are a result of daytime measurements only, not incorporating carbon loss due to night–time respiration.

The fate of the benthic micro–organism production will be different to that of phytoplankton. The former are largely restricted to intertidal or shallow subtidal regions (much on the sand banks at the mouth of Flat Rock Creek) and DIC produced by respiration will have an enhanced chance of being lost from SWA by diffusion into the atmosphere.

The seagrass bed outside the mouth of SWA, as suggested by its rate of primary production (0.12 t C day^{-1}) relative to that within SWA (0.25 t C day^{-1}), is of obvious potential importance to SWA. Fig. 5 shows that most of the carbon fixed by the seagrasses is released either as sediment detritus or as water column DOC. Some of the DOC will undoubtedly be moved off the seagrass beds to SWA by tidal action, say one–half. Since the seagrass beds are 30 ha in area, a total of 0.04 – 0.09 t C is released from the seagrass beds each day. This implies an annual inflow of the same order as the DOC pool within SWA. Since mean DOC mass within SWA does not vary widely (53% C.V.), and since we assume no direct losses to Port Hacking due to tidal exchange during dry–weather conditions, we must conclude that the introduced DOC is cycled within SWA before being removed from SWA in some other form. If we assume that the total DOC : PDOC ratio of 4.0 : 2.6 for the flow from phytoplankton to DOC (Fig. 2) also applies to the flow from DOC to microheterotrophs, then a net flow from DOC of 4.5 mg C m^{-3} day^{-1} may be calculated. A bay–wide estimate equals 0.04 t C day^{-1}, sufficient to process the smaller of the two estimates of the DOC generated by the seagrass beds. Less than 0.01 t C day^{-1} flows from SWA phytoplankton to DOC.

The influence of the detrital mass produced by the seagrasses was studied by Kirkman *et al.* (1979). From their studies they postulate that the seagrass bed releases reactive phosphate through microbiological remineralization of the detritus. The phosphate enters the water column either by diffusion or by a tidal–flow mechanism. The high–phosphate water is then flushed off the seagrass beds into SWA once tidal movement begins. The effects of this contribution will, of course, already be reflected in the budget parameters as a determinant of the average rate – and its variability – of primary production of phytoplankton.

We now turn to looking for major paths of carbon flow within SWA. Fig. 2 shows that the largest flows observed in the water column are from DIC to phytoplankton (40 ± 30 mg C m^{-3} day^{-1}), from phytoplankton to intermediate sized zooplankton (34 – 50 mg C m^{-3} day^{-1}), and from water column detritus to sediment detritus (35 mg C m^{-3} day^{-1}). This suggests that the main path of carbon flow in the water column is from DIC to phytoplankton to intermediate sized zooplankton. Both phytoplankton and intermediate sized zooplankton will have flows to detritus, but the available data suggest that the largest flow will be from intermediate sized zooplankton: if the mid–range rate of the flow from phytoplankton to intermediate sized zooplankton is used, the phytoplankton compartment is unbalanced by an outflow excess of 6 mg C m^{-3} day^{-1}. But the variation on the flow rates, as known, suggests the system is capable of dramatic changes from instance to instance.

The detritus which falls onto the sediments of the shallow regions (0 – 10 m) is either rapidly and completely processed or transported to the sediments of the deeper regions (> 10 m), as suggested by Rainer's (1980) observation that the sediments of the shallow regions largely consist of clean sand and shell fragments. Fig. 3 suggests that carbon cycling rates in the sediment subsystem are insufficient for using the carbon deposited by fallout plus seagrass litter but we do not know the applicability of our estimate of the settling rate of water column detritus to these shallow waters.

In the deeper regions the water column fallout of detritus is correspondingly larger (~350 mg C m^{-2} day^{-1} for the average 10–m depth). The rate of accumulation of the detritus into the deeper sediments (600 mg C m^{-2} day^{-1}) is large enough to account for this rate of detrital fallout. But our estimate of this accumulation rate is not good (see Section 5) and we have suggested that it may be overestimated.

The pathway just sketched seems to encapsulate the majority of the carbon flowing during dry–weather conditions, but one can see from Fig. 2 a secondary pathway involving microheterotrophs. DOC is generated from the phytoplankton and, as already discussed, as inflow from the seagrass beds. The DOC is converted into living POC by microheterotrophs but, interestingly, the living POC is not transferred up the food chain but cycled back to water column DIC (Fig. 2 shows a larger respiratory outflow from microheterotrophs than an inflow from DOC.) Hence the main functional significance of this flow path seems to be to inhibit a buildup of DOC in the water column. This conclusion is at odds with that of Wiebe & Smith (1977): they mistakenly multiplied the correct rate constant by PDOC mass, rather than POC mass, to calculate the respiration rate of the microheterotrophs – see Eq. 3 of Fig. 1.

7.3 Implications of the available information

Even though knowledge about carbon flow in SWA is incomplete, we believe it still provides a useful contribution to discussions of research priorities, both experimental and theoretical, and can serve as a useful input into management decisions regarding development impacts on SWA.

From hindsight, there exist a few glaring areas in which further experimental effort would bring substantial gains to our understanding. The extensive chlorophyll *a* data will only reach its full usefulness when an adequate C/Chl *a* ratio, ± variance, for SWA is obtained. Likewise the C/ATP ratio for SWA is needed to get more reliable mass estimates for water column detritus and microheterotrophs, and a C/DW ratio is needed to confirm our epifaunal estimate. Such studies would require innovative work for the development of a general procedure for estimating these ratios (Banse, 1977) before its application to SWA.

As is clear from Section 3, work remains to be done on the definition of sufficiently homogeneous compartments. Some steps in this direction have been made in this paper by the definition of subcompartments, following the aggregations actually used in experiments, as opposed to those agreed upon at the beginning of the study (Parker *et al.*, 1983). The intermediate sized zooplankton subcompartment remains a heterogeneous grouping of a variety of species, but Griffiths (1983) has made an interesting beginning to a useful further subdivision. Another problem concerns the benthos. The variety of species of epifauna and macrobenthos makes it difficult to visualize a useful aggregation of species. Hence, Colquhoun–Kerr (1977) opted to study only two prominent species of the epifauna in the intertidal–subtidal region.

A problem of more theoretical interest concerns the variance estimates calculated for mean compartment masses and flow rates. They were estimated to help in data interpretation in the face of much missing, or else preliminary, data. While they seem to be quite useful for this task, reliable estimates are difficult to obtain and the best form of variance estimate is not clear. We have aimed to present estimates equivalent to the Total MS of an AOV with data from various time and space co–ordinates. This idea was not always attained or, when attained, was at the expense of covering some trend in the data (*e.g.* phytoplankton primary production had temporal and spatial trends). Fortunately, temporal and spatial trends in compartment masses and flow rates were not very strong in SWA (see Table 4), but it is not obvious how one would calculate reasonable variance estimates where such trends occur. A further problem: how does one estimate variances of compound quantities when the components are not independent?

A surprising result that arose from the variance estimates was the dominance of the "local effects" variance component (Table 4). Within–cell replication was small, with some cells empty, but the consistency of the result suggests that this conclusion is not an artifact of inadequate data. And it is relevant to the degree to which understanding of carbon flow within SWA is possible. It suggests that no reasonably small spatial elements can be defined such that the majority of the variability is between, rather than within, the elements and thus that high C.V.'s cannot be avoided. (On the other hand, had stations been defined in terms of the structure (hydrodynamics) of the water column at the time of sampling, rather than fixed in space, the variation may have been less. It is an interesting research project, fundamental to the design of any sampling program.)

Provided we are aware of the degree to which variations from the expected response can occur within SWA, one should not be inhibited from using the information in this paper to make decisions regarding the consequences of developments affecting SWA. One of the advantages of a budget as a synthesis of data on carbon flow is that it is capable of being summarized in a readily understood, transparent form (Table 3 and Figs 2 – 5). This advantage is especially important in comparison with many dynamic ecosystem models, where the processes underlying a particular simulation may appear beyond comprehension.

It is possible to devise development–oriented questions regarding SWA and to use the (incomplete) budget presented here to provide insight, but referees of earlier drafts, where we did have such examples, have shown that at our current state of knowledge about SWA much discussion and argument will be involved in reaching any concensus on SWA response to some development. Suffice it to say that our experience has shown that an incomplete budget can make a useful contribution to the argument, but presents only one of a number of worthwhile viewpoints.

ACKNOWLEDGEMENTS

Several people have made useful suggestions to the work outlined in this paper and to this synthesis. We wish to formally acknowledge the contributions of P.R. Benyon, P.J. Sands, J.J. Wilkes, M. Tomczak, R.B. Humphries, and R. McMurtrie.

The credits and responsibilities for this synthesis are: W.R. Cuff initiated the paper, assembled the information, and retains overall responsibility for the conclusions. R.E. Sinclair took the main responsibility for obtaining information from the computer data base. The other authors took responsibility for the accuracy of information in their areas and lent freely of their extensive knowledge of SWA.

REFERENCES

Albani, A.D., Rickwood, P.C., Tayton, J.W., Johnson, B.D.: Geological aspects of the Port Hacking estuary. In: W.R. Cuff and M. Tomczak jr, eds *Synthesis and Modelling of Intermittent Estuaries*. Berlin, Heidelberg, New York: Springer (1983)

Allen, K.R.: Introduction to the Port Hacking Estuary Project. In: W.R. Cuff and M. Tomczak jr, eds *Synthesis and Modelling of Intermittent Estuaries* Berlin, Heidelberg, New York: Springer (1983)

Banse, K.: Determining the carbon–to–chlorophyll ratio of natural phytoplankton. *Marine Biology (Berlin)* **41**, 199–212 (1977)

Bell, J.D., Burchmore, J.J., Pollard, D.A.: Feeding ecology of three sympatric species of leatherjackets (Pisces: Monocanthidae) from a *Posidonia* seagrass habitat in New South Wales. *Australian Journal of Marine and Freshwater Research* **29**, 631–643 (1978a)

Bell, J.D., Burchmore, J.J., and Pollard, D.A.: Feeding ecology of a scorpaenid fish, the fortescue *Centropogon australis*, from a *Posidonia* seagrass habitat in New South Wales. *Australian Journal of Marine and Freshwater Research,* **29**, 175–185 (1978b)

Bulleid, N.C.: The nutrient cycle of an intermittently stratified estuary. In: W.R. Cuff and M. Tomczak jr, eds *Synthesis and Modelling of Intermittent Estuaries*. Berlin, Heidelberg, New York: Springer (1983)

Colquhoun–Kerr, J.S.: Carbon flux through the South West Arm populations of *Crassostrea commercialis* and *Trichomya hirsuta*. *CSIRO Division of Fisheries and Oceanography Report* **79** (1977)

CSIRO: *Estuarine Project Progress Report 1974–1976*. Sydney: CSIRO Division of Fisheries and Oceanography (1976)

Fenchel, T.M.: The ecology of micro– and meiobenthos. *Annual Review of Ecology and Systematics* **9**, 99–121 (1978)

Giles, M.S.: Primary production of benthic micro–organisms in South West Arm, Port Hacking, New South Wales. In: W.R. Cuff and M. Tomczak jr, eds *Synthesis and Modelling of Intermittent Estuaries*. Berlin, Heidelberg, New York: Springer (1983)

Godfrey, J.S.: Tidal flushing and vertical diffusion in South West Arm, Port Hacking. In: W.R. Cuff and M. Tomczak jr, eds *Synthesis and Modelling of Intermittent Estuaries*. Berlin, Heidelberg, New York: Springer (1983)

Godfrey, J.S., Parslow, J.: Description and preliminary theory of circulation in Port Hacking estuary. *CSIRO Division of Fisheries and Oceanography Report* **67** (1976)

Griffiths, F.B.: Zooplankton community structure and succession in South West Arm, Port Hacking. In: W.R. Cuff and M. Tomczak jr, eds *Synthesis and Modelling of Intermittent Estuaries*. Berlin, Heidelberg, New York: Springer (1983)

Griffiths, F.B.: Zooplankton community structure and succession in South West Arm, Port Hacking. In: W.R. Cuff and M. Tomczak jr, eds *Synthesis and Modelling of Intermittent Estuaries*. Berlin, Heidelberg, New York: Springer (1983)

Griffiths, F.B., Caperon, J.: Description and use of an improved method for determining estuarine zooplankton grazing rates on phytoplankton. *Marine Biology (Berlin)* **54**, 301–309 (1979)

Jitts, H.R.: A twin six litre water sampler. *Limnology and Oceanography* **9**, 452 (1964)

Kirkman, H., Reid, D.D.: A study of the role of a seagrass *Posidonia australis* in the carbon budget of an estuary. *Aquatic Botany* **7**, 173–183 (1979)

Kirkman, H., Griffiths, F.B., Parker, R.R.: The release of reactive phosphate by a *Posidonia australis* seagrass community. *Aquatic Botany* **6**, 329–337 (1979)

Parker, R.R.: Some ecological effects of rainfall on the protoplankton of South West Arm. In: W.R. Cuff and M. Tomczak jr, eds *Synthesis and Modelling of Intermittent Estuaries*. Berlin, Heidelberg, New York: Springer (1983)

Parker, R.R., Rochford, D.J., Tranter, D.J.: History and organization of the Port Hacking Estuary Project. In: W.R. Cuff and M. Tomczak jr, eds *Synthesis and Modelling of Intermittent Estuaries*. Berlin, Heidelberg, New York: Springer (1983)

Parsons, T.R., Takahashi, M.: *Biological Oceanographic Processes*. Oxford: Pergamon Press (1973)

Rainer, S.F.: The benthic biotopes of South West Arm, Port Hacking, N.S.W., 1975. *CSIRO Division of Fisheries and Oceanography Report* **109** (1980)

Rochford, D.J.: Sediment trapping of nutrients in Australian estuaries. *CSIRO Division of Fisheries and Oceanography Report* **61** (1974)

Scott, B.D.: Phytoplankton distribution and light attenuation in Port Hacking estuary. *Austalian Journal of Marine and Freshwater Research* **29**, 31–44 (1978a)

Scott, B.D.: Nutrient cycling and primary production in Port Hacking, New South Wales. *Austalian Journal of Marine and Freshwater Research* **29**, 803–815 (1978b)

Scott, B.D.: Seasonal variations of phytoplankton production in an estuary in relation to coastal water movements. *Austalian Journal of Marine and Freshwater Research* **30**, 449–461 (1979)

Scott, B.D.: Diurnal and other variations of photosynthesis versus irradiance curves for phytoplankton in Port Hacking. *CSIRO Division of Fisheries and Oceanography Report* **140** (1980)

Scott, B.D.: Phytoplankton distribution and production in Port Hacking estuary, and an empirical model for estimating daily primary production. In: W.R. Cuff and M. Tomczak jr, eds *Synthesis and Modelling of Intermittent Estuaries*. Berlin, Heidelberg, New York: Springer (1983)

Sinclair, R.E., Cuff, W.R., Parker, R.R.: Ecosystem modelling of South West Arm, Port Hacking. In: W.R. Cuff and M. Tomczak jr, eds *Synthesis and Modelling of Intermittent Estuaries*. Berlin, Heidelberg, New York:'Springer (1983)

Smith, D.F., Higgins, H.W.: An interspecies regulatory control of dissolved organic carbon production by phytoplankton and incorporation by microheterotrophs. In: M.W. Loutit and J.A.R. Miles eds. *Microbial Ecology*. Berlin: Springer (1978)

Smith, D.F., Bulleid, N.C., Campbell, R., Higgins, H.W., Rowe, F., Tranter, D.J., Tranter, H.: Marine food–web analysis: an experimental study of demersal zooplankton using isotopically labelled prey species. *Marine Biology (Berlin)* **54**, 49–59 (1979)

Snedecor, G.W., Cochran, W.G.: *Statistical Methods. Sixth Edition.* Ames: The Iowa State University Press (1967)

Sokal, R.R., Rohlf, F.J.: *Biometry.* San Francisco: W.H. Freeman and Company (1969)

Spencer, J.W.: Sydney solar tables. Tables for solar position and radiation for Sydney (Latitude 34°S) in SI units. *CSIRO Division of Building Research Technical Paper (Second Series)* **8** (1975)

Tafe, D.J., Griffiths, F.B.: Seasonal abundance, geographical distribution and feeding types of the copepod species dominant in Port Hacking, New South Wales. In: W.R. Cuff and M. Tomczak jr, eds *Synthesis and Modelling of Intermittent Estuaries* Berlin, Heidelberg, New York: Springer (1983)

Tranter, D.J., Kennedy, G.: Size–specific respiration rate of Port Hacking zooplankton. In: W.R. Cuff and M. Tomczak jr, eds *Synthesis and Modelling of Intermittent Estuaries*. Berlin, Heidelberg, New York: Springer (1983)

Vaudrey, D.J., Griffiths, F.B., Sinclair, R.E.: Data base for the Port Hacking Estuary Project: Parameters, monitoring procedure, and management system. In: W.R. Cuff and M. Tomczak jr, eds *Synthesis and Modelling of Intermittent Estuaries* Berlin, Heidelberg, New York: Springer (1983)

Wiebe, W.J., Smith, D.F.: Direct measurement of dissolved organic carbon release by phytoplankton and incorporation by microheterotrophs. *Marine Biology (Berlin)* **42**, 213–223 (1977)

Williams, W.T.: *Pattern Analysis in Agricultural Science* Melbourne: CSIRO and Elsevier (1976)

Wood, E.J.F.: Studies in microbial ecology of the Australasian region. V. Microbiology of some Australian estuaries. *Nova Hedwigia* **8**, 461–527 (1964)

Synthesis and Modelling of Intermittent Estuaries
(W.R. Cuff and M. Tomczak jr. eds) Berlin, Heidelberg,
New York: Springer (1983), pp. 233–258.

An Evaluation of the Dynamic Information
for South West Arm, Port Hacking

Wilfred R. Cuff

CSIRO Division of Computing Research

P.O. Box 1800, Canberra, A.C.T. 2601, Australia

Present address: Maritimes Forest Research Centre

Canadian Forestry Service

P.O. Box 4000

Fredericton, New Brunswick, E3B 5P7

Canada

Summary. A multidisciplinary study of the structure and dynamics of a small (~78 ha) Australian marine embayment (South West Arm of Port Hacking, New South Wales) was conducted during 1973–1978. Some chemical and biological studies were conducted specifically to obtain data for incorporation into a dynamic model of the flow' of carbon through South West Arm. This paper represents an attempt to synthesize that dynamic information: the functions and parameters were based on studies in South West Arm, rather than on the literature. This attempt to study an incomplete set of data of the sort required to make a dynamic model of the flow of carbon (or other chemical species) through the environment is novel, or at least such an attempt has not yet been reported in the literature. The study shows that synthesis of dynamic information can be usefully done, using a procedure analogous to that used in synthesizing an incomplete set of static information in a budget. The value of such a synthesis is that it organizes currently available information, and hence should make a useful contribution to further planning for research or ecosystem management.

Key words: carbon, dynamic modelling, synthesis, Port Hacking, South West Arm

1. INTRODUCTION

A multidisciplinary study of the structure and dynamics of a small marine embayment (South West Arm of Port Hacking, New South Wales, Australia) was conducted during 1973–1978. Compatible data were obtained by studying biological and chemical processes in terms of the flow of carbon. The initial aim was to construct a predictive, dynamic model of carbon flow through South West Arm (SWA) on the basis of data collected during the study (CSIRO, 1976).

This aim was not realized. Once it was accepted that the effort required to build such a model would be exorbitant, attention was turned to synthesizing the data that had been collected. A first attempt was reported in Cuff *et al.* (1983); a static model was constructed and shown to be capable of effectively using a wide variety of types of static and dynamic information available for SWA.

This paper is concerned only with that information from SWA that is suitable to the construction of a dynamic model of carbon flow. The experiments on which this information is based took cognizance of the literature, of course, but generally speaking this information is not supplemented by data from other ecosystems (as reported in the literature). While such extensions represent common procedure in most ecosystem–modelling efforts (see, for example, Kremer & Nixon, 1978), I believe that our goal is not best served by mixing local data with data of unknown applicability to SWA from other ecosystems. Such a mixture would denigrate our relatively reliable data which, incidentally, were obtained at significant cost. (The bias of this paper is, perhaps, an unwarranted rejection of current ecological theory and in following it too rigidly one risks discovering what is already well established – and at a large cost. A balance is needed but in this paper I try to redress the balance from unwarranted belief in inadequate ecological theory.)

The greatest value of dynamic models lies in their ability to clarify the logical consequences of using particular component functions and parameters. But reliable predictions require a complete specification of all components: feedback loops can enable a large variety of types of model behaviour dependent on the structure of the missing functions. A completely developed model remains a long–term goal but perhaps for now it may be useful simply to study in detail the available dynamic information for SWA. I assume that such a synthesis of incomplete dynamic information can, by organizing what is already known about SWA, make a useful contribution to future planning both for research and ecosystem management. Wisdom must be exercised in the use of incomplete information, of course, but one should not ignore available information just because it presents a limited viewpoint.

The form taken by this study is a novel one since, to my knowledge, no attempt has previously been made to evaluate systematically an incomplete data set, not supplemented from the literature, to make a predictive, dynamic ecosystem model. I begin with a summary of the available information.

2. THE INFORMATION AVAILABLE FOR DYNAMIC MODELLING

2.1 Experiments conducted for functional identification

It is always desirable in the construction of dynamic models that experiments be designed specifically to determine the functional relationships between the rate of flow of a material between compartments and relevant state/forcing variables. The final product is, of course, the estimated set of parameter values for the appropriate function. As part of the study of SWA, five experimental programs were set up specifically for functional identification and parameter estimation: Griffiths & Caperon (1979) studied the state dependency of the rate of carbon flow from phytoplankton to intermediate sized zooplankton ($> 124 \, \mu m$ but not ctenophores and large medusae); Wiebe & Smith (1977) studied the state dependency of the rate of carbon flow from photosynthetically derived dissolved organic carbon (PDOC) to "microheterotrophs" to dissolved inorganic carbon (DIC) (microheterotrophs being bacteria attached to detritus or clumped to themselves to form particles of $106 - 124 \, \mu m$); Caperon & Smith (1978) studied the effect of DIC limitation on the rate of carbon flow from DIC to phytoplankton; Tranter & Kennedy (1983) studied the effect of organism size on the flow rate from heterotrophs ($25 - 800 \, \mu m$) to DIC; and Giles (1983) studied the effect of irradiance on the photosynthetic rate of benthic micro–organisms. In each study the authors assumed a particular model on theoretical grounds and fitted the model to their data to estimate the parameters and, thereby, to check on the suitability of the model for describing the data.

2.1.1 Grazing of intermediate sized zooplankton on phytoplankton. Griffiths & Caperon (1979) followed Holling (1959) and Rashevsky (1959) in writing the grazing rate ($-dD/dt$) of zooplankton biomass (z) feeding on phytoplankton (D) as follows

$$dD/dt = - v \, D \, z \, / \, (b+D) \tag{1}$$

where v and b are constants and time, t, has units of one half–hour. The variable z was defined by Griffiths & Caperon (1979) as the ratio of the zooplankton density in the experiment and the zooplankton density in SWA at the time and position of sample collection.

Griffiths & Caperon (1979) narrowed their investigation to those $D << b$ to simplify the governing equation to

$$dD/dt = - K D \qquad (2)$$

where

$$K = v z / b = kz \qquad (3)$$

Only relationships between dD/dt and D were dealt with experimentally, the assumed partition of K being used to obtain estimates of $v/b = k$.

They conducted three experiments. In the first they obtained a regression–based estimate of $k = 0.14$ day^{-1}, assuming zooplankton to feed for 24 h each day, for a relative zooplankton density of 46 times ambient. In the second they obtained a regression–based estimate of 0.12 day^{-1}, and further showed the linear model to be an adequate approximation within a phytoplankton density range of at least 80 – 320 mg carbon (C) m^{-3}, for a relative zooplankton density of 28 times ambient. In the third experiment they did not estimate k from a regression of dD/dt on D but estimated dD/dt separately for each of the component zooplankton species, as the increase in radioactivity. They summed these estimates and divided by D, in units of radioactivity, to estimate k. (In this experiment the zooplankton were kept at ambient densities: $z = 1$.) Two subsamples gave estimates of k of 0.28 and 0.42 respectively. These estimates of k are about 2 – 3 times higher than those estimated from the first two experiments. The authors thought that the higher zooplankton densities in the first two experiments may have inhibited zooplankton feeding.

2.1.2 Carbon flow from DOC to microheterotrophs to DIC. Wiebe & Smith (1977) followed classical procedures in compartmental modelling (Berman & Schoenfeld, 1956; Jacquez, 1972; Smith, 1974) in assuming the flows to be donor dependent. Experiments were devised such that carbon flowed from the PDOC compartment, Q_1, to the microheterotroph compartment, Q_2, to the DIC compartment, Q_3, and nowhere else. The descriptive equations were written as

$$dQ_1/dt = - \lambda_{21} Q_1 \qquad (4)$$
$$dQ_2/dt = \lambda_{21} Q_1 - \lambda_{32} Q_2 \qquad (5)$$
$$dQ_3/dt = \lambda_{32} Q_2 \qquad (6)$$

where λ_{ij} represents the fractional turnover rate from the jth compartment to the ith compartment and t has units of ks.

From four experiments they obtained a mean value of λ_{21} of $0.0278 \pm 4.8\%$ ks^{-1}, equivalent to 2.40 day^{-1}; and of λ_{32} of $0.00091 \pm 17\%$ ks^{-1}, or 0.08 day^{-1}. Separate estimates are given for light and dark conditions, for each of experiments conducted on 18 July 1975 and 8 August 1975. For the former date, λ_{21} (λ_{32}) equalled 0.0243 ± 6.5% (0.00097 ± 19%) for light conditions and 0.0285 ± 6.2% (0.00124 ± 16.1%) for dark conditions. For the latter date, λ_{21} (λ_{32}) equalled 0.0276 ± 2.7% (0.00063 ± 16.5%) for light conditions and 0.0308 ± 3.9% (0.00078 ± 15.8%) for dark conditions. The estimates of percent variation about the mean parameter estimates are taken from Table 2 of Wiebe & Smith (1977). The authors do not explain what these estimates represent but they probably represent standard errors of estimate generated by the fitting routine (SAAM 25, Mathematical Research Branch, National Institute of Arthritis and Metabolic Diseases, National Institutes of Health, Bethesda, Md. 20013, USA).

2.1.3 Carbon flow from DIC to phytoplankton, as a function of DIC concentration. Caperon & Smith (1978) fitted the Michaelis–Menten equation to experimental data relating the carbon fixation rate of algae to the total inorganic carbon concentration. The equation fits the data well and parameter estimates (± std. error) are given (their Table 1) for 3 groups of species and 3 individual species. DIC concentrations resulting in half–maximal fixation rates in the range 4.3 – 5.3 g C m^{-3} for mixed phytoplankton species and 0.8 – 2.8 g C m^{-3} for the axenic algal cultures. DIC concentration in SWA was estimated to be 23.9 ± 1.2 (std. dev.) g C m^{-3} (see Cuff et al., 1983).

2.1.4 Respiration rate of heterotrophs, as a function of the weight of the heterotroph organism. Tranter & Kennedy (1983) followed established practice in representing the specific respiration rate (R) as a power function of the weight (W) of the individual heterotroph, to give

$$R = a \, W^{-b} \tag{7}$$

where a (= 0.857) and b (= −0.306) are constants (r^2 = 0.62) and W has units of µg, R of µg–atom O$_2$/mg dry weight (DW)/h.

The coefficient of proportionality, a, may be compared to that found by Ikeda (1970, 1974): 0.224 (r^2 = 0.86). Ikeda (1970, 1974) allowed the field–collected organisms time to adapt to experimental conditions before measuring what was probably the basal metabolic rate. Tranter & Kennedy (1983) aimed to estimate the rate occurring in the field (the "active" rate) and so conducted experiments as soon as possible (4 – 5 h) after the samples were collected.

2.1.5 Primary production rate of benthic micro–organisms, as a function of light. Giles (1983) fitted three equations, found useful for describing phytoplankton production by Jassby & Platt (1976), to experimental data relating the primary production rate of benthic micro–organisms (per m^2) to illuminance. He tested the models and chose the following one for obtaining monthly and annual integrated productivities:

$$P = m\alpha I \, / \, (\alpha^2 + m^2 I^2)^{\frac{1}{2}} \qquad\qquad\qquad (8)$$

where P = productivity, I = illuminance, m and α are constants.

For the sand banks at the mouth of Flat Rock Creek – see Fig. 1 of Vaudrey *et al.* (1983) – Giles (1983) obtained an estimate (± std. error) of α of 3.1 ± 0.19 and of m of 0.011 ± 0.0015; for the seagrass beds outside the mouth of SWA, he obtained an α estimate of 1.7 ± 0.14 and an estimate of m of 0.01 ± 0.0017. Thus α, the saturating value of P with increasing I, varies within regions of (or near to) SWA.

Productivity (P) measurements suggested by Eq. 8 were normalized for chlorophyll *a* content before being used to estimate the α and m parameters. (The significance of this step, suggested by Platt *et al.* (1977), is clarified in Section 2.2.1.) This allowed Giles (1983) to incorporate time series of chlorophyll *a* concentrations and of light measurements at the sediment surface into Eq. 8 to obtain a time series of P (see Table 4 of Giles, 1983).

2.2 Other data from South West Arm amenable to functional identification

2.2.1 Primary production rate of phytoplankton, as a function of light and chlorophyll *a* concentration. Following studies by Scott (1978a, 1978b, 1979a, 1979b, 1983), there exists a comprehensive set of data relating to the distribution, abundance, and production of SWA phytoplankton.

Scott (1979b, 1983) considered the usefulness of light–curve based modelling to the prediction of the rate of primary production. Emphasis was placed on the daily, depth–integrated rate and the initial plan was to adopt the procedures developed by Fee (1973) and Jitts *et al.* (1976). These authors estimated the attenuation of light with depth, which they assumed to be constant between days; then estimated daily the parameters of the light curve; recorded daily the pattern of surface irradiance; and used this information to calculate the instantaneous photosynthetic rate at each depth throughout each day, and thence the integral rate. Jitts *et al.* (1976) point out that where characteristic light curves (only minor changes in parameters over time) occur, this method can be used with a time series of surface irradiance data to estimate a time series of depth–integrated rates.

Large variation in the maximal photosynthetic rate (p_{max}) from various light curves from SWA made the adoption of this modelling technique inappropriate. So Scott (1979b, 1983) devised an "empirical model", in the sense of Platt *et al.* (1977): "to describe, with more or less precision, the essential characteristics of the available data relating to two or more variables ... with little regard to the internal mechanism of the system". Scott's model represents a convenient summary of the relationship between the *in situ* photosynthetic rate (at 1 and 2 m depths) as a function of a measure of phytoplankton mass and total daily solar irradiance at the appropriate depth. In a plot of modelled against observed photosynthetic rates, Scott (1983) found an r^2 of 80% (n=155). This model form represents one way of dealing with variation in p_{max} but, conceptually at least, there are other options; one alternative, *viz.* an extension of the mechanistic, "rational model" (Platt *et al.*, 1977) approach, has obvious advantages and so is developed here.

Most of the variation in p_{max} is thought to be due to changes in the phytoplankton mass (Scott, pers. comm.); and Bannister (1974a) showed how p_{max} (as determined in a nutrient–saturated environment) can be written as an explicit function of chlorophyll *a* (c) and a new parameter which is independent of temperature. For Steele's (1962) light curve it has been shown by Bannister (1974a, Eq. 11) that

$$p_{max} = 12 \; \phi_{max} \; I_{max} \; k_c \; c \; / \; e \tag{9}$$

where ϕ_{max} is the maximum quantum yield, being the maximum value of the ratio of the photosynthetic rate (in moles carbon incorporated per unit time) and the rate (in einsteins absorbed per unit time) at which light is absorbed by phytoplankton. I_{max} is the irradiance at which p_{max} occurs, and k_c is the extinction coefficient of light with depth per unit of chlorophyll *a* .

The new parameter is, of course, ϕ_{max}. Bannister (1974a) put forward three arguments to the effect that the value of ϕ_{max} observed in nutrient–saturated laboratory conditions (0.10 moles oxygen evolved per einstein absorbed) implies a value of 0.06 moles carbon incorporated per einstein absorbed for most lakes, accurate within a factor smaller than 1.5. Smith (1980) notes that the three arguments used by Bannister (1974a) to reduce the laboratory value of ϕ_{max} may apply generally in nature. But Platt *et al.* (1977) caution that ϕ_{max} may be more variable than Bannister (1974a) suggests; they suggest that c may be separated from p_{max} by setting out the light curves as the photosynthetic rate per unit of phytoplankton mass. (The good theoretical sense of this suggestion is apparent from Eq. 9 above; such a plot would, in fact, provide an alternative way of estimating ϕ_{max} empirically.) But Scott's light curves were not set out with a normalized ordinate, and since Platt *et al.* (1977) do not give convincing evidence for the variability of ϕ_{max}, Bannister's (1974a) method for writing p_{max} as an explicit function of c is used here.

Development of the "integral model" begins with choosing an appropriate equation for describing the light curve and I chose Steele's (1962) function because inhibition of the instantaneous photosynthetic rate occcured at high light intensities (see Scott, 1979b). Thus the photosynthetic rate may be written as

$$p = p_{max} \; (I/I_{max}) \; \exp \; (1 - I/I_{max}) \tag{10}$$

where p_{max} and I_{max} are as defined above.

Using this light curve, the depth (from D_1 to D_2) and time (day length) integrated production, Π, may be derived, following the method used by Bannister (1974a), as

$$\Pi = (p_{max} \, e \, \lambda \, / \, \epsilon)$$
$$\int_{-\frac{1}{2}}^{+\frac{1}{2}} [\exp (- T_2 \, I_0 \, / \, I_{max}) - \exp (- T_1 \, I_0 \, / \, I_{max})] \, dt' \tag{11}$$

where p_{max} and I_{max} are as defined above; λ is time in days between sunrise and sunset; ϵ is the extinction coefficient of light with depth; I_0 is the incident subsurface illumination; T_1 and T_2 are the fraction of the incident illumination I_0 transmitted to depths D_1 and D_2 respectively; and t' is a dimensionless time of day such that $t' = -1/2$ at sunrise, zero at noon, and $+1/2$ at sunset (Vollenweider, 1965).

This equation may be made an explicit function of chlorophyll *a* concentration by substituting for p_{max} as given in Eq. 9 and by expanding the extinction coefficient. Scott (1978a) showed that in SWA the extinction coefficient ϵ can be written as a linear function of chlorophyll *a* concentration (c), as

$$\epsilon = k_c \, c + k_w \tag{12}$$

where k_c is as defined above and k_w is the non–phytoplankton component of light absorption.

Substituting Eq. 9 & 12 into Eq. 11 yields the fully expanded model of daily production

$$\Pi = (12 \, \phi_{max} \, I_{max} \, k_c \, c \, \lambda \, / \, (k_c \, c + k_w))$$
$$\int_{-\frac{1}{2}}^{+\frac{1}{2}} [\exp (- T_2 \, I_0 \, / \, I_{max}) - \exp (- T_1 \, I_0 \, / \, I_{max})] \, dt' \tag{13}$$

All parameters needed to evaluate the equation for Π are estimable for SWA. From light curves given in Appendix B (2 – 7) of Scott (1979b) and from chlorophyll *a* data obtained from Scott (unpublished), I used Eq. 9 to estimate a mean ($\pm$ std. dev.) $\phi_{max} = 0.01 \pm 0.004$ moles C einst^{-1} absorbed. This value is about one sixth that suggested by Bannister (1974a) for nutrient–saturated conditions, probably because primary production in SWA is nutrient–limited. I suspect that it was not ϕ_{max} that was estimated, but a nutrient–limited version of ϕ_{max} , say Φ_{max}; the p_{max} of Eq. 9 was not estimated from Scott's (1979b) light curve data but probably P_{max}, defined as

$$P_{max} = p_{max} \, (N \, / \, (K + N)) \tag{14}$$

where N is the concentration of the (relevant but unspecified) nutrient and K is the half–saturation constant for N in the Michaelis–Menten expression. Bannister (1974a) suggested that his predicted value of Φ_{max} would be correct within a factor of 1.5; the SWA data support this contention, with the estimate of P_{max} correct to a factor of about 2.

An estimated value of I_{max} = 318 ± 79 exaquanta m^{-2} s^{-1} was also calculated from Scott's (1979b) data; this value may be compared to that of 300 reported by Scott (1979a). (In Eq. 13, I_{max} must have units of einst m^{-2} day^{-1} outside the integral and J m^{-2} s^{-1} inside.) The conversion from quanta to joules has been considered by Morel & Smith (1974); the ratio of quanta to joules, or watt–seconds, (Q : W) is a well behaved function of chlorophyll a concentration, with a recommended value of 2.65 ± 0.13 in waters with a chlorophyll a concentration of > 1 mg m^{-3}. To convert from joules to einsteins, I first converted from joules to calories, and then calculated visible solar energy to be 48% of total (direct plus diffuse) radiation. (Monteith (1973) suggests a value of 50% but Scott (1979b) measured (in air) a visible exaquanta : total joules ratio of 1.33 which could be divided by the above–surface, total exaquanta : total joules value given by Morel & Smith (1974) to get a better estimate.) Finally, I used the conversion to einst m^{-2} day^{-1} visible from cal (total) cm^{-2} min^{-1} as given in Bannister (1974a), although I modified it slightly since his calculation assumed visible energy is only 43% of the total. These conversions yielded I_{max} estimates of 47.8 einst m^{-2} day^{-1} and 120 J m^{-2} s^{-1}.

The parameters of the extinction coefficient are given in Scott (1978a) as

$$\epsilon = k_c c + k_w = 0.048 c + 0.220 \tag{15}$$

with c units of mg Chl a m^{-3}, k_c of m^2 mg^{-1} Chl a, and k_w of m^{-1}. Chlorophyll a concentrations (c) over time were measured by Scott at various positions throughout SWA and these data were supplied by him. T_1 and T_2 are calculated from ϵ, assuming Beer's Law to apply, as

$$T_i = \exp(-\epsilon D_i) \tag{16}$$

The remaining parameters, λ and I_0, were calculated from tables covering the Sydney region (Spencer, 1975). A cosine function adequately described the seasonal variation in daylength, to give

$$\lambda = 0.507 + 0.093 \cos(2 \pi (T + 9) / 365) \tag{17}$$

(r^2=1.00) where T is day number from 1 January.

The incident subsurface illumination was modelled as a function of the time of day by Vollenweider (1965) as

$$I_0 = 0.5 I_{0 \, max} (1 + \cos 2\pi t') \tag{18}$$

where $I_{0 \, max}$ is the maximum illumination at noon. Total incident "above ground" illumination values (at noon) for sunny days (Spencer, 1975) were modelled as a cosine function, to give (r^2=0.99)

$$855 + 283 \cos (2 \pi (T + 12) / 365)$$

I assumed a 10% reflective loss at the water surface (see p. 17 of Bannister, 1974b) and, as above, took 48% of total solar radiation to be within 400 – 700 nm. This illumination was then reduced by a factor of 0.7 to account for average cloud cover. This value (std. dev. $= 0.27$; n $= 331$) was estimated by comparing the surface irradiance values calculated as described herein with values of total irradiance given in Appendix G of Scott (1979b). Then $I_{0\ max}$ may be written as

$$I_{0\ max} = (0.9)(0.48)(0.7)(855 + 283 \cos (2 \pi (T + 12) / 365)) \tag{19}$$

The results of a simulation of the 0 – 1 m photosynthetic rates for 1975 (g C m^{-2} day^{-1}, converted to mg C m^{-3} day^{-1}) are compared to the *in situ* rates at 1 m depth (Scott, unpublished) in Fig. 1 (day 1 $=$ 1 January). The model does quite well at matching the observed data (r^2 $= 0.68$; n $= 76$); this correspondence may be compared to the r^2 $= 0.8$ obtained by Scott (1983) for his empirical model (n $= 155$). The linear regression highlighted the inability of the model to match the very low observed values (y–intercept was $+28$ mg C m^{-3} day^{-1}).

The fact that the model mimics the annual trend best is in agreement with Scott's (1979a) view that the long–term variation in primary production is mainly related to variation in light. The residuals (model minus data) are shown in Fig. 2 and, except for the very large data values (below –50 in Fig. 2), no temporal trend is apparent. The main weakness of this model is its inability to deal with short–term variability, and this weakness is considered in the next section.

3. AN EVALUATION OF THE AVAILABLE INFORMATION

3.1 Specific functions

In this section an attempt will be made to identify strengths and weaknesses of each of the available functions. Where important deficiencies are uncovered, further attempts will be made to try to find ways of bridging them.

3.1.1 Grazing of intermediate sized zooplankton on phytoplankton. The work of Griffiths & Caperon (1979) was split between developing a methodology for directly estimating the amount of natural, mixed phytoplankton population mass consumed by the zooplankton community and applying this methodology to SWA. The latter aspect is hence not as complete as one may like.

The experimentally determined model of Eq. 2 is a function of relative zooplankton density. Such a dimensionless zooplankton mass was useful experimentally because it circumvented the need to estimate absolute zooplankton mass but it implies that zooplankton are not limiting: that the grazing rate per unit of phytoplankton mass, determined at or adjusted to ambient zooplankton mass, is independent of the absolute zooplankton density. This assumption might be appropriate if mean zooplankton mass in SWA had a low coefficient of variation (C.V.), but Cuff *et al.* (1983) found a C.V. of 83%, too high to support this suggestion. It is obviously desirable to have dD/dt as a function of absolute zooplankton density as well as absolute phytoplankton density.

I assume the following model to be an adequate representation of the data for phytoplankton densities between 80 – 320 mg C m^{-3} and for zooplankton densities (absolute, Z) of an unknown range:

$$dD/dt = -\kappa \, Z \, D \qquad\qquad\qquad (20)$$

The term κZ of this model can be estimated from the experimental results of Griffiths & Caperon (1979); in their second experiment κZ can be shown to be equal to 3.37 by using the value of k given above and writing D (and Z by implication) in units of g C m^{-3} and t in units of day (24 h). Experiment 2 was conducted on 6 November 1975. Data base records of zooplankton mass are available for 3 and 10 November 1975 (Vaudrey *et al.*, 1983), and each record suggests a DW mass of 75 mg m^{-3}. Using a C/DW ratio of 0.27 (see Tranter & Kennedy, 1983), a C mass estimate of 20.3 mg C m^{-3} is obtained. Since the experimental zooplankton density was 28 times ambient, this gives a $\kappa = 5.9$ (g C m^{-3} day)$^{-1}$.

An estimate of κ is also possible for Experiment 3 of Griffiths & Caperon (1979). The experiment was conducted on 25 November 1975; data base (Vaudrey *et al.*, 1983) records suggest a zooplankton mass of 30 mg DW m^{-3}, or 8.1 mg C m^{-3}, on 24 November 1975. Ambient zooplankton densities were used in this experiment (z = 1). Hence k = 0.28 (see discussion relating to Eq. 2 & 3) implies $\kappa = 34.6$; k = 0.42 implies $\kappa = 51.9$.

If Eq. 20 is taken to be an adequate model of zooplankton grazing in SWA, then average phytoplankton and zooplankton masses (Cuff *et al.*, 1983) can be used to calculate expected bay–wide grazing rates on phytoplankton. A value of $\kappa = 5.9$ yields a flow rate of 24 mg C m^{-3} day^{-1}, $\kappa = 35$ yields a rate of 143, and $\kappa = 52$ yields a rate of 212. The static flow rates calculated in Cuff *et al.* (1983) using Eq. 2 and parameter values for Griffiths & Caperon's (1979) third experiment range from 34 – 50 mg C m^{-3} day^{-1}. The data base records show that zooplankton mass was variable around 25 November 1975, suggesting that the value of zooplankton mass on 24 November 1975 as used in the calculations was an inadequate estimate of the mass

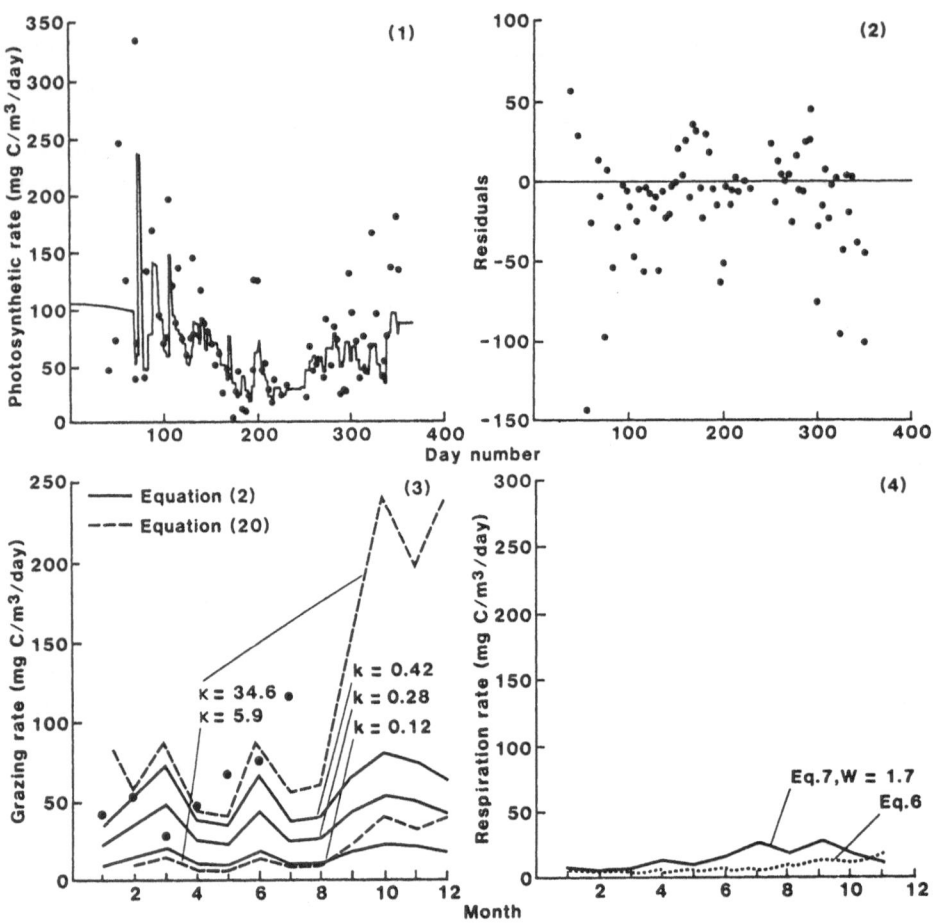

Fig. 1: A simulation of the *in situ* photosynthetic rate of South West Arm phytoplankton in the 0 – 1 m
depth, assuming average cloud cover and average nutrient limitation effects. The *in situ* rates are
also plotted.

Fig. 2: The residuals (modelled minus data) for the simulation illustrated in Fig. 1.

Fig. 3: Predicted grazing rates of intermediate sized zooplankton on phytoplankton, according to two models
described as Eq. 2 (k= various parameter values; solid lines) and as Eq. 20 (κ = various parameter
values; dotted lines) in the text. Month 1 represents June 1975. Average monthly *in situ* primary
production rates are also plotted.

Fig. 4: Predicted respiration rates for intermediate sized zooplankton (Eq. 7; W=1.7) and microheterotrophs
(Eq. 6). Month 1 represents July 1975.

on the next day. The best estimate of κ is thus 5.9, although this value underestimates the static flow rate estimates somewhat.

An additional ambiguity regarding the appropriate model for calculating the grazing rate of zooplankton arises from results presented in Fig. 1 of Griffiths & Caperon (1979). The model assumed by them (Eq. 2 & 3) suggests that if the relative zooplankton density (z) is doubled, then the grazing rate (dD/dt) will be doubled. Their Fig. 1 illustrates the grazing rates from two experiments with z = 23 and z = 46; the relationship between dD/dt and D was linear and independent of z. Since experimental zooplankton masses were much larger than the ambient densities, this result is surprising and may be an artifact: Griffiths (pers. comm.) views his first experiment primarily in terms of illustrating aspects of their experimental technique.

The potential significance of each hypothesis (grazing rate as a function of D alone; of D and Z) can be evaluated by using a time series of D and Z data from the data base to calculate the annual time series of grazing rates. Z and D were not always collected at the same days and, furthermore, there is a lot of noise in the data. Hence average, monthly values were used, with data available for Stn C (see Vaudrey et al., 1983) for the period June 1975 (Month 1 in Fig. 3) to May 1976 (Month 12 in Fig. 3). (This procedure can in theory produce a biassed estimate of average grazing rate.) Chlorophyll a data (C/Chl a = 60) from the 0 – 2 m column were combined with the average–column values available for Z. The two models with parameters derived from Griffiths & Caperon's (1979) second experiment (k = 0.12; κ = 5.9) show similar grazing rates. The grazing rates from the model of Eq. 20 with the parameter value from the first subsample of Experiment 3 (κ = 34.6) give much larger rates than the other models. It is interesting to compare these rates to the net rates of average, *in situ* primary production (also plotted) since it is generally impossible for zooplankton to graze on phytoplankton at a higher rate than the net primary production rate of phytoplankton. Except for Month 3 all models are feasible, although, since the C.V.'s of the primary production rates average 50%, some of the topmost models seem unlikely. But the Month 3 datum eliminates all models except Eq. 2 with k = 0.12 and Eq. 20 with κ = 5.9 from further consideration.

3.1.2 Carbon flow from DOC to microheterotrophs to DIC. In the model of Wiebe & Smith (1977) the flow rate from DOC to microheterotrophs applies only to a flow from photosynthetically derived DOC (PDOC). The flow rate from total DOC to microheterotrophs is unknown and as the concentration of PDOC in SWA is only about 0.1% that of total DOC (see Cuff *et al.*, 1983), use of total DOC instead of PDOC in the model would give unreasonable results.

A second, minor limitation of their study concerns the variability of the parameters, spatially and temporally, in SWA. Wiebe & Smith (1977) conducted experiments only on 18 July 1975 and 8 August 1975, with samples apparently collected from the same station, for both light and dark conditions. The two parameters do not vary substantially due to either factor, but the data are too limited to draw firm conclusions about the natural variability of the parameters.

It is not possible to calculate a time series of the PDOC utilization rate since the time variation in PDOC is unknown. But, since time series data are available for microheterotrophs from July 1975 (Month 1 in Fig. 4) to May 1976 (Month 11 in Fig. 4) for Stn C, it is possible to estimate the time variation of the respiration rate of microheterotrophs (see Fig. 4 – Eq. 6). No strong temporal trend is apparent.

3.1.3 Carbon flow from DIC to phytoplankton, as a function of DIC concentration.
This experiment of Caperon & Smith (1978) is of limited significance to the model developed here because the estimated DIC concentrations at half–maximal fixation rates are far below the lowest DIC value recorded in SWA.

3.1.4 Respiration rate of heterotrophs, as a function of the weight of the heterotroph organism.
Since a time series of intermediate sized zooplankton data is available on the data base (Vaudrey et al., 1983), Tranter & Kennedy's (1983) model was used to calculate a time series of active respiration rates from July 1975 through May 1976 (see Fig. 4 – Eq. 7). The active respiration rate of intermediate sized zooplankton seems about the same order as the (probably basal) rate of microheterotrophs (see Fig. 4 – Eq. 6). I assume a basal rate because, although Wiebe & Smith (1977) do not mention when their experiments were conducted relative to the collection times, their illustrations show that experiments, once begun, were terminated only after ~9 h (their Fig. 1) or ~6 h (their Fig. 2). The most important data in estimating the parameter value in the increasing exponential function (Eq. 6) for the respiration rate of microheterotrophs occur at the later times.

3.1.5 Primary production rate of benthic micro–organisms, as a function of light.
The estimation of a time series of primary production rates of benthic micro–organisms, and its comparison to *in situ* estimates was detailed in Giles (1983), and summarized in Section 2.1.5 above. His approach is sound, but from a "rational" modelling viewpoint (Platt *et al.*, 1977), it would have been desirable had Giles (1983) used modelled, rather than observed, values of illuminance at the sediment surface. To this end Eqs 16, 17, 18, and 19 would have been useful – assuming that exposed intertidal regions would be exposed to incident "above ground" illumination. But a time series of tide heights over the stations sampled by Giles (1983) would also be required. This information is available but numerical estimates of productivity should change only slightly and so I choose simply to mention this option.

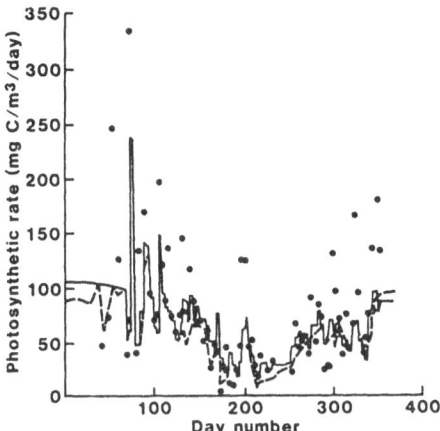

Fig. 5: A simulation of the *in situ* photosynthetic rate of South West Arm phytoplankton in the 0 – 1 m
depth, with the inclusion of average (solid line) and actual daily (broken line) irradiances and average
nutrient values. The *in situ* rates are also plotted.

3.1.6 Primary production rate of phytoplankton, as a function of light and chlorophyll *a* concentration.

The model developed above fails to mimic the exceptionally high photosynthetic rates which occur occasionally throughout the year, but especially before about day 150 and after about day 250.

The available data allow the assumption relating to the average effect of cloud cover to be relaxed. Irradiance data for most of 1975 are available in Appendix G of Scott (1979b) and in Fig. 5 a simulation of the model described above with light reduced to 70% (solid line) is compared to the same model with observed light values per day (broken line); variations in light from the average cloud pattern do not explain the unexpectedly high *in situ* rates, but do occasionally account for unexpectedly low values (to about day 200). Scott (1983) showed (his Fig. 5) that the irradiance observed at the Cronulla laboratory can vary from the irradiance at a place the same distance away as SWA, but this variation is unlikely to change the conclusion about the relative significance of the actual light data.

Scott (1979a) suggested that the short–term variations in primary production are mainly related to nutrient concentration; the exceptionally high values which the models described above failed to reproduce, are in fact those producing the short–term variability in primary production. It thus seems potentially valuable to try to include a nutrient term in the model. Nutrient data are available on the data base for a set of depths but the response of the photosynthetic rate of SWA phytoplankton to varying nutrient concentrations is unknown. To investigate the possible role of nitrates and phosphates, I assumed the Michaelis–Menten equation and took half–saturation constants from di Toro *et al.* (1971): 10 μg P l^{-1} and 50 μg N l^{-1}. Use of mean nutrient concentrations in SWA allowed calculation of the average percentage reduction attributable to each of nitrates and phosphates. Phosphates, on average, reduce primary production by about 45% and nitrates by about 75% of the nutrient–saturated environment. Thus, on average, nitrates appear to be the main limiting nutrient in SWA; nitrates are commonly assumed to be the limiting nutrient in marine waters (Scott, 1979a).

To obtain an estimate of the degree to which nitrates do account for the unexpectedly high *in situ* photosynthetic rates (see Fig. 1), I used a $\phi_{max} = 0.06$ (as suggested by Bannister, 1974a) and the nitrate values on the data base with the Michaelis–Menten expression to simulate the photosynthetic rate. Surface nitrate values are available for 1975 only about every fortnight. These values were used to set up a nitrate time series, where nitrate concentrations between measured values were assumed equal to the last record; the variation was too great (see Section 2.2 of Cuff *et al.*, 1983) to make interpolation attractive. Half–saturation coefficients vary widely in the literature, from 1.0 to 100μg N l^{-1} according to di Toro *et al.* (1971), although Kremer & Nixon (1978) suggest a typical range of 14 – 42 μg N l^{-1} (1 – 3 μg-atom N l^{-1}). Fig. 6 shows the results of a simulation with a half–saturation constant of 35 μg N l^{-1}. A plot of the residuals (modelled minus observed) is given in Fig. 7. It is clear that the addition of even relatively inadequate information about nitrates adds considerably to the goodness of the model, at least to day 230. A correlation of modelled and observed does not, however, reflect the improvement ($r^2 = 0.53$ for data to day 250 compared to 0.68 for average nutrient limitation). A linear regression suggests results contrary to those produced by the model reflecting average nutrient limitation; the current model underestimates the low primary production rate (y–intercept was –21 mg C m^{-3} day^{-1}).

The large predicted values on day 76 and 91, and to a lesser degree on days 198 and 202 (see Fig. 7), are above the observed values. An attempt to match the modelled and observed points on these days by varying the half–saturation constant (from 35 to 100) resulted in a perfect match at days 198 and 202, although the model result was still too high at day 76. Apart from the exceptionally high *in situ* values, however, the trend of the model was considerably below that of the data; a half–saturation constant for N of 100 is too large for SWA.

Fig. 7 shows that the model lies within the central region of the data only to day 250. After that the model lies consistently below the data. The reason for this is unclear but it is instructive to compare Fig. 1 which shows average nutrient conditions with Fig. 6 which shows actual nutrient data; it is relevant that during this period there were especially frequent nitrate observations and they all indicated low concentrations. The average–nutrient simulation corresponds well to the majority of the data. This suggests a shift in community structure with the new community having a greater tolerance to low–nutrient conditions (low half–saturation constant). It is

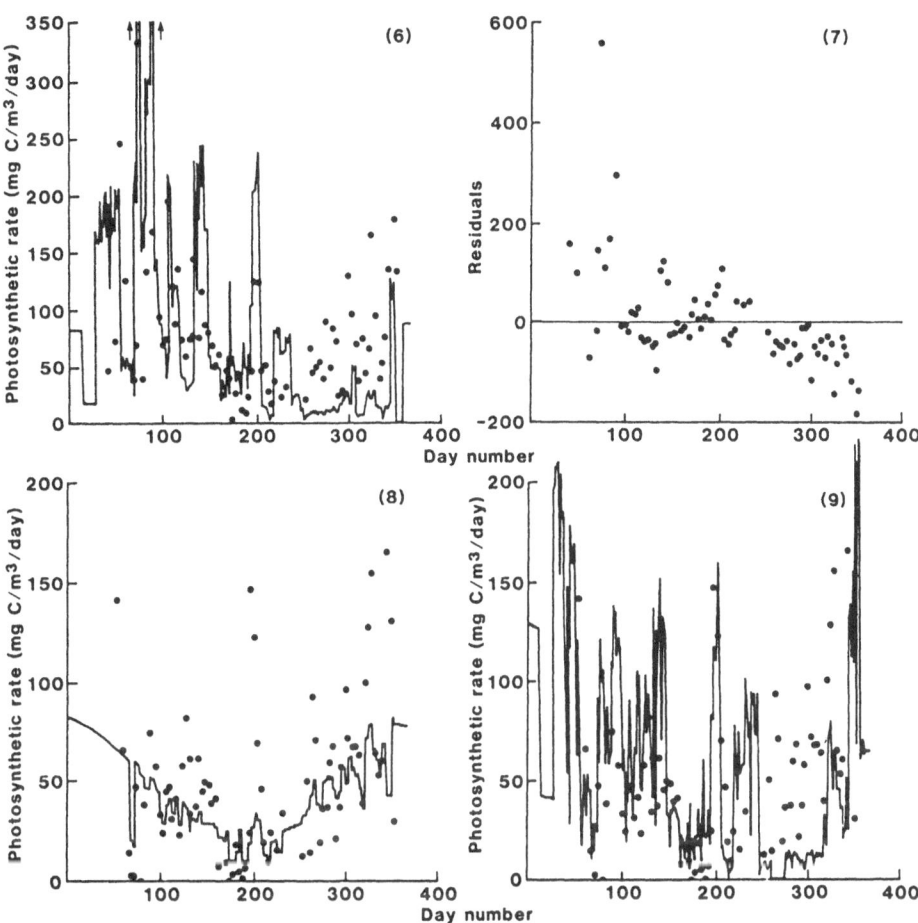

Fig. 6: A simulation of the *in situ* photosynthetic rate of South West Arm phytoplankton in the 0 – 1 m depth, with the inclusion of actual daily irradiances and actual nutrient concentrations, the latter modelled by the Michaelis–Menten expression. See text for more details.

Fig. 7: The residuals (modelled minus data) for the simulation illustrated in Fig. 6.

Fig. 8: A simulation of the *in situ* photosynthetic rate of South West Arm phytoplankton in the 5 – 6 m depth, assuming average cloud cover and average nutrient limitation effects. The chlorophyll concentration at 6 m was assumed in calculating the extinction coefficient. The *in situ* rates are also plotted.

Fig. 9: A simulation of the *in situ* photsynthetic rate of South West Arm phytoplankton in the 5 – 6 m depth, with the inclusion of actual daily irradiances and actual nutrient values. The *in situ* rates are also plotted.

known (Scott, 1979a) that at times hydrological conditions result in the introduction and entrapment of marine water and this period may reflect this process. An intriguing observation that obscures this picture further is that on day 349 there was a production of 180 mg C m^{-3} day^{-1} and on 351 a production of 134 but on these days zero nitrate concentrations were measured. This observation suggests that the regeneration rate of nitrates just balanced the assimilation rate by phytoplankton – an example of this may be found in Scavia (1979, p. 1341).

Finally, to test the model developed for the 0 – 1 m depth, I used the model unaltered to simulate the 5 – 6 m deep column (nitrate data available at this depth). The results are shown in Fig. 8 and show, in relation to the data (6 m), many of the same strengths and weaknesses as applied to the 0 – 1 m depth.

The correlation, modelled v. observed, was lower than the average–effects simulation for the 1 m depth ($r^2 = 0.50$ cf 0.68). This may have occurred because in calculating the extinction coefficient (Eq. 15) I used the 6–m deep chlorophyll a concentration. A weighted average of the available data (at 1, 2, 3, 4, and 6 m) would be more correct. The inclusion of nitrate data in the model produced the results shown in Fig. 9. Unlike the analogous 0 – 1 m simulations, this model does not under–estimate at low primary production rates (y–intercept was +31 mg C m^{-3} day^{-1}) and the correlation was poor ($r^2=0.28$). Nevertheless, I believe that Fig. 9 shows that the model captures the essence of the variability and that the correspondence would come good with a less crude incorporation of the nitrate data.

3.2 Network of functions

In this section an attempt will be made to deal with more than one function at a time, aiming at the construction of a dynamic model encompassing the network of functions available for SWA.

The functions presently identified are shown (solid lines) in Fig. 10. There is one feedback loop, via DIC, but since it does not limit the rate of primary production of phytoplankton in SWA (Section 3.1.3), the loop is effectively open. All functions flowing from phytoplankton and from PDOC are state–dependent and with possibly one exception all are donor controlled; this means that any primary production or DOC input (e.g. from the seagrass beds outside the mouth of SWA; see Cuff et al., 1983) is passed onto DIC. The exception is the joint dependency for the flow from phytoplankton to zooplankton.

One function, describing the rate of primary production of benthic micro–organisms as a function of light, is not shown in Fig. 10. This flow, from sediment (or near–sediment) DIC to micro–organisms (see Cuff et al., 1983), was omitted because we have no dynamic information on its relationship with other functions; i.e. it is not yet part of a network.

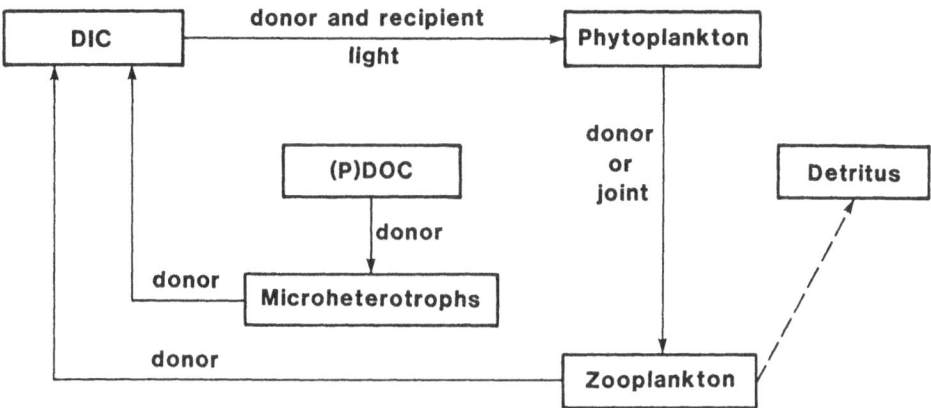

Fig. 10: The functions, and their state/forcing ·variable dependencies, that have been identified for South West Arm. State dependencies are either donor, recipient, or jointly dependent. The dotted line indicates a function assumed in some of the modelling described in the text.

A model describing the network shown in Fig. 10 is readily set up from Eqs 2 (or 20), 4, 5, 6, 7, and 13. (The equation discussed in Section 2.1.3 could be added for completeness.) A simulation of this set of simultaneous equations, using the average light and nutrient concentration for primary production in the 0 – 1 m depth range, reflected the incompleteness of the information. For a constant PDOC concentration (set at the average concentration of 1.2 mg C m⁻³), the microheterotroph compartment adjusted downward until the constant inflow was balanced by an equal outflow rate. The simple dynamics of the microheterotroph–containing pathway are clear from inspection of Eqs 4 – 6. An interesting deduction from these equations is that microheterotrophs are in balance only at a constant PDOC concentration of 7.67 mg C m⁻³, or about 6 times the known concentration in SWA. Such high concentrations – or even higher ones – may occur during the rising tide periods when DOC is being transported from the seagrass beds to SWA. This would suggest a cycle in microheterotroph population growth corresponding to the tidal cycle. The complication of active/basal respiration rates may not be of significance here since each rate would probably change by the same factor.

The phytoplankton–containing path, being the work of a number of separate studies, showed less interpretable behaviour in simulation. The phytoplankton and zooplankton compartments are of main interest since DIC in the model acts simply as a reservoir for respiratory carbon. With initial conditions set at average compartment masses (see Cuff *et al.*, 1983), a simulation resulted in high frequency oscillations in the zooplankton and phytoplankton compartments, with the cycles out of phase. These oscillations decreased with time so that by the third year of simulation the pattern had stabilized to that shown in Fig. 11 but oscillations continued through the

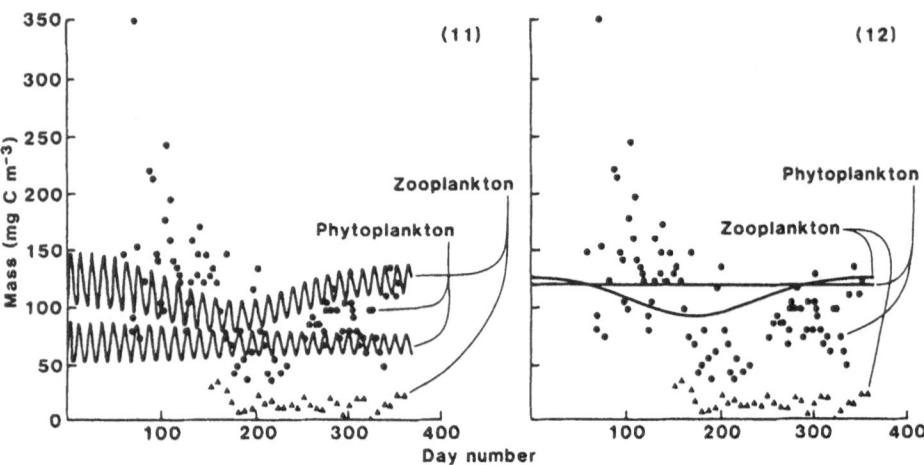

Fig. 11: A simulation of phytoplankton and intermediate sized zooplankton carbon masses using the model
described in the text and illustrated in Fig. 10. The measured phytoplankton (0 - 1 m) (designated
by dots) and intermediate sized zooplankton (designated by triangles) rates are also plotted.

Fig. 12: A simulation of phytoplankton and intermediate sized zooplankton carbon masses using a model
including the flow from zooplankton to detritus (see Fig. 10). The measured phytoplankton (0 - 1 m)
(designated by dots) and intermediate sized zooplankton (designated by triangles) rates are also
plotted.

end of the sixth year, with an amplitude in phytoplankton mass of about 5 mg C m^{-3}.
The simulated phytoplankton mass is of the correct order (chlorophyll (C/Chl a =
60) and zooplankton measurements also plotted) but zooplankton mass is much higher
than the average 24 mg C m^{-3} observed (Cuff et $al.$, 1983). Zooplankton showed an
annual cycle but phytoplankton mass stayed relatively constant over the year
(although the data show an annual cycle). This simulation assumes the grazing rate to
be controlled as by Eq. 20, κ = 5.9. A simulation with Eq. 2, k = 0.12, was also run;
phytoplankton rose quickly to ~15 000 mg C m^{-3} and zooplankton to ~5 000 before
stabilizing. No high frequency oscillations were shown and both compartments showed
annual cycles and the compartments cycled in phase. Since Eq. 2 does not involve
zooplankton mass, the high predicted phytoplankton mass suggests that either the
model is wrong or some unknown flow from phytoplankton exists. The discrepancy
between observed and modelled phytoplankton mass suggests the former explanation
to be the most likely. The joint–dependency model is thus the most likely of the two
models.

This is as far as available data lets one go but I wish to go one step beyond the
data. Cuff et $al.$ (1983) suggest a large flow from water column detritus to sediment
detritus of roughly the same size as the flow from phytoplankton to zooplankton. The
flow to water column detritus (dashed line in Fig. 10) will be donor dependent and if
it is assumed to be linear as well then one may estimate the rate constant from the

budget information. If one assumes that 35 mg C m^{-3} day^{-1} flows from zooplankton to water column detritus a simulation shows phytoplankton mass to stabilize at a constant value of 320 mg C m^{-3} and zooplankton to an annual cycle of 90 – 125 mg C m^{-3}. If one reduces the flow to only 7 mg C m^{-3} day^{-1}, then the same qualitative behaviour occurs but phytoplankton mass stabilizes at 117 mg C m^{-3}, compared to the mean value of 120 mg C m^{-3} (or 2 mg Chl a m^{-3}) calculated by Scott (1979a). But zooplankton mass does not change; it is still too high. The pattern is illustrated in Fig. 12. Some additional experimental work is required to understand the phytoplankton/zooplankton dynamics in SWA.

4. DISCUSSION

It is obvious from the analyses presented in this paper that we are far from a complete understanding of the dynamics of carbon flow within SWA. Complete understanding of the dynamics of even such a small (78 ha) estuary will obviously require a long–term and costly project. But I believe that progress has been made towards this goal, and that the available dynamic information represents an advance on the static information presented by Cuff *et al.* (1983). The former information gave a partial understanding of the paths through which, and at what rates, carbon flows within SWA. The dynamic information, where available, provides an understanding of the ways in which controlling factors influence these rates and hence allows the following question to be put: "What would happen if such and such were done to SWA?"

Useful information about SWA can come simply from a consideration of the individual functions. Perhaps the most useful function presently available is that describing photosynthesis, as developed from Scott's extensive data. For example, it is clear that primary production can be substantially increased by an increase in nitrate concentration and a fairly reliable estimate of the magnitude of such growth by any suggested increase in nitrate concentration is possible. (Further, there would seem to be immediate pay–off from experiments aimed at a mechanistic model of nitrogen intake and the dependence of productivity on the amount of nitrogen in phytoplankton.) Depth– and turbidity–related influences on primary production can be explored. Griffiths & Caperon's (1979) investigation of the grazing rate of intermediate sized zooplankton, along with the analyses of their model presented in this paper, suggests a degree of variability in the percentage of the phytoplankton that zooplankton consume (see Fig. 3) and a degree of uncertainty in model form and parameter values that should serve to make one careful about making dogmatic predictions about matters relating to the grazing rate. Any development–related changes of SWA that could affect the grazing rate of intermediate sized zooplankton should obviously be accompanied by a monitoring program to look for significant changes in phytoplankton or intermediate sized zooplankton masses.

Useful information about SWA can also follow from a consideration of the dynamics of the two pathways of carbon flow illustrated in Fig. 10. The discussion of the microheterotroph–containing pathway in the previous section suggests – as did an analysis of the budget data (Cuff *et al.*, 1983) – that microheterotrophs exist in SWA primarily to prevent a DOC buildup. If the DOC input from the seagrass beds was somehow removed, then, other things being equal, the microheterotroph population would decline and if we could measure PDOC in SWA, it would be possible to estimate the new steady state mass of the microheterotroph community. The other pathway, containing the phytoplankton and zooplankton compartments, is less directly useful because the phytoplankton mass depends, through the jointly dependent grazing rate of zooplankton on phytoplankton, on the zooplankton mass; the factors controlling zooplankton mass are unfortunately still largely unknown. One might still wish to accept as useful the model used to generate the most realistic looking simulation (Fig. 12) and study the effects on relative phytoplankton and zooplankton mass of interesting "what if" questions.

It is of general relevance that the incomplete set of dynamic information for SWA can be used to obtain useful – if conditional – insights into how SWA works. (I do not wish to overstate the usefulness of our current information about SWA: most of the functions could use more study, and much is unknown. On the other hand, I do not wish to overlook the knowledge about SWA that has been obtained.) This conclusion is of significance because published studies of material or energy flow through ecosystems have consistently supplemented locally collected and reliable data with information from the scientific literature and with guesstimates of questionable reliability; the assumption is that incomplete sets of dynamic information cannot be used effectively in practical situations. The essence of the argument appears to be that unknown information can change conclusions drawn on the basis of incomplete information. An example of this occurred in SWA: some unknown flows from zooplankton limited, through the jointly dependent grazing rate, our ability to predict phytoplankton mass. Also, feedback loops can, if unknown, have major effects on conclusions drawn on the basis of their assumed absence. While this argument is sound, it is of less importance when the goal is to synthesize the dynamic information that is available for the ecosystem of interest, and to get the best answers possible to those questions of interest that the data can be used to answer. If the answers prove incorrect, then one must accept that the reliability of predictions always increases with the completeness of the information available. I am not naively suggesting that a very incomplete dynamic model can provide reliable quantitative, or even qualitative, insights. Rather I believe that all information that one has should be considered; the weight placed on the information must of course be related to the degree of reliability (or completeness) of the information. If the data cannot answer all questions, that is simply the result of our current knowledge and the answer must wait until relevant data become available.

Although little use was made of non–SWA literature in supplementing local information, it is interesting to note that all functions studied here were developed elsewhere. Project participants simply chose likely functions, verified that they were appropriate, and estimated the parameters. Thus non–SWA experience was used but only as it was shown to be of proven relevance. In this way our work is constrained by the current state of theoretical knowledge.

An example of such a constraint concerns the use of the Michaelis–Menten expression for ascertaining the effect of nutrient concentration on phytoplankton growth. The parameters of the Michaelis–Menten expression are known to vary with the nutrition and stage of the life cycle of the phytoplankton species, and Harris (1980) even suggests that the expression may never be appropriate for use with natural populations. Yet we use the expression here, to describe the nutrient–growth kinetics of a phytoplankton community that changes in species composition. Also, the parameters of the other functions studied here are known to vary. Obviously, in using the synthesis of this paper one must be aware of the model's dependence on the general level of ecological understanding.

Synthesis of an incomplete set of dynamic information has followed essentially the same format as synthesis of an incomplete set of static information. In the static case the component information consists of average compartment masses and average flow rates between compartments, as known; in the dynamic case the component information consists of functions and parameters describing the flow rates between compartments, as known. In both the paper synthesizing the static information (Cuff et al., 1983) and this paper, synthesizing the dynamic information, I began by setting out and evaluating the available information. Then I used a block diagram to summarize what is known: Figs 2 – 5 in Cuff et al. (1983) and Fig. 10 in this paper. In the static case the diagram is essentially the model. Since average compartment masses and average flow rates do not change with new information (relating to previously unmeasured compartment sizes and flow rates) as it may be discovered, the static model provides a useful summary on how carbon flows through the ecosystem. The block diagram in this paper, although useful in itself, does not represent the final form of the dynamic model because this information allows one to postulate various "what if" questions and through analysis of the model provide an answer.

One aspect of synthesizing dynamic information that proved useful, as well as providing much fun, was the ability to compare the predictions to what is known, either as qualitative conclusions or as modelled v. observed plots.

ACKNOWLEDGEMENTS

In this study I have tried to balance modelling and experimental aspects: thanks go to R.E. Sinclair for help with some modelling aspects and to R.R. Parker for help with some experimental aspects. The experimental staff involved in the Port Hacking Estuary Project have been very helpful and I thank specifically F.B. Griffiths, D.J. Tranter, B.D. Scott, M.S. Giles and D.F. Smith for discussions regarding their specific studies. I am obliged to R.B. Humphries and R. McMurtrie for reviewing the manuscript and J.J. Wilkes for much help of various sorts.

REFERENCES

Bannister, T.T.: Production equations in terms of chlorophyll concentration, quantum yield, and upper limit to production. *Limnology and Oceanography* **19**, 1–12 (1974a)

Bannister, T.T.: A general theory of steady state phytoplankton growth in a nutrient saturated mixed layer. *Limnology and Oceanography* **19**, 13–30 (1974b)

Berman, M., Schoenfeld, R.: Invariants in experimental data on linear kinetics and the formulation of models. *Journal of Applied Physics* **27**, 1361–1370 (1956)

Caperon, J., Smith, D.F.: Photosynthetic rates of marine algae as a function of inorganic carbon concentration. *Limnology and Oceanography* **23**, 704–708 (1978)

CSIRO: *Estuarine Project Progress Report 1974–1976*. Sydney: CSIRO Division of Fisheries and Oceanography (1976)

Cuff, W.R., Sinclair, R.E., Parker, R.R., Tranter, D.J., Bulleid, N.C., Giles, M.S., Godfrey, J.S., Griffiths, F.B., Higgins, H.W., Kirkman, H., Rainer, S.F., Scott, B.D.: A carbon budget for South West Arm, Port Hacking. In: W.R. Cuff and M. Tomczak jr, eds *Synthesis and Modelling of Intermittent Estuaries*. Berlin, Heidelberg, New York: Springer (1983)

di Toro, D., O'Connor, D.J., Thomann, R.V.: A dynamic model of the phytoplankton population in the Sacramento–San Joaquin Delta. In: *Nonequilibrium Systems in Natural Water Chemistry. Advances in Chemistry Series, No. 106*. American Chemical Society (1971)

Fee, E.J.: A numerical model for determining integral primary production and its application to Lake Michigan. *Journal of the Fisheries Research Board of Canada* **30**, 1447–1468 (1973)

Giles, M.S.: Primary production of benthic micro–organisms in South West Arm, Port Hacking, New South Wales. In: W.R. Cuff and M. Tomczak jr, eds *Synthesis and Modelling of Intermittent Estuaries*. Berlin, Heidelberg, New York: Springer (1983)

Griffiths, F.B., Caperon, J.: Description and use of an improved method for determining estuarine zooplankton grazing rates on phytoplankton. *Marine Biology (Berlin)* **54**, 301–309 (1979)

Harris, G.P.: Temporal and spatial scales in phytoplankton ecology: Mechanisms, methods, models, and management. *Canadian Journal of Fisheries and Aquatic Sciences* **37**, 877–900 (1980)

Holling, C.S.: Some characteristics of simple types of predation and parasitism. *Canadian Entomologist* **91**, 385–398 (1959)

Ikeda, T.: Relationship between respiration rate and body size in marine plankton animals as a function of the temperature of habitat. *Bulletin of the Faculty of Fisheries Hokkaido University* **21**, 91–112 (1970)

Ikeda, T.: Nutritional ecology of marine zooplankton. *Memoirs of the Faculty of Fisheries Hokkaido University* **22**, 1–97 (1974)

Jacquez, J.A.: *Compartmental Analysis in Biology and Medicine: Kinetics of Distribution of Tracer-labeled Materials*. Amsterdam: Elsevier (1972)

Jassby, A.D., Platt, T.: Mathematical formulation of the relationship between photosynthesis and light for phytoplankton. *Limnology and Oceanography* **21**, 540–547 (1976)

Jitts, H.R., Morel, A., Saijo, Y.: The relation of oceanic primary production to available photosynthetic irradiance. *Australian Journal of Marine and Freshwater Research* **27**, 441–454 (1976)

Kremer, J.N., Nixon, S.W.: *A Coastal Marine Ecosystem, Simulation and Analysis*. Berlin: Springer-Verlag (1978)

Monteith, J.L.: *Principles of Environmental Physics*. London: Edward Arnold (1973)

Morel, A., Smith, R.C.: Relation between total quanta and total energy for aquatic photosynthesis. *Limnology and Oceanography* **19**, 591–600 (1974)

Platt, T., Denman, K.L., Jassby, A.D.: Modelling the productivity of phytoplankton. In: E.D. Goldberg, J.N. McCave, J.J. O'Brien and J.H. Steele, eds. *The Sea, Volume 6*. New York: John Wiley & Sons (1977)

Rashevsky, N.: Some remarks on the mathematical theory of nutrition of fishes. *Bulletin of Mathematical Biophysics* **21**, 161–183 (1959)

Scavia, D.: Examination of phosphorus cycling and control of phytoplankton dynamics in Lake Ontario with an ecological model. *Journal of the Fisheries Research Board of Canada* **36**, 1336–1346 (1979)

Scott, B.D.: Phytoplankton distribution and light attenuation in Port Hacking estuary. *Australian Journal of Marine and Freshwater Research* **29**, 31–44 (1978a)

Scott, B.D.: Nutrient cycling and primary production in Port Hacking, New South Wales. *Australian Journal of Marine and Freshwater Research* **29**, 803–815 (1978b)

Scott, B.D.: Seasonal variations of phytoplankton production in an estuary in relation to coastal water movements. *Australian Journal of Marine and Freshwater Research* **30**, 449–461 (1979a)

Scott, B.D.: *The Relation of Irradiance to Photosynthesis in Marine Phytoplankton*. M.Sc. Thesis, University of New South Wales (1979b)

Scott, B.D.: Phytoplankton distribution and production in Port Hacking estuary, and an empirical model for estimating daily primary production. In: W.R. Cuff and M. Tomczak jr, eds *Synthesis and Modelling of Intermittent Estuaries*. Berlin, Heidelberg, New York: Springer (1983)

Smith, D.F.: Quantitative analysis of the functional relationships existing between ecosystem components. I. Analysis of the linear intercomponent mass transfers. *Oecologia (Berlin)* **16**, 97–106 (1974)

Smith, R.A.: The theoretical basis for estimating phytoplankton production and specific growth rate from chlorophyll, light and temperature data. *Ecological Modelling* **10**, 243–264 (1980)

Spencer, J.W.: Sydney solar tables. *CSIRO Division of Building Research, Technical Paper (Second Series)* **8** (1975)

Steele, J.H.: Environmental control of photosynthesis in the sea. *Limnology and Oceanography* **7**, 137–150 (1962)

Tranter, D.J., Kennedy, G.: Size-specific respiration rate of Port Hacking zooplankton. In: W.R. Cuff and M. Tomczak jr, eds *Synthesis and Modelling of Intermittent Estuaries*. Berlin, Heidelberg, New York: Springer (1983)

Vaudrey, D.J., Griffiths, F.B., Sinclair, R.E.: Data base for the Port Hacking Estuary Project: Parameters, monitoring procedure, and management system. In: W.R. Cuff and M. Tomczak jr, eds *Synthesis and Modelling of Intermittent Estuaries*. Berlin, Heidelberg, New York: Springer (1983)

Vollenweider, R.A.: Calculation models of photosynthesis–depth curves and some implications regarding day rate estimates in primary production measurements. In: C.R. Goldman ed., Primary Productivity in Aquatic Environments. *Memorie dell'Istituto Italiano di Idrobiologia Dott Marco de Marchi* **18** (**suppl.**), 425–457 (1965)

Wiebe, W.J., Smith, D.F.: Direct measurement of dissolved organic carbon release by phytoplankton and incorporation by microheterotrophs. *Marine Biology (Berlin)* **42**, 213–223 (1977)

Synthesis and Modelling of Intermittent Estuaries
(W.R. Cuff and M. Tomczak jr. eds) Berlin, Heidelberg,
New York: Springer (1983), pp. 259–271.

Ecosystem Modelling of South West Arm, Port Hacking

Richard E. Sinclair[†], Wilfred R. Cuff[†§], Robert R. Parker[‡]

[†] CSIRO Division of Computing Research
P.O. Box 1800, Canberra, A.C.T. 2601, Australia

§Present address: Maritimes Forest Research Centre
Canadian Forestry Service
P.O. Box 4000
Fredericton, New Brunswick, E3B 5P7
Canada

[‡] Division of Fisheries Research
CSIRO Marine Laboratories
P.O. Box 21, Cronulla, N.S.W. 2230, Australia

Summary. A major aim of the Port Hacking Estuary Project was to produce an ecosystem model of a small marine embayment (South West Arm) in the Estuary. This paper describes the modelling efforts of the Project and puts them in perspective.

Key words: ecosystem modelling, Port Hacking, South West Arm

1. INTRODUCTION

From mid–1973 to 1978 scientists from CSIRO Division of Fisheries and Oceanography and from several other organizations conducted a study aimed at understanding the principles underlying the structure and dynamics of Australian estuarine systems. The participating scientists chose to focus their studies on a small marine embayment called South West Arm (SWA).

The major aim of the study was the construction of a predictive, dynamic ecosystem model (of carbon flow) for SWA, with data collected from SWA as suggested by the model. With hindsight, this seems like an ambitious task but it must be remembered that SWA was chosen because it is small (~78 ha), unpolluted, and close to Divisional staff – supposedly an ideal estuarine ecosystem to test the ability of a group of scientists to build a detailed and reliable model for a particular ecosystem.

The objective was not attained. But a number of models were constructed and a gradual evolution of ideas about ecosystem modelling occurred. Our experiences are described in this paper.

2. SYSTEM AND FEATURES

2.1 Study site

South West Arm (Fig. 1 of Vaudrey *et al.*, 1983) is a marine embayment off Port Hacking (N.S.W., Australia), located about 25 km south of Sydney and about 5 km by water from the CSIRO Marine Laboratories at Cronulla. The channel entering SWA from Port Hacking varies in depth from 2 to 4 m while the depth of SWA is from 18 to 20 m. For more details see Cuff *et al.* (1983).

2.2 Estuarine and marine states

South West Arm has a tendency to assume one of two states. Most of the time there is only a small freshwater inflow and a uniform salinity–depth profile (Fig. 2 of Bulleid, 1983). We refer to this condition as the marine state. Tidal circulation during this state simply involves the export and import of the top 1 to 2 m (the tidal range).

The second, estuarine, state occurs after rain when there is an influx of fresh water. As the surface water layer freshens following rainstorms, a pycnocline (*i.e.* the region of rapid change of density with depth) develops, intensifies, and moves towards the surface. Fresh water on the surface flows out on the ebb tide, mixing in the seaward channel with marine water. The returning, or flooding, tide is denser than the surface water and as it enters the basin it sinks, entraining some water, until it reaches a depth where it merges with water of equal density. Over a number of tidal cycles the pycnocline weakens and moves downwards until, with complete removal of low–salinity surface water, the incoming tidal water enters as described above for the marine state.

As detailed in a number of papers (see Cuff *et al.* 1983), the lower water column can become stagnant, not mixing with surface waters, during the estuarine state. Due to respiratory activities of micro–organisms, this results in irregular periods of de–oxygenation of the lower water column and sediments, and regeneration of nutrients (Fig. 2 of Bulleid, 1983). The end of the estuarine state allows a renewed entry of nutrients into the surface waters resulting in phytoplankton blooms (described in more detail in Scott, 1983).

3. MODELLING CONSIDERATIONS

3.1 Reasons for the ecosystem model

The development of an ecosystem model was set as a goal early in the Port Hacking Estuary Project (Allen, 1983; Parker *et al.*, 1983). The model was seen as a way of co–ordinating the efforts of the various scientists and disciplines involved, and as a framework for combining and evaluating the knowledge obtained (CSIRO, 1976). It was also seen early in the Project that the distinctive feature of SWA (the estuarine state leading to the development of de–oxygenation) resulted from a complex interaction between physical and biological processes and could be profitably studied using an ecosystem approach.

3.2 Objectives of the ecosystem model

An ideal ecosystem model would be comprehensive and possess reliable predictive capabilities. It would produce good agreement with the monitoring results; for example, good agreement with a temporal (or spatial) series of, say, phytoplankton mass. It would be useful in a management role by indicating the system's response to a wide variety of changed conditions. In specific terms, an ecosystem model of SWA would be able to match the observed behaviour after rain and simulate de–oxygenated conditions whenever they occur.

3.3 Requirements of the ecosystem model

This section considers the information, structure, and assumptions that are required if an ecosystem model of SWA is to meet the general and specific requirements listed above.

One requirement is the setting out of homogeneous spatial regions. Monitoring results have shown that, because of its relatively small area, it is possible to regard SWA as horizontally homogeneous. However, because of the importance of the depth–related effects of the estuarine water circulation pattern and of light attenuation and photosynthesis, it is necessary to consider a depth–stratified model. It is then possible to separate the ecosystem into a chemical biological compartment model and an independently operating water circulation model.

The requirements for a dynamic, depth–stratified, carbon flow model – representing the chemical and biological processes in terms of the flow of carbon between compartments – are: that there is an adequate number of compartments which are defined so that

each member (*e.g.* a species complex) shows a sufficiently similar type of dynamic behaviour;
all significant flows between compartments are included and expressed as believable functions of all relevant state and forcing variables;
the depth–stratification interval gives a reasonable balance between cost and accuracy; and
other important non–carbon entities are available as required: *e.g.* oxygen changes following stratification of the water column; nutrients as they affect primary production.

A water circulation model is used to provide an · information input to the depth–stratified carbon flow model. The requirement for this model is that it adequately reproduces the tidal inflow and outflow of SWA, especially the depth–related features of the flows during estuarine conditions. This could be achieved with a sophisticated hydrological model but the approach used in this Project was to build a much simpler model, which is described in Cuff *et al.* (1978), and summarized in Section 4.2.

Each model also requires inputs or information from outside SWA. The carbon flow model requires light, water temperature, data on influx of nutrients, and flows into compartments (outside values). The water circulation model requires temperature and salinity values for Port Hacking and rainfall values for the catchment area.

4. MODELLING ATTEMPTS

4.1 Dynamic modelling of carbon flow

The initial suggestion to model was provided by D.F. Smith and he provided the modelling direction from 1973 to mid–1975. His expertise lay in the area of radioactive tracer analysis (e.g. Wiebe & Smith, 1977; Smith et al., 1979), and linear (Smith, 1974a, 1975) and non–linear (Smith, 1974b, 1975) compartmental modelling (Berman & Schoenfeld, 1959; Jacquez, 1972). The (mostly) linear–modelling philosophy of compartmental analysis/modelling was applied, without discussion, to the task of modelling carbon flow within SWA. It played a more central and restrictive role in future (modelling and non–modelling) developments than the Project participants realized at the time, in terms of dictating what input the model needed from the experimentalists in order to provide "an explicit statement of the work that was required so that individuals could continuously evaluate their progress and re–allocate their efforts" (CSIRO, 1976).

But to go back to the beginning, interaction matrices and flow diagrams were proposed and discussed at length (final forms in Tables 1 and 2 of CSIRO, 1976; Parker et al., 1982). SWA was idealized to a uniform 10 m depth and the chemical and biological entities were idealized into sediment and water column sub–systems, each with

Dissolved inorganic carbon (DIC),
Dissolved organic carbon (DOC),
Detritus,
Autotrophs, and
Heterotrophs.

In the sediment sub–system there were both aerobic and anaerobic heterotroph compartments.

Methods were soon proposed to determine the parameters in functions formulated for the transfers between compartments. In practice, much early work of the Project involved the development of experimental techniques which were needed to be able to provide the information required for the model. Work proceeded apparently unaltered from the initial plan until mid–1975 when one of us (R.E. Sinclair) was invited to join the group to bolster the modelling efforts. At that point no model had been constructed: a final interaction matrix and flow diagram were available and parameters were thought to be forthcoming from the experimental work in progress. Based on this information, R.E. Sinclair and D.F. Smith implemented in ACSL (Mitchell & Gauthier, Assoc., 1975) the model, using dummy parameter values. R.E. Sinclair then took over the sole responsibility for the model; D.F. Smith put subsequent efforts into radioactive–tracer experiments (*e.g.* Smith & Wiebe, 1976; Wiebe & Smith, 1977; Smith *et al.*, 1979). The parameters required for the implementation of this model (SWAMP: SWA Modelling Program) did not arrive, for reasons discussed in Cuff (1983b) and Tomczak (1983).

By late 1976, W.R. Cuff had joined the CSIRO Division of Computing Research and was asked to assist R.E. Sinclair with modelling aspects of the Project. (It was felt by the Division that Cuff's biological experience could be a useful complement to Sinclair's control–theory experience.) But, by this time, research programs were well underway and there was little chance of influencing their direction by modelling arguments. So attention was turned to the problem of *synthesizing* the information that had been – and was being – collected. From this point onwards R.E. Sinclair, W.R. Cuff, and R.R. Parker (Co–ordinator of the Project) worked as a team, with only specific contributions from other members of the Project, at least until the production of this book.

While we were having great difficulties getting off the ground, the literature continued to document ecosystem models of various estuaries, lakes, and so on. Nearly all were found to contain complicated functions to represent the various flow rates. They required a large number of parameters, which seemingly were available to them. Yet our Project did not seem to be producing even the required, smaller, set of parameters. A closer inspection of the published models revealed that they used only a small amount of reliable, local dynamic information; they apparently found in the literature other ecosystems similar enough to their own to use as the source of most parameter values. The remaining values were, seemingly, adequately obtained as "guesstimates", especially when combined with "tuning" (*i.e.* fitting outputs of the model to observed results) to get to the best estimates.

This approach was not adopted for obtaining the required parameter values for SWAMP for two reasons. Firstly, the idea at the beginning of the Project was to use a model to guide and co–ordinate the data collection from SWA. Functions were pre–defined (based on Smith's 1974a, 1974b, and 1975 papers) and participants set out to estimate average compartment sizes and average transfer rates (fluxes). Secondly, enthusiasm within the scientific community for the literature–based type of ecosystem model had changed from 1973 to 1977; scepticism had emerged. To summarize the state of affairs in the Port Hacking Estuary Project in early 1977: the simple, basically linear, compartmental model followed to that point was seen as too simple–minded by a sizable proportion of the participants, but the literature–based approach to ecosystem modelling seemed (to the modellers at least) to have only tangential relevance to any *particular* ecosystem. The experimental ramification of this confusion and disenchantment over modelling meant that in early 1977 there was a variety of experimental programs in progress: some experimentalists continued to concentrate on the originally defined measurements of compartment sizes and transfer rates; some turned to (state–dependent) functional identification and parameter estimation; and yet others simply followed the conventions of their experimental disciplines to try to understand what was going on in SWA.

In trying to find a way into the problem of synthesizing this bewildering array of information, we thought that since budgetary information is easier to obtain (directly by experimentation or indirectly by reduction of dynamic information; see Cuff *et al.*, 1983) than dynamic information, it would be useful if a budget could be the first step leading to a dynamic model. By looking back to earlier ecosystem modelling methods (*e.g.* see articles in Patten, 1971), we remembered that this approach had once been the standard. The first ecosystem modellers by and large obtained parameter estimates for their linear models by thinking in terms of average compartment sizes and average flow rates, *i.e.* by mentally constructing a budget. (A characteristic feature of many of these earlier models, however, was the cursory attention given to the budgetary contribution of their dynamic models. The role of the budget was implied, not stressed.) Since the initial model for SWA was also conceived as being essentially linear, so also our Project had, in fact, been working on nothing more mysterious than a carbon budget. Needless to say, this fact was not at all clear to the Project participants, as records of the vigorous debate at group meetings show. It seems to us that much useful de–mystification of dynamic modelling would occur if experimentalists *clearly* understood that (at the simplest level) dynamic models are immediate descendants of budgets.

In obtaining a linear dynamic model from budgetary information, the flow rates are assumed to be linear functions of the states. Then one must know whether the flow is donor or recipient dependent (this information is often available and reliable) to estimate the rate constants. A set of differential equations for SWA was derived in this way and a carbon flow model developed and published in a set of Conference Proceedings (Cuff *et al.*, 1978). At this stage we also built a water circulation model (Section 4.2) and combined it with the carbon flow model discussed here to produce an ecosystem model (see Section 4.3).

Although this approach seems to have merit, it did not lead us to further developments: by this time the Project had been stopped and, besides, even our budget was in total not all that reliable. (In constructing the budget we mixed reliable data with information of unknown reliability for SWA to get a "complete" budget.) So we returned to the budget and eliminated the unreliable data. This (despite its gaps) highlighted the major flow path of carbon through the ecosystem. We could then, by invoking the assumption that some compartments are on average balanced, obtain all required flow rates for the major flow path. This allowed the building of a dynamic model of carbon flow (Cuff *et al.*, 1980) which was more believable than the previous model.

Our more recent work has been in two directions. In one paper (Cuff *et al.*, 1983) we made our most serious effort to date to identify and co–ordinate the reliable data for a carbon budget. There are still many details missing and more observational work will be required to obtain this information. In a companion paper (Cuff, 1983a), the available and reliable dynamic information for SWA was evaluated and the best possible dynamic model of carbon flow was defined.

The most reliable information relates to phytoplankton (Scott, 1978a, 1978b, 1979a, 1979b, 1980, 1983). We have some dynamic information regarding a major outflow from the phytoplankton compartment (Griffiths & Caperon, 1979) and while the possibility of building an "ecosystem model" of phytoplankton alone comes to mind, simulations showed that the nature of the zooplankton grazing rate function (jointly dependently on autotroph and heterotroph densities) combined with the uncertainty concerning the zooplankton compartment makes this option out of reach – although not that far away. This could be a useful next direction for further research in SWA, since the only known feedback to phytoplankton is due to the limitation of DIC, but this loop will never limit primary production in the field.

We ended our efforts to dynamically model carbon flow at this point. The evolution of our ideas has gone from collection of SWA data for obtaining parameter values for a mostly linear compartmental model, to explicit reliance on budgetary information – whole budget (Cuff *et al.*, 1978) to major flow path (Cuff *et al.*, 1980) – to the use of only that dynamic information available for SWA. In fact, our most extensive and reliable information relates to the phytoplankton compartment and so, in a sense, we have argued ourselves back to one compartment! It is an interesting and surprising evolution, but at least we now feel that we (soon) could have a firm base on which to build a dynamic model of carbon flow.

4.2 Dynamic modelling of water circulation

The water circulation model is based on the simple concept of a wedge of water entering SWA, sinking if necessary, to find water of similar density and merging at this depth. The model simulates the water column in terms of 20 "layers" each of 1 m depth. The volume of the layers decreases with depth according to measured values of cross–sectional area. Horizontal homogeneity of SWA is assumed, as observed by Godfrey & Parslow (1976).

In the model the ebb tide is simulated by removing the top layer. This is mixed with an equal volume of seawater (mixing ratios of 0.62 to 1.5 volumes of seawater to 1 volume of surface water were calculated from data of Godfrey & Parslow, 1976) and the temperature and salinity of the mixed water are calculated. The density (ρ, kg l^{-1}) of the mixed water is calculated from its temperature and salinity, using the standard formulae (Lafond, 1951) for shallow water, but it is expressed in terms of "density anomaly", σ_t which is related to ρ by

$$\sigma_t = 1000 * (\rho - 1) \tag{1}$$

The mixed water, of density anomaly σ_t, and of volume equal to that of the top layer, enters SWA on the flood tide and sinks until it merges with water at the depth where its density is intermediate between the densities of the two adjacent layers. As the incoming water sinks it entrains a proportion (currently 5%) of each layer it passes. The temperature, salinity, and density of the incoming water are thus modified at each layer. The temperature and salinity of the incoming water are then averaged with those of the two adjacent layers. The layers above the incoming level are moved up and their temperature and salinity values are modified by the small amounts of "excess" water from below. This excess results from the fact that the surface layer is larger than the layer it merges into. Some mixing between adjacent layers occurs from the incoming level down to the bottom.

The above details are sufficient to simulate both the marine and estuarine states. The marine state is simulated as described above, but with density differences caused only by possible temperature differences between the water inside and outside SWA. The estuarine state is initiated by adding fresh water to the surface layer at the beginning of both the ebb and flood tides. The amount of fresh water added is determined by using the daily rainfall averaged between Cronulla and Audley. These two sites span SWA (see Fig. 1 of Albani *et al.*, 1983). The rainfall is multiplied by a factor (currently set to 10) to take account of the area of the SWA catchment, and is spread over three tidal cycles.

The model was implemented in FORTRAN and contour plots of temperature, salinity, and density (over depth and time) were produced. An example of the density contour output of the model is given in Fig. 1a. The corresponding observed density contour is given in Fig. 1b. This run (from 29 September (day 0) 1976 to 8 December 1976) simulated marine conditions from 29 September to 1 October (day 2), when a rainstorm resulted in an influx of fresh water. The decay of the estuarine conditions was evident to 15 October (day 16), when the effect of another pulse of fresh water can be seen. Continuing general rainfall maintained estuarine conditions until it ceased and the system reverted to the marine condition by 3 December (day 65).

The observed behaviour was reproduced reasonably well although low surface densities were not achieved (not visible on Fig. 1) and the non–uniform density profile (Fig. 1b) was not maintained towards the end of the simulated period (Fig. 1a).

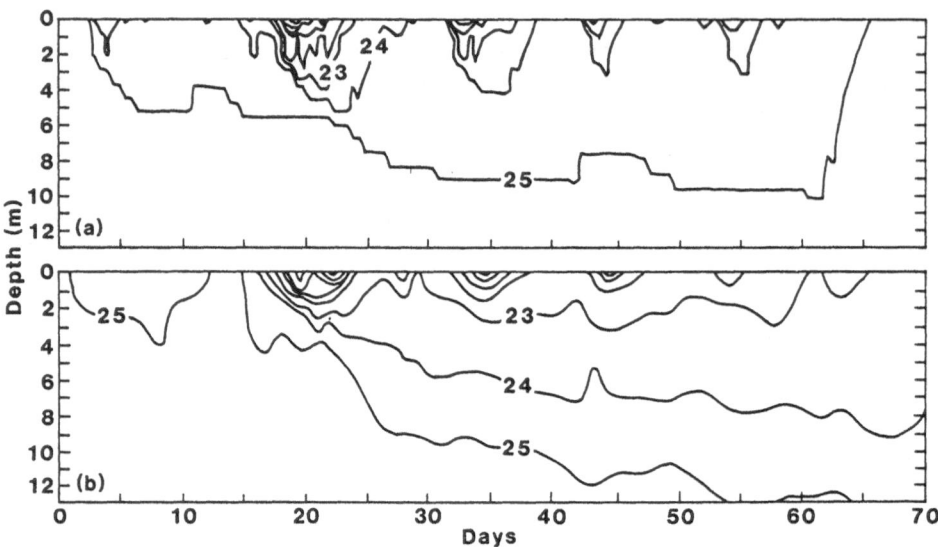

Fig. 1: An example of output from the water circulation model: modelled (a) and observed (b) density σ_t for 29 September to 8 December 1976. From Cuff *et al.* (1978).

4.3 Dynamic ecosystem modelling

In our early modelling efforts (Cuff *et al.*, 1978) thought was given to methods of combining a carbon flow model with a water circulation model, in order to investigate the depth–related processes of SWA. The solution arrived at appears to be the best and does not depend on the degree of complexity of either model.

4.3.1 The marine state – The marine state is simulated as a simple extension of the carbon flow model to include depth stratification. The water column sub–system is subdivided into 7 layers each of 2 m depth, with a final layer from 14 to 20 m. The model is defined by 46 simultaneous differential equations, being the sum of the 5 water column equations for each of the 8 layers and the 6 sediment equations. Exchanges between layers are represented by reciprocal exchange terms to simulate the effects of turbulence (mixing and settling) and diffusion. In the absence of data, we arbitrarily set the exchange terms to allow rapid exchange between the layers.

4.3.2 The estuarine state – During the estuarine state there is a non–uniform density profile in SWA. Thus the water circulation model was run independently from the time of freshwater input, through the estuarine to the marine state, to generate daily the following information: the strength and position of the pycnocline and the depth at which the incoming wedge of water merged.

To accommodate this information the ecosystem model as described above was extended as follows. The incoming wedge was incorporated at the appropriate level thereby affecting the characteristics of the water at that depth both directly, and indirectly by turbulence. The turbulence was modelled by setting the exchange coefficients to high values near the wedge and decreasing them with distance. The rising exchange coefficients were set to zero below a strong pycnocline although some mixing was allowed when a pycnocline formed near the surface and the incoming wedge was below it.

4.3.3 Comment – Although this modelling will not find immediate use in SWA, it does serve to describe one approach to modelling an intermittent estuary and suggests the cost and effort that would be involved in building a reliable ecosystem model of such a system.

5. DISCUSSION

This paper has provided a history of our efforts in building a reliable ecosystem model. We did build an ecosystem model but did not feel we could trust that part relating to carbon flow. But we know the direction in which to proceed. By concentrating on available and trustworthy dynamic carbon flow information (for SWA, this more–or–less means phytoplankton information), we seem to have found a reliable kernel from which growth is possible.

The story speaks for itself. Perhaps it is useful to repeat a statement of our feeling as expressed in our 1978 paper:
"The development of an estuarine ecosystem model within a multidisciplinary framework is not a task for the faint–hearted,... ."

REFERENCES

Albani, A.D., Rickwood, P.C., Tayton, J.W., Johnson, B.D.: Geological aspects of the Port Hacking estuary. In: W.R. Cuff and M. Tomczak jr, eds *Synthesis and Modelling of Intermittent Estuaries*. Berlin, Heidelberg, New York: Springer (1983)

Allen, K.R.: Introduction to the Port Hacking Estuary Project. In: W.R. Cuff and M. Tomczak jr, eds *Synthesis and Modelling of Intermittent Estuaries*. Berlin, Heidelberg, New York: Springer (1983)

Berman, M., Schoenfeld, R.: Invariants in experimental data on linear kinetics and the formulation of models. *Journal of Applied Physics* **27**, 1361–1370 (1959)

Bulleid, N.C.: The nutrient cycle of an intermittently stratified estuary. In: W.R. Cuff and M. Tomczak jr, eds *Synthesis and Modelling of Intermittent Estuaries*. Berlin, Heidelberg, New York: Springer (1983)

CSIRO: *Estuarine Project Progress Report 1974–1976*. Sydney: CSIRO Division of Fisheries and Oceanography (1976)

Cuff, W.R.: An evaluation of the dynamic information for South West Arm, Port Hacking. In: W.R. Cuff and M. Tomczak jr, eds *Synthesis and Modelling of Intermittent Estuaries*. Berlin, Heidelberg, New York: Springer (1983a)

Cuff, W.R.: An evaluation of the Port Hacking Estuary Project from the viewpoint of applied science. In: W.R. Cuff and M. Tomczak jr, eds *Synthesis and Modelling of Intermittent Estuaries*. Berlin, Heidelberg, New York: Springer (1983b)

Cuff, W., Sinclair, R., Parker, R.R.: The development of an ecosystem model for South West Arm (Port Hacking, N.S.W.). In: P. Whitehead, ed. *Simulation Modelling Techniques and Applications. Proceedings of SIMSIG–78 Simulation Conference*. Canberra: Australian National University (1978)

Cuff, W., Sinclair, R., Parker, R.R.: Carbon flow within the South West Arm of Port Hacking (N.S.W., Australia). In: P.A. Trudinger, M.R. Walter and B.J. Ralph, editorial committee. *Biogeochemistry of Ancient and Modern Environments*. Canberra: Australian Academy of Science (1980)

Cuff, W.R., Sinclair, R.E., Parker, R.R., Tranter, D.J., Bulleid, N.C., Giles, M.S., Godfrey, J.S., Griffiths, F.B., Higgins, H.W., Kirkman, H., Rainer, S.F., Scott, B.D.: A carbon budget for South West Arm, Port Hacking. In: W.R. Cuff and M. Tomczak jr, eds *Synthesis and Modelling of Intermittent Estuaries*. Berlin, Heidelberg, New York: Springer (1983)

Godfrey, J.S., Parslow, J.: Description and preliminary theory of circulation in Port Hacking estuary. *CSIRO Division of Fisheries and Oceanography Report* **67** (1976)

Griffiths, F.B., Caperon, J.: Description and use of an improved method for determining estuarine zooplankton grazing rates on phytoplankton. *Marine Biology (Berlin)* **54**, 301–309 (1979)

Jacquez, J.A.: *Compartmental Analysis in Biology and Medicine: Kinetics of Distribution of Tracer-labelled Materials*. Amsterdam: Elsevier (1972)

Lafond, E.C.: Processing oceanographic data. *U.S. Navy Hydrographic Office, Washington, D.C., H.O. Publication* **614** (1951)

Mitchell and Gauthier, Assoc.: *Advanced Continuous Simulation Language (ACSL). User Guide/Reference Manual*. 2nd ed. Concord, Mass.: Mitchell and Gauthier, Assoc., Inc. (1975)

Parker, R.R., Rochford, D.J., Tranter, D.J.: History and organization of the Port Hacking Estuary Project. In: W.R. Cuff and M. Tomczak jr, eds *Synthesis and Modelling of Intermittent Estuaries*. Berlin, Heidelberg, New York: Springer (1983)

Patten, B.C.: *Systems Analysis and Simulation in Ecology, Vol. I*. New York: Academic (1971)

Scott, B.D.: Phytoplankton distribution and light attenuation in Port Hacking estuary. *Australian Journal of Marine and Freshwater Research* **29**, 31–44 (1978a)

Scott, B.D.: Nutrient cycling and primary production in Port Hacking, New South Wales. *Australian Journal of Marine and Freshwater Research* **29**, 803–815 (1978b)

Scott, B.D.: Seasonal variations of phytoplankton production in an estuary in relation to coastal water movements. *Australian Journal of Marine and Freshwater Research* **30**, 449–461 (1979a)

Scott, B.D.: *The Relation of Irradiance to Photosynthesis in Marine Phytoplankton*. M.Sc. Thesis, University of New South Wales (1979b)

Scott, B.D.: Diurnal and other variations of photosynthesis versus irradiance curves for phytoplankton in Port Hacking. *CSIRO Division of Fisheries and Oceanography Report* **140** (1980)

Scott, B.D.: Phytoplankton distribution and production in Port Hacking estuary, and an empirical model for estimating daily primary production. In: W.R. Cuff and M. Tomczak jr, eds *Synthesis and Modelling of Intermittent Estuaries*. Berlin, Heidelberg, New York: Springer (1983)

Smith, D.F.: Quantitative analysis of the functional relationships existing between ecosystem components. I. Analysis of the linear intercomponent mass transfers. *Oecologia (Berlin)* **16**, 97–106 (1974a)

Smith, D.F.: Quantitative analysis of the functional relationships existing between ecosystem components. II. Analysis of non–linear relationships. *Oecologia (Berlin)* **16**, 107–117 (1974b)

Smith, D.F.: Quantitative analysis of the functional relationships existing between ecosystem components. III. Analysis of ecosystem stability. *Oecologia (Berlin)* **21**, 17–29 (1975)

Smith, D.F., Wiebe, W.J.: Constant release of photosynthate from marine phytoplankton. *Applied and Environmental Microbiology* **32**, 75–79 (1976)

Smith, D.F., Bulleid, N.C., Campbell, R., Higgins, H.W., Rowe, F., Tranter, D.J., Tranter, H.: Marine food–web analysis: an experimental study of demersal zooplankton using isotopically labelled prey species. *Marine Biology (Berlin)* **54**, 49–59 (1979)

Tomczak, M., jr.: Some conclusions from the Port Hacking Estuary Project. In: W.R. Cuff and M. Tomczak jr, eds *Synthesis and Modelling of Intermittent Estuaries*. Berlin, Heidelberg, New York: Springer (1983)

Vaudrey, D.J., Griffiths, F.B., Sinclair, R.E.: Data base for the Port Hacking Estuary Project: Parameters, monitoring procedure, and management system. In: W.R. Cuff and M. Tomczak jr, eds *Synthesis and Modelling of Intermittent Estuaries*. Berlin, Heidelberg, New York: Springer (1983)

Wiebe, W.J., Smith, D.F.: Direct measurement of dissolved organic carbon release by phytoplankton and incorporation by microheterotrophs. *Marine Biology (Berlin)* **42**, 213–223 (1977)

Synthesis and Modelling of Intermittent Estuaries
(W.R. Cuff and M. Tomczak jr. eds) Berlin, Heidelberg,
New York: Springer (1983), pp. 273–292.

An Evaluation of the Port Hacking Estuary Project
from the Viewpoint of Applied Science

Wilfred R. Cuff

CSIRO Division of Computing Research

P.O. Box 1800, Canberra, A.C.T. 2601, Australia

Present address: Maritimes Forest Research Centre

Canadian Forestry Service

P.O. Box 4000

Fredericton, New Brunswick, E3B 5P7

Canada

Summary. The Port Hacking Estuary Project, a model–guided study of the flow of carbon through a small Australian estuary, is reviewed from the viewpoint of applied science. The Project did not reach its goal of constructing a predictive dynamic model of carbon flow in the South West Arm of Port Hacking and key ambiguities in project design and execution that inhibited progress are identified. It is suggested that the model structure chosen to be compatible with time and manpower constraints did not allow sufficient mechanistic contribution to attract the support of the experimental participants. It is a general characteristic of ecosystem modelling to date that a complete (self–consistent) set of data is assumed before the model is constructed. Since no group, including our Project, has been able to collect such a complete set of data from the specific ecosystem, recourse to the literature is regularly made to complete the model. This means most of the data are of unknown reliability, and this fact implicitly demeans the (often hard won) data from the specific ecosystem; this, in turn, leads to decreasing interest in the model from the experimental participants and the group eventually splits apart. To avoid this scenario, I suggest more emphasis on the data that has been collected from the specific ecosystem: in particular, periodic syntheses of the available data set not supplemented by data of unknown reliability.

Key words: ecosystem, applied science, environmental science, synthesis, modelling, Port Hacking, South West Arm

1. INTRODUCTION

It seems appropriate to state at the outset of this evaluation that the Port Hacking Estuary Project was a difficult – and often frustrating – scientific endeavour. The broad, conceptual aim of the project was ambitious: to gain "increased understanding of the principles underlying the structure and dynamics of estuarine systems" (Allen, 1983). Even the specific representation of this broad statement of intent is, with hindsight, daunting: to model mathematically through to reliable predictions "the flow of carbon into, within, and out of South West Arm" (Parker et al., 1983).

Yet the Port Hacking Estuary Project was neither unique nor exceptionally ambitious in its objectives, there having been a number of such multidisciplinary projects set up in the last decade or so. Early examples are from the US–IBP (International Biological Program). These projects, first funded in 1968, were expected to progress from the initiation of field studies to complete ecosystem analyses in 4 – 5 years. In an evaluation of three of the five studies, Mitchell et al. (1976) note: "Ecosystem models were one of the central goals for each of the biome programs. Originally the models were intended to be general enough to deal with large regions, precise enough to allow for meaningful applications in management decisions, and realistic enough to add new insights to theoretical ecology. These overly ambitious goals were not achieved." Descriptions of the models relating to the Eastern Deciduous Forest Biome (EDFB) and to the Grassland Biome (GB) have been published: EDFB – Park et al., 1974; GB – Innis, 1978.

The lack of the expected level of success of the biome models has not dampened the rate of development of other ecosystem models. Examples include a carbon–flow model for a coastal Georgia salt marsh (Wiegert et al., 1975); a carbon–flow model for the Bristol Channel and Severn Estuary (Longhurst, 1978); a water, carbon, and energy–flow model for a coniferous forest in Oregon (Swartzman, 1979); and a comprehensive model for Lake Ontario (Scavia, 1980). The last model includes components of the phosphorus, nitrogen, silicon, and carbon cycles, dissolved oxygen, and particulate sediment and pore water dynamics.

The fact that the Port Hacking Estuary Project is a member of an ongoing genre does not change the fact that it did not achieve its goal of making a predictive, dynamic model of carbon flow in South West Arm. Why? Was there something unique either about the Project or about South West Arm that prevented the expected progress? Or should one look at the whole class of model–guided studies? These questions are addressed in this paper from the viewpoint of applied science. Tomczak (1983) attempts a similar examination from the viewpoint of basic science.

2. WHAT IS "APPLIED SCIENCE"?

While the terms "applied" and "basic" are often used in describing research, there are many shades of meaning. Some would equate university–based research with basic (*i.e.* "academic") science; and industry and government–based research with applied science. Others would hold that all long–term research is basic and all short–term research is applied. Others would say that all research conducted for the purpose of advancing the fund of scientific knowledge is basic (*i.e.* "pure"); all other research is applied. Yet others would contend that all research is applied, with variation only in the time to application.

For the purpose at hand, I choose another variant, *viz.* basic research as that whose prime aim is to contribute to the development of theory, whether that research is conducted in industry, government, or university; and whether it is short or long–term research. A characteristic of basic research is thus the search for generality. I further choose applied research as that solely concerned with a particular problem, whether that research is conducted in industry, government or university; and whether it is short or long–term research. From these definitions, it is clear that applied science may use theory, but only as it is of *demonstrable* relevance to the particular problem. Otherwise innovative research programs will be set up to obtain the advances needed for the particular problem. Equally, basic science may use many of the results from applied science for formulating and testing theories.

3. THE PORT HACKING ESTUARY PROJECT

3.1 Study location

Port Hacking is a small subtropical estuary on the east coast of Australia about 30 km south of Sydney city centre. The Project was centred on the South West Arm (SWA) of Port Hacking. The catchment of SWA lies entirely within a National Park; no industrial activity affects the estuary, and low–density urbanization extends only along the northern side. All this makes Port Hacking a relatively unpolluted estuary, despite its proximity to a city of several million people.

3.2 Key ambiguities

Allen (1983) discussed the factors leading to the Port Hacking Estuary Project; Parker *et al.* (1983) discussed the organization of the Project; and Sinclair *et al.* (1983) discussed the evolution of the modelling activities. These papers suggest certain key ambiguities in the initiation and execution of the Project that I wish to discuss in some detail here. In doing so I rely heavily on an extensive set of minutes and memoranda; I did not join the Project until late 1976.

3.2.1 What was the reason for making the carbon flow model? – Allen (1983) made clear how, in the early 1970's, it was possible to see environmental problems developing in Australian estuaries and mentioned a few *ad hoc* studies that had been set up to deal with development–related issues. He suggested that these studies could benefit by having available to them a general pool of knowledge of the principles underlying the structure and dynamics of (Australian) estuarine systems.

This idea had its origins in a staff meeting (with a member of the CSIRO Executive in attendance) held on 30 – 31 July 1973 to discuss a possible project on estuarine research. The meeting was wide ranging, with reviews of Australian and international estuarine research and a discussion of major pressures likely to affect the Australian estuarine environment, and with a review of ongoing CSIRO projects. During the discussion of major pressures likely to affect the Australian estuarine environment, it was suggested that research in an estuary project should be directed towards an understanding of broad principles, *e.g.* food webs, energetics. The idea was to study a broad range of interacting features and "to weld these into a comprehensive enough model to determine general principles". In subsequent discussion relating to the desirability of intensive or extensive estuarine studies, the former type was favoured because in order to "build a conceptual model of an estuary one had to thoroughly investigate one system and experiment until the model could predict the behaviour of that system". In other words, the search for principles was to be conducted by thoroughly understanding one ecosystem, to the level of prediction.

On 4 October 1973, K.R. Allen (Chief of the CSIRO Division of Fisheries and Oceanography) suggested in a memorandum that, on the basis of the July meeting, he felt that there was a real need within Australia for the development of *basic* research on estuaries and called a meeting to discuss a proposition that the Division "should embark immediately on a preliminary program of estuarine studies, based mainly on Port Hacking, and to extend over one or two years during which time the foundations for a more comprehensive and long term program of estuarine studies could be established". Port Hacking was suggested both for convenience of location and because it has both temperate and tropical (*e.g.* mangroves) characteristics, and has some anaerobic basins. Port Hacking was also seen as an unpolluted representative of other, polluted, estuaries in Sydney, so the studies would potentially have practical relevance.

In a meeting (9 October 1973) called by K.R. Allen it was agreed to mount such a preliminary program. Suggestions were invited regarding the identification of specific questions for the preliminary Project and an Interim Co–ordinator (D.F. Smith) appointed, pending the search for a permanent appointee. (For a description of the role of the Co–ordinator see Parker *et al.*, 1983.) Various suggestions were subsequently put forward and it was generally agreed in a meeting held on 19 October 1973 that three projects were of general interest:

> i) what is the nature of the flow of energy and materials in an estuarine system and what factors modify this flow?
> ii) what are the mechanisms of the oxygenation cycle in an estuary and how important is this cycle to organisms?
> iii) the water budget of Port Hacking.

Although the decision was not unanimous, it was agreed that "it would be useful to construct a model of mass transport and energy flow through the estuarine ecosystem. Building such a model would help in identifying the specific questions we need to ask, suggest experiments (and point out inherent technical difficulties) to be undertaken, and establish the roles individuals might take during various phases of the study". The model was referred to as a "Preliminary Model" and it was said "that model building was useful as an initial approach to define the problems to be studied". It was agreed that South West Arm should be the area of interest, but with effort devoted to other areas of Port Hacking as required.

In their suggestions regarding the specific questions to be addressed, participants had suggested a wide variety of chemical and physical properties to be traced into, within, and out of SWA: heat, fresh water, heavy metals, nutrients, and oxygen. In a meeting held on 13 November 1973, carbon was suggested as the "best central element in which to express the model", given its association with many biological processes and its close relationship with biomass and energy. This suggestion was greeted by general, but not unanimous, acceptance by those present at the meeting.

Thus the carbon flow model was underway, although there are suggestions in the minutes that the reason for making the carbon flow model were never fully accepted. For example, minutes of a meeting held on 10 December 1973 record discussion on whether the objective of the Project was to make "a working carbon model" or to "understand how the ecosystem functions", with the model as a useful tool for obtaining such knowledge. Minutes from a meeting held on 12 December 1973 record: "There was discussion – again – about the utility of the detailed model which had been prepared."

I do not think that principles underlying the structure and dynamics of Australian estuaries necessarily follow from the construction of a predictive model of carbon flow through SWA. Not that I know how to design a project to uncover "principles" (although Tomczak (1983) makes some useful suggestions towards this goal) but I think that the ambiguity inherent in that broad objective was to plague the Project throughout its life. From the viewpoint of applied science, however, the Project makes good sense. In the environmental area many ecosystems are being studied to understand the flow of various undesirable chemical species (often toxicants or stimulants). Success in such applied biogeochemical studies has been patchy in the large and polluted estuaries usually studied and it is of interest to know if a hand-picked, small estuary can be dynamically modelled to give detailed, reliable prediction.

To conclude: the broad objective of searching for principles and the specific aim of making a carbon flow model are not necessarily compatible, and if attempted will probably lead to at least an undercurrent of confusion and argument. It is interesting, however, that the Port Hacking Estuary Project shared this ambiguity with the US–IBP studies (Mitchell *et al.*, 1976), and was probably characteristic of thinking in the early 1970's. From the applied viewpoint, however, it would still be noteworthy if some group could make a demonstrably predictive dynamic model of some (however small and simple) natural ecosystem. (Although I refer to a "predictive dynamic model" as an absolute quantity, I wish to imply that proven predictive ability of current ecosystem models is very limited and well below the uses to which they are put. It would be noteworthy if some group could make a dynamic model of a natural ecosystem that could correctly predict the response of the ecosystem to the usual range of man-made disturbances.)

3.2.2 What prevented the construction of a predictive dynamic model of carbon flow through South West Arm? –

Even though not all participants were always convinced of the appropriateness of the Project, many were convinced and much effort was expended to that aim. Once again, I use the minutes and memoranda (the "records") of the Project, this time to see what factors impeded progress towards a predictive dynamic model of carbon flow.

The group was fortunate in having available an expert in multidisciplinary modelling (P.R. Benyon), both in terms of a recent paper on the subject (Benyon, 1972) and in discussions with him held during a meeting on 26 October 1973. An offer of collaboration in making the model was extended from the Systems Analysis and Simulation Group of the CSIRO Division of Computing Research but the offer was declined because it was felt the modelling expertise within the group was sufficient to the task. Accessable and competent statistical help was available from the beginning of the Project.

The first step (October 1973) the group undertook was to formulate a "conceptual model", being a verbal/diagrammatic model of how the physical, chemical, and biological features of estuaries (especially SWA) interact (see illustrations in Parker *et al.*, 1983). Various conceptual models were put forward and from these models a "composite carbon model" produced during the meeting on 13 November 1973. The composite model identified the state variables, or compartments, as inorganic carbon, reduced carbon, oxygen, detrital carbon, microalgae, macroalgae, seagrasses, mangroves, bacteria, fungi, protozoa, worms, crustaceans, molluscs, echinoderms, ascidians, and fish. The manner in which carbon flowed among the compartments was not discussed at this meeting, and there is a suggestion in the minutes that the Interim Co-ordinator drew up a diagram of the expected flows among the compartments.

In this meeting it was agreed that the model should deal with two spatial elements: the shallow water column and sediments that do not undergo de-oxygenation; and the deep water column and sediments that do. The Interim Co-ordinator requested (15 November 1973) that participants indicate on a flow diagram that had been produced those compartment masses and transfer rates that would change during de-oxygenation of the central basin of SWA. A model was thus evolving, with 17 compartments in each of two spatial elements and two substantially different types of system behaviour (dependent upon oxygen availability). The model first produced was extended by the 19 November 1973 meeting to 21 compartments, 137 flows between compartments (dependent on various factors, mainly state variables), and 39 source/sink terms. The faunal compartments differed from those first proposed in being based on functional categories (zooplankton, zoobenthos, and nekton) subdivided by size (macro, meso, meio/micro) and feeding types (carnivore, herbivore, detritivore; grazer, browser, filter feeder), as for example macro (filter feeding) zoobenthos. By 29 November 1973, a list of 117 transfer rates had been identified as potentially dependent on the oxygenated or de-oxygenated state of the lower water column and sediments of SWA.

A "top down" approach to modelling was to be adopted throughout: the model would suggest what was to be measured and if inadequate staff were available, the model would be simplified. Priorities for data gathering would be set as deemed appropriate to the current version of the model. It was decided at the meeting held on 29 November 1973 to form "working groups" to decide on the "courses of action to be taken to obtain information on each project". The following 16 working groups were formed: survey, compartment composition, compartment size, mass transport, circulation, stratification, gaseous diffusion, oxygen regime, nutrient regime, primary production and *in situ* light, respiration rates, transfer rates, detrital regime, secondary production, behaviour, and diversity regime. There were 24 scientists involved in the Project at that time.

It was not long (7 December 1973) before the participants realized that available effort was insufficient even to obtain measurements of compartment size and the model got its first simplification. The simplification was based on feasible experimental methodology, as described by Parker *et al.* (1983). The simplified model contained 13 compartments and required 51 transfer–rate determinations.

Attention was now turned to obtaining the data suggested by the model. One of the first observational programs set up was a survey planned to establish the main biotic features of SWA as a basis for making detailed studies of the selected features. Subsequent detailed studies centred around the measurement of compartment composition and compartment size, and later of "transfer rates" of carbon between compartments.

A recurrent theme raised during the continuing meetings was: how does one disaggregate an ecosystem into suitable experimental units (with implied studies), without going beyond reasonable manpower over an acceptable project life?

Numerous discussions were held on identifying and scheduling experimental techniques; on simplification of the model to allow fewer tasks, or on expansion to allow more attainable tasks; and on finding an acceptable organizational structure. An appropriate caution was put foreward in a meeting held on 10 December 1973; a participant "foresaw considerable difficulty in relating measurements to one another, in building the model, and that there should be a continuing attempt to compare and assemble these as the program progressed". Another participant added that "there was a moral that it is easier to assemble a few large 'pieces' than many small ones". As will be shown below, these cautionary suggestions were quite prophetic.

Some examples of the modelling and organizational decisions made during this period are given here. The feeding–type categorization for compartment definition was removed, apparently for lack of manpower. Working groups were reduced from 16 to 4: survey, compartment size, transfer rates, controlling factors. A further reduction in compartment numbers occurred on 6 February 1974; on the other hand, there was evidence of a strengthening of the concepts of a water–column and sediment sub–systems. New working groups were again (25 January 1974) formed, each with a Convener, to "produce the numerical values for each transfer rate and each compartment size associated with his working group's sub–system". Working groups were: water column, microbenthos, macrofauna, and macrophytes. (These working groups lasted throughout the rest of the Project.)

It is significant that these perceived tasks are typical of a static (budget) model, not a dynamic model. One early record (12 December 1973) makes it clear that, at least initially, modelling direction was towards a dynamic model of carbon flow: not only were relevant flows identified but so were their dependencies (usually being state variable dependencies but some forcing variable dependencies as well). Yet in much subsequent discussion the distinction was often clouded; for example, the expressions "carbon model" and "carbon budget" were used thoughout the Project with no apparent difference in meaning. Why a stronger distinction was not made between the two types of models is not clear but I would guess that three factors were involved: i) some people did not know the difference between static and dynamic models; ii) the intention to obtain from SWA *all* data required by the model in a relatively short time period suggested a rather primitive data set; and iii) the modelling expert at the time was experienced mainly in (biomedical) compartmental modelling (Berman & Schoenfeld, 1956; Jacquez, 1972) and was probably used to thinking in terms of models with linear transfer rates, dependent only on the donor state variable. The parameter values for such functions can be estimated from a knowledge of average compartment size and average transfer rates.

It is relevant that the modelling expert at the time wrote on 27 March 1974 that the project "will involve constructing a hybrid carbon/oxygen model of the estuarine system and writing a program to permit the simulation of the model on the CYBER 76 Network". A discussion of the model in the Project's progress report (CSIRO, 1976) confirms continuing thought in the direction of dynamic modelling. Since no recorded discussion took place on the appropriate model structure, it is likely that the modelling expert's experience in linear, donor–dependent models explains the type of data required by the model. It also shows that the Project had a strong, if unstated, impetus towards a specific type of predefined model structure.

A significant turn of events occurred around March 1974, when the first reference was made (in Working Group notes) to use of the literature in obtaining the required average compartment sizes and average transfer rates. A complete set of literature values were requested (by the Interim Co–ordinator) from the Working Groups by 22 April 1974 and many were, in fact, obtained from the literature.

The permanent Co–ordinator (R.R. Parker) arrived on 24 April 1974 and D.F. Smith relinquished his role as Interim Co–ordinator but retained the role of modelling expert. The remainder of 1974 saw data being collected both from SWA and from the literature. The arrival of a new participant, W.J. Wiebe, saw another version (4 February 1975) of the model, this time with 11 compartments, 33 transfer rates, and 12 source/sink terms. (This model is similar to that illustrated in Fig. 3 of Parker *et al.* (1983); both contain the same compartments but they have slightly differing flows.) A memorandum regarding the new model notes: "This model has been constructed with the view of defining the compartments and fluxes in such a manner that they can all be measured by techniques and with the manpower that we currently have available. This has given us a model for which field data acquisition can begin immediately". (It must be pointed out, however, that the model was not much different from the previous version, in that both models only required compartment sizes and average transfer rates.) Methods were suggested for obtaining the compartment masses, the mass transfer rates, the allochthonous inputs, and the intercompartmental transfer rates; in only one case was recourse to the literature suggested. (It is relevant to note that this version of the model was produced 1 year and 4 months after the Project began: the expected life of the Project was to be 1 – 2 years!)

At this stage one would have expected to see the data that had been collected being used in obtaining parameter values for the model. Minutes of a meeting held on 20 May 1975 show that a substantial amount of data had been collected. There was a general feeling expressed that the currently available data suggested the desirability of more meetings to allow "prolonged and involved discussion" of those specific projects which were well advanced. Programming assistance was requested (12 February 1975) and given. And, finally, mention was made of the construction – rather than the conceptualization – of a model: "programming for SWAMP (South West Arm Modelling Program) is nearly complete".

A Modelling Workshop, originally proposed during a March 1975 meeting, was held on 18 – 19 August 1975. Various speakers, mainly mathematical statisticians from around Australia, talked about models they had constructed. Yet another model for SWA was suggested, this one having only 5 compartments and 7 intercompartmental transfer rates. The ecosystem was visualized in terms of CO_2, autotrophs, heterotrophs, and a Dissolved Organic Carbon (DOC) – Particulate Organic Carbon (POC) compartment. Sunlight, nutrients, temperature, and salinity were assumed to affect the primary production rate of autotrophs but all other fluxes were state dependencies only. There appeared to be no awareness on the part of the invited speakers that SWA has both sediments and water column; that there are exchanges with the ocean and input from catchment; that it would be very difficult indeed to deal experimentally with a "heterotroph"; and that it is quite inappropriate to include DOC and POC in one compartment. The Workshop was apparently intended to reassure the participants about the usefulness of modelling but there is no evidence that it had any positive effect.

Time passed without the appearance of even a first model incorporating parameter values derived from the data that had been collected. A memorandum of 12 September 1975 then noted that R.E. Sinclair of the Systems Analysis and Simulation Group of the CSIRO Division of Computing Research (P.R. Benyon's group – see reference above) was now "working with D.F. Smith to complete and test the model". A request was sent from the Co-ordinator to the participants: pass on "compartment sizes and rates of flux. Please help him in any way you can and when he asks for some values you can give him data from this project, literature values, or your best guess. The objective now is to get the model working. When it is working you will see the need, or lack thereof, for futher sophistication or precision".

On February 1976 a person was assigned to begin "data collation on phytoplankton respiration for the carbon model" and after that to get data on the chlorophyll : carbon conversion ratio. In their initial efforts, R.E. Sinclair and D.F.

Smith wrote an ACSL (Mitchell and Gauthier, Assoc., 1975) computer program, based on the hypothesized Smith–Wiebe model of 4 February 1975 and using the functions proposed by D.F. Smith in a "day long course in modelling techniques" (CSIRO, 1976). R.E. Sinclair gave a talk on 21 April 1976 to explain this model and to suggest the need for parameter values for SWA.

The data values were not forthcoming. First, we had the situation where the data were apparently available and awaiting the writing of a computer program to use for simulations of the model; now we had the situation where there was a computer program but there were no data forthcoming. This contradiction is probably explained in terms of frustration of the participants at this time. I think there were two important contributing factors. One is that the model had been *talked* about for 2 years (the expected maximum life of the Project) and data had been collected for some time but without incorporation in the "model". And, in fact, new versions of the model kept appearing at regular intervals, quite independent of the data that had been collected. Patience was simply wearing thin. The other factor is of more conceptual interest: the envisaged model required a primitive data set (average compartments and average transfer rates). A number of participants felt that a model based on such limited experimental understanding was not going to give reliable predictions and collection of data for such a model was a waste of their time. This feeling strengthened as the difficulty of obtaining many of the transfer rates became clear and as the likelihood of the existence of high natural variation on these rates became apparent. (It is interesting that the envisaged model was eventually built and discarded because it didn't lead anywhere; see review in Sinclair et al., 1983.)

The records of the Ecosystem Group (see Parker et al., 1983) show periodic mention of the model and its data needs but the impetus for model–guided activity seemed to have slowed and people were simply finishing the projects they had begun. (Minutes of meetings of the "Estuarine Programme" – "Port Hacking Project" had stopped with a meeting held on 20 May 1975.)

Thus although much experimental and modelling experience was now available, the will to proceed of the group as a whole was lacking. The remaining records suggest a realization that the Project was not going to be successful, and a response to this realization. The remaining period was characterized by personal animosities; by a suggestion that the study of carbon flow in Port Hacking was not the way to uncover principles of the structure and dynamics of estuaries and perhaps an extensive, not an intensive, program would be more useful; and by a comment to the effect that the model had not proved to be useful, as first hoped, in defining reasonable experimental questions to be asked.

I came into the Project on 24 November 1976 to see what R.E. Sinclair and I could do together with this Project. (D.F. Smith had already withdrawn from modelling considerations.) We went through the records compiled by the Project and talked with individual participants and got a set of data values. We put these into the computer program already constructed and talked about the model at a meeting held on 24 February 1977. But we could offer no suggestions as to how to make such a model produce reliable predictions in the short term. This pessimistic view, combined with a general tiredness of talking about models, suggested that there was no value in trying to proceed with the project as had been conceived and so we turned our attention to *synthesizing what was known about SWA*. No further model–guided studies of SWA were conducted. The subsequent modelling effort, divorced from the experimental efforts, is explained in Sinclair *et al.* (1983).

It is interesting that in this meeting another "conceptual model" was produced. Also it was agreed that the general goal of the Ecosystem Group was to "understand both the energy flow in ecosystems and the response of species and their diversity, in relation to natural and man–made environmental change". Both steps have their parallels in 1973 work; the group had gone full cycle.

The Project was formally terminated in October 1977.

To conclude: there are apparent in this historical summary certain features that re–appear and were probably responsible for the troubles experienced by the Project. The main problem, I think, stemmed from the Group's original idea that it would obtain all the data required for the ecosystem model. This demanded very simple functions for each of the many flows, in turn demanded by the need to give each compartment experimental validity. But, although these features made it conceptually possible to obtain all required parameter values, it also suggested that the resultant model would be so simple as to be unrealistic – and not worth the (at times considerable) effort involved in obtaining the various average transfer rates. This is leaving aside the problem that some participants envisaged of writing a scientific paper on the basis of one number, however much trouble was involved in getting that number. A second problem was that while data for the model arrived progressively over the duration of the project, the model did not develop in concert with the data. This ambiguity gave the impression of data collection to no end. The attractive idea of evaluating current understanding by using the model was thus never realized. A third problem revolved about the difficulty of visualizing an ecosystem, as reflected in the continually changing model. This is a problem shared by all ecosystem workers, and an adequate definition has still to arrive.

3.3 Evaluation

The 1973–1978 project cost about 2 million unadjusted Australian dollars (Table 1).

What are the benefits? A simple answer is about 40 articles – or a relatively cheap $Aust. 50 000 per article. Since the goal of building a predictive dynamic model and formulating principles was not reached, a more satisfactory answer to this question must require a detailed look at the studies that were conducted.

Table 1.
Expenditure on the Port Hacking Estuary Project.
(Direct costs only).

Year	Expenditure ($ Aust.)		
1973/74	Salaries	89 000	
	Operating	71 000	
	TOTALS	160 000	160 000
1974/75	Salaries	177 000	
	Operating	177 000	
	TOTALS	354 000	514 000
1975/76	Salaries	238 000	
	Operating	120 000	
	TOTALS	358 000	872 000
1976/77	Salaries	353 000	
	Operating	232 000	
	TOTALS	585 000	1 457 000
1977/78	Salaries	356 000	
	Operating	188 000	
	TOTALS	544 000	2 001 000

The lack of a single generally accepted functional form resulted in a variety of functions being identified. Wiebe & Smith (1977) pursued the linear compartmental modelling formalism advanced by D.F. Smith. Griffiths & Caperon (1979) envisaged a non–linear grazing rate of zooplankton but, for experimental convenience, narrowed the domain of the model to get a linear, donor–dependent function. Tranter & Kennedy (1983) used the non–linear function typically used in the literature. Scott (1978a, 1978b, 1979a, 1979b) ignored mathematical modelling and collected data and drew conclusions in the established way for phytoplankton research. These data were subsequently used to develop an empirical model (Scott, 1983) and a non–linear, rational model (Cuff, 1983).

In those areas where there is not a tradition of dynamic modelling the participants were amenable to the model's requirement for data typical of a static model and produced as much budgetary information as time and techniques allowed: i.e. a complete seagrass carbon budget (Kirkman & Reid, 1979), and a reliable estimate of the annual average primary production rate of benthic micro–organisms (Giles, 1983).

In some fields, techniques for obtaining average transfer rates are not available and participants in these specialities had to devote much of their effort to the development of techniques (Bulleid, 1977, 1978).

Lastly, in some fields it was difficult simply to define a useful compartment. Griffiths (1983) addressed this problem for zooplankton. Colquhoun–Kerr (1977) found the problem of aggregating benthic species too difficult and simply chose two prominent intertidal benthic species; he left unanswered the question of the relevance of the estimated transfer rates to "benthic heterotrophs" (Table 2 of CSIRO, 1976).

The most reliable and comprehensive information that was obtained was in those specialities most well developed: e.g. hydrology (Godfrey & Parslow, 1976; Godfrey, 1983) and primary production of phytoplankton (Scott, 1978a, 1978b, 1979a, 1979b, 1983). Budgetary studies were generally well and comprehensively developed whereas dynamic information tended to be incompletely developed: compare the carbon budget of seagrasses (Kirkman & Reid, 1979) to a discussion of our dynamic information (Cuff, 1983).

On a broader scale, this book analyses the Port Hacking Estuary Project as a member of the genre and presents a new idea relating to synthesis of incomplete sets of static and dynamic information. Projects of this genre being set up now will have scientific and organizational problems and the experiences described in this book should help to anticipate and avoid some.

Thus it is clear that many results, at a variety of levels, have come from the Port Hacking Estuary Project. All studies were aimed at the original goal, with the diversity of approaches related to the stage of development of the fields of study. Only time will tell if together the studies were worth the cost of the Project. If our experiences help to clarify thoughts about the scientific and organizational directions needed for successful programs, then I think the money will have been well spent.

To finish this evaluation, it may be interesting to consider if significant progress would have been made if the Project had been funded for another year, at an additional cost of about $Aust. 500 000 ? This question is, of course, impossible to answer with complete reliability but it is useful once again to consider the published papers resulting from the Project. The only experimental study of water circulation was reported in 1976 (Godfrey & Parslow, 1976). Monitoring of physical and chemical parameters was begun in 1974 (Vaudrey et al., 1983). The extensive studies on phytoplankton (Scott, 1978a, 1978b, 1979a, 1979b) were all based on observations conducted mainly in 1975 and those on seagrasses (Kirkman & Reid, 1979) on observations conducted mainly in 1974 and 1975. Experiments on zooplankton grazing (Griffiths & Caperon, 1979) and work on microheterotrophs (Wiebe & Smith, 1977) was also conducted in 1975. These examples are sufficient to show that it took about a year to gear up from the initial discussions and then a few years to collate and analyse the data and write up. In a multidisciplinary study an obvious next stage would be a thinking out of progress with the next round of observation and experiment in mind. As shown in Section 3.2.2 above this did not happen as part of the expected development of the Project. Thus there seems little evidence to suggest substantial progress would have been achieved for the next $Aust. 500 000. The termination of the observational and experimental aspects of the Project allowed the rather considerable, and diverse, data set to be co–ordinated and synthesized. This synthesis was reported by Cuff et al. (1983) and Cuff (1983) and shows that a lot of useful information exists about SWA and thus about similar marine embayments, thus going part way to the original broad objective. The termination has also allowed us to see what impeded progress in the Project, and to put forward suggestions on how to deal with some of the problems we faced, and which others will presumably encounter.

4. WHICH CHARACTERISTICS GENERALIZE?

One of the more unexpected characteristics of the Port Hacking Estuary Project was the long interval between the beginning of data availability (early 1975) and the use of that data in model construction. The records do not explain why this was so; in early 1977 we used the SWA and literature data collected by the participants during this period to construct the intended model. To do this, however, we had to use balancing considerations to estimate some missing parameter values. And I think that this step illustrates the reason why the model was not built earlier: there is a general view (implicitly expressed) in the modelling literature that all data must be available before model construction begins. The difficulty here is that it is very difficult to keep a model–guided multidisciplinary study together for the required time (usually years) until all required data come in and modelling begins. Probably the project will break itself apart before all the data is collected.

When we realized that modelling had to begin but we did not have all required data we went to the literature to get an idea of what to do. A brief look at the literature suggested that we alone had difficulty in obtaining the required parameter estimates. It seemed that we may have erred in opting for linear, donor–dependent functions, thereby promoting disbelief and discontent among the participants; the published models generally contained complex non–linear functions, presumably with much more mechanism. This may have given the experimental participants of these studies more freedom to explore the processes operating in their areas of expertise, and not forced them to be low–level "data gatherers" for the model.

If this assumption is true then one should be able to trace in the published models the experimental origins and deduce the reliability of these estimates. However, in too many cases the parameter values are explained by publications in obscure journals, in unpublished reports, or simply as unpublished data or personal communication. When it is possible to check the reference source, one finds too often that the datum is for a different species or a different locality (and not uncommonly a different continent). Such a data source may make sense if one is constructing a theory but from the viewpoint of applied science, the subject of this paper, it does not make sense unless it can be demonstrated that the different species and different localities are relevant to the ecosystem of specific interest. And most of the published models *do* refer to specific ecosystems, usually mentioned in the titles of the papers; often the text notes that the purpose of the model is to "synthesize" the data available for the named ecosystem.

The tendency of the published models to opt for a complete model – incorporating components of unknown unreliability – is most apparent with the static (budget) models. A characteristic of budgets is that fairly secure predictions can be made in the face of missing data; in dynamic models omitted feedback loops can seriously affect the model's predictions. Hence there should be less incentive to make "guesstimates" in static models. Yet it is not at all difficult to find budgets based on what appear to be unreliable estimates of compartment sizes and transfer rates. In fact, the only two budgets I know of that actually admit to missing information are by Woodmansee *et al.* (1978) and Valiela & Teal (1979a). In fact, Valiela & Teal (1979b), by concentrating only on input and ouput rates of nitrogen to the Great Sippewissett Marsh, managed to get all the required data. They presented "the first complete nitrogen budget for a marine ecosystem where each of the various inputs and outputs from a salt marsh has been accounted for".

A recent review of dynamic ecosystem modelling (Swartzman, 1979) confirms suspicions that only a very small fraction of data incorporated is reliable information from the ecosystem of specific interest. Much information simply has to be guessed. Hence Margalef (1973) seems to have been correct in his stated belief: "It does not pay to build a complicated model for its own sake, and, in order to implement it, go through a wild chase after data that are often inexistent and even unobtainable and that, after all, may turn [out] to be irrelevant. Perhaps this expression reflects the frustration of the ecologist, unable to provide data to fit the holes of almost any model. Anyways many an ecologist believes that a better prediction can be obtained by rapid intellectual appraisal and elaboration of a complex situation, than after a tiresome computation based on unreliable measurements and on constants that may be only 'uneducated guesses'. "

So it seems that our Project was not alone in having difficulty getting the data required by the model. In our perusal of the literature, we noted the direction taken by the published models (described well in Kremer & Nixon, 1978), but could not see the relevance, at this stage, to applied science. Articles such as that by Simons & Lam (1980) point to some serious limitations. Some recent progress has been made in this type of modelling, as *e.g.* Fedra (1980); Hornberger & Spear (1980); Spear & Hornberger (1980), but the work is too new to see the implications of these advances.

In moving in another direction, we started from the bias that it is desirable to synthesize that data from South West Arm, not making recourse to the literature or "guesstimates" for missing data as had been done so often before (and by us in our early modelling efforts; see Sinclair *et al.*, 1983). Rather than aiming for completeness we aimed for reliability of components. We also believed that *any model–guided multidisciplinary study will fail (for very human reasons) unless synthesis can be an ongoing process, beginning as soon as the first data are collected and moving in a visible manner along with the data collection.* The results of our approach to synthesis are illustrated by Cuff *et al.* (1983) and Cuff (1983). As the project continues the synthesized information can modify the conceptual model and allow both deductive and inductive processes to operate. This approach has its parallel in the development of computer software, where it is known as "top down design, bottom up implementation".

Our experience also suggests that it might be worth considering the construction of a budget before attempting a dynamic model. I suspect that if the participants of our Project had been aware that they were simply building a budget there would have been much less argument, although it is open to debate as to whether a budget of carbon in SWA would uncover general principles. The advantage of first constructing a budget is that it should help to clarify the main features of the ecosystem (*viz.* compartment definitions that are internally homogeneous and measurable; the flows that exist, their magnitude, and their variability; inputs to and outputs from the ecosystem; transient and steady state behaviours of the ecosystem). Such preliminaries can be invaluable to the more difficult and long–term studies involved in making a dynamic model.

As far as I can see these two suggestions (and particularly the former), carried out systematically within a particular model–guided study, should eventually show to what degree it is possible to make an ecosystem model of a natural ecosystem. The information contained in this book suggests that the job will be long and expensive. Nevertheless, given the problems experienced by the Port Hacking Estuary Project, it is surprising to be able to end this evaluation on an optimistic note.

ACKNOWLEDGEMENTS

I thank A.C. Heron, R.R. Parker, and R.E. Sinclair for supplying me with the minutes and memoranda that enabled this evaluation. I also thank M. Tomczak jr., R.B. Humphries, and J.J. Wilkes for helpful discussions.

REFERENCES

Allen, K.R.: Introduction to the Port Hacking Estuary Project. In: W.R. Cuff and M. Tomczak jr, eds *Synthesis and Modelling of Intermittent Estuaries* Berlin, Heidelberg, New York: Springer (1983)

Benyon, P.R.: Computer modelling and interdisciplinary teams. *Search* 3, 250–256 (1972)

Berman, M., Schoenfeld, R.: Invariants in experimental data on linear kinetics and the formulation of models. *Journal of Applied Physics* 27, 1361–1370 (1956)

Bulleid, N.C.: Adenosine triphosphate analysis in marine ecology: a review and manual. *CSIRO Division of Fisheries and Oceanography Report* 75 (1977)

Bulleid, N.C.: An improved method for the extraction of adenosine triphosphate from marine sediment and seawater. *Limnology and Oceanography* 23, 174–178 (1978)

Colquhoun–Kerr, J.S.: Carbon flux through the South West Arm populations of *Crassostrea commercialis* and *Trichomya hirsuta*. *CSIRO Division of Fisheries and Oceanography Report* 79 (1977)

CSIRO: *Estuarine Project Progress Report 1974–1976*. Sydney: CSIRO Division of Fisheries and Oceanography (1976)

Cuff, W.R., Sinclair, R.E., Parker, R.R., Tranter, D.J., Bulleid, N.C., Giles, M.S., Godfrey, J.S., Griffiths, F.B., Higgins, H.W., Kirkman, H., Rainer, S.F., Scott, B.D.: A carbon budget for South West Arm, Port Hacking. In: W.R. Cuff and M. Tomczak jr, eds *Synthesis and Modelling of Intermittent Estuaries*. Berlin, Heidelberg, New York: Springer (1983)

Cuff, W.R.: An evaluation of the dynamic information for South West Arm, Port Hacking. In: W.R. Cuff and M. Tomczak jr, eds *Synthesis and Modelling of Intermittent Estuaries*. Berlin, Heidelberg, New York: Springer (1983)

Fedra, K.: Mathematical modelling – a management tool for aquatic ecosystems? *Helgoländer Wissenschaftliche Meeresuntersuchungen* 34, 221–235 (1980)

Giles, M.S.: Primary production of benthic micro–organisms in South West Arm, Port Hacking, New South Wales. In: W.R. Cuff and M. Tomczak jr, eds *Synthesis and Modelling of Intermittent Estuaries.* Berlin, Heidelberg, New York: Springer (1983)

Godfrey, J.S.: Tidal flushing and vertical diffusion in South West Arm, Port Hacking. In: W.R. Cuff and M. Tomczak jr, eds *Synthesis and Modelling of Intermittent Estuaries.* Berlin, Heidelberg, New York: Springer (1983)

Godfrey, J.S., Parslow, J.: Description and preliminary theory of circulation in Port Hacking estuary. *CSIRO Division of Fisheries and Oceanography Report* 67 (1976)

Griffiths, F.B.: Zooplankton community structure and succession in South West Arm, Port Hacking. In: W.R. Cuff and M. Tomczak jr, eds *Synthesis and Modelling of Intermittent Estuaries.* Berlin, Heidelberg, New York: Springer (1983)

Griffiths, F.B., Caperon, J.: Description and use of an improved method for determining estuarine zooplankton grazing rates on phytoplankton. *Marine Biology (Berlin)* 54, 301–309 (1979)

Hornberger, G.M., Spear, R.C.: Eutrophication in Peel Inlet – I. The problem – defining behavior and a mathematical model for the phosphorus scenario. *Water Research* 14, 29–42 (1980)

Innis, G.S. (Ed): *Grassland Simulation Model.* Ecological Studies, 26. New York: Springer (1978)

Jacquez, J.A.: *Compartmental Analysis in Biology and Medicine: Kinetics of Distribution of Tracer–labeled Materials.* Amsterdam: Elsevier (1972)

Kirkman, H., Reid, D.D.: A study of the role of a seagrass *Posidonia australis* in the carbon budget of an estuary. *Aquatic Botany* 7, 173–183 (1979)

Kremer, J.N., Nixon, S.W.: *A Coastal Marine Ecosystem, Simulation and Analysis.* Berlin: Springer (1978)

Longhurst, A.R.: Ecological models in estuarine management. *Ocean Management* 4, 287–302 (1978)

Margalef, R.: Some critical remarks on the usual approaches to ecological modelling. *Investigación Pesquera* 37, 621–640 (1973)

Mitchell and Gauthier, Assoc.: *Advanced Continuous Simulation Language (ACSL). User Guide/Reference Manual. 2nd ed.* Concord, Mass.: Mitchell and Gauthier, Assoc., Inc. (1975)

Mitchell, R., Mayer, R.A., Downhower, J.: An evaluation of three biome programs. *Science* 192, 859–865 (1976)

Park, R.A., O'Neill, R.V., Bloomfield, J.A., and 22 others: A generalized model for simulating lake ecosystems. *Simulation* 23, 33–56 (1974)

Parker, R.R., Rochford, D.J., Tranter, D.J.: History and organization of the Port Hacking Estuary Project. In: W.R. Cuff and M. Tomczak jr, eds *Synthesis and Modelling of Intermittent Estuaries.* Berlin, Heidelberg, New York: Springer (1983)

Scavia, D.: An ecological model of Lake Ontario. *Ecological Modelling* 8, 49–78 (1980)

Scott, B.D.: Phytoplankton distribution and light attenuation in Port Hacking estuary. *Australian Journal of Marine and Freshwater Research* 29, 31–44 (1978a)

Scott, B.D.: Nutrient cycling and primary production in Port Hacking, New South Wales. *Australian Journal of Marine and Freshwater Research* 29, 803–815 (1978b)

Scott, B.D.: Seasonal variations of phytoplankton production in an estuary in relation to coastal water movements. *Australian Journal of Marine and Freshwater Research* 30, 449–461 (1979a)

Scott, B.D.: *The Relation of Irradiance to Photosynthesis in Marine Phytoplankton.* M.Sc. Thesis, University of New South Wales (1979b)

Scott, B.D.: Phytoplankton distribution and production in Port Hacking estuary, and an empirical model for estimating daily primary production. In: W.R. Cuff and M. Tomczak jr, eds *Synthesis and Modelling of Intermittent Estuaries.* Berlin, Heidelberg, New York: Springer (1983)

Simons, T.J., Lam, D.C.L.: Some limitations of water quality models for large lakes: a case study of Lake Ontario. *Water Resources Research* 16, 105–116 (1980)

Sinclair, R.E., Cuff, W.R., Parker, R.R.: Ecosystem modelling of South West Arm, Port Hacking. In: W.R. Cuff and M. Tomczak jr, eds *Synthesis and Modelling of Intermittent Estuaries*. Berlin, Heidelberg, New York: Springer (1983)

Spear, R.C., Hornberger, G.M.: Eutrophication in Peel Inlet II. Identification of critical uncertainties via generalized sensitivity analysis. *Water Research* **14**, 43–49 (1980)

Swartzman, G.L.: Simulation modelling of material and energy flow through an ecosystem: methods and documentation. *Ecological Modelling* **7**, 55–81 (1979)

Tomczak, M., jr.: Some conclusions from the Port Hacking Estuary Project. In: W.R. Cuff and M. Tomczak jr, eds *Synthesis and Modelling of Intermittent Estuaries*. Berlin, Heidelberg, New York: Springer (1983)

Tranter, D.J., Kennedy, G.: Size–specific respiration rate of Port Hacking zooplankton. In: W.R. Cuff and M. Tomczak jr, eds *Synthesis and Modelling of Intermittent Estuaries*. Berlin, Heidelberg, New York: Springer (1983)

Valiela, I., Teal, J.M.: Inputs, outputs and interconversions of nitrogen in a salt marsh ecosystem. In: R.L. Jefferies and A.J. Davy, eds. *Ecological Processes in Coastal Environments*. Oxford: Blackwell (1979a)

Valiela, I., Teal, J.M.: The nitrogen budget of a salt marsh ecosystem. *Nature (London)* **280**, 652–656 (1979b)

Vaudrey, D.J., Griffiths, F.B., Sinclair, R.E.: Data base for the Port Hacking Estuary Project: Parameters, monitoring procedure, and management system. In: W.R. Cuff and M. Tomczak jr, eds *Synthesis and Modelling of Intermittent Estuaries*. Berlin, Heidelberg, New York: Springer (1983)

Wiebe, W.J., Smith, D.F.: Direct measurement of dissolved organic carbon release by phytoplankton and incorporation by microheterotrophs. *Marine Biology (Berlin)* **42**, 213–223 (1977)

Wiegert, R.G., Christian, R.R., Gallagher, J.H., Hall, J.R., Jones, R.D.H., Wetzel, R.L.: A preliminary ecosystem model of coastal Georgia *Spartina* marsh. In: L.E. Cronin, ed. *Estuarine Research Volume I, Chemistry, Biology, and the Estuarine System*. New York: Academic Press (1975)

Woodmansee, R.G., Dodd, J.L., Bowman, R.A., Clark, F.E., Dickinson, C.E.: Nitrogen budget of a shortgrass prairie ecosystem. *Oecologia (Berlin)* **34**, 363–376 (1978)

Synthesis and Modelling of Intermittent Estuaries
(W.R. Cuff and M. Tomczak jr. eds) Berlin, Heidelberg,
New York: Springer (1983), pp. 293–302.

Some Conclusions from the Port Hacking Estuary Project

Matthias Tomczak jr.

Division of Oceanography
CSIRO Marine Laboratories
P.O. Box 21, Cronulla, N.S.W. 2230, Australia

Summary. The Port Hacking experiment, an interdisciplinary five year study of an east Australian estuary guided by a model of carbon flow, is reviewed as an application of systems analysis to marine ecology. It is argued that the experiment was of the basic research type, irrespective of statements by participants at the start of the project. It is observed that the numerical model suffered from insufficient input data, a situation which is shown to be common with models of the basic research type. A discussion of some general characteristic features of research programs with strong mathematical input leads to a recommendation to weaken the link between numerical model and field program and let both disciplines develop along their own lines, with weak interaction through a joint project.

Key words: ecosystems, mathematical analysis, field programs, planning, Port Hacking, South West Arm

1. INTRODUCTION

Management and planning decisions rely on the ability to forecast. The observed deterioration of estuarine and coastal marine ecosystems made management of estuaries an urgent task and resulted in efforts to model these ecosystems as a basis for reliable forecasts. Modelling can be done in a variety of ways, but for quantitative predictions the choice is limited to a few mathematical techniques and to their application in systems analysis. A decade ago, systems analysis was seen as "the wave of the future" (Odum, 1971) by ecologists; and scientists, engineers, and administrators alike expressed the belief that it held the key to mutual understanding between the researcher eager to be "helping to make the world a better place" (Clymer, 1972) and the administrator whose "managerial data requirements are far less demanding than those required for the full scientific understanding" (Ketchum, 1972).

In the following years some ecosystem models for estuarine and coastal marine areas were developed (di Toro *et al.*, 1977; Legovic *et al.*, 1977; Kremer & Nixon, 1978). Although they relied to a large extent on available information, such modelling projects initiated a number of interdisciplinary field programs for the collection of missing input data. The underlying concept of using systems analysis as a guide for data aquisition and evaluation has also been suggested for the open ocean (Walsh, 1972), but the space and time scales involved in its implementation pose severe problems.

Space and time scales in model–guided estuarine and coastal projects were generally considered managable, yet success of estuarine and coastal ecosystem models was rather limited, causing disappointment and, in more instance than one, bitter frustration. Rather than help to understand existing data and throw some light on the behaviour of the ecosystem, the models tended to call for more and more experimental information before they could be put to predictive use. Sometimes, the resulting field experiments could be described as "a wild chase after data that are often inexistent and even unobtainable and that, after all, may turn [out] to be irrelevant" (Margalef, 1973).

Today, the process of conceptualization and implementation of ecological simulation models is well documented (Swartzman, 1979), but at least for marine ecosystems it seems fair to state that a model which adequately describes an ecosystem and is used for its day–to–day management does not exist. Not surprisingly, managing authorities have doubts about the effectiveness of interdisciplinary ecosystem modelling activities, and ecosystem modelling has fallen back in the race for public funding (it has never had a chance with private funding). Healthy and logical as the concerns may be, they can only be judged from a critical evaluation of a marine ecosystem study, which seems to be still missing. The CSIRO study of South West Arm (SWA) in the Port Hacking estuary offers an opportunity for such an evaluation from which some conclusions for the direction of future work could be drawn.

2. EVALUATION OF THE PORT HACKING ESTUARY PROJECT

Port Hacking is a small sub–tropical estuary on the east coast of Australia about 30 km south of Sydney city centre. It is next to the laboratory of the (then) Division of Fisheries and Oceanography of the Commonwealth Scientific and Industrial Research Organization (CSIRO) which during 1973–1978 performed an interdisciplinary program centred around a model of the flow of carbon through SWA of Port Hacking (Parker *et al.*, 1983.) The watershed of SWA, and most of the watershed of Port Hacking, lies entirely within a National Park; no industrial activity affects the estuary, and low–density urbanization extends only along the northern side. All this makes Port Hacking a relatively unpolluted estuary, despite its proximity to a city of several million people.

Although there is evidence in CSIRO progress reports and other unpublished material that the carbon model was seen by some participants as a convenient way to structure and organize a *descriptive* investigation of the estuary, the main reasoning behind adopting a systems analysis approach was the need for reliable prediction: "Questions which might be asked are, for example, 'How will the estuarine ecosystem respond if a major portion of the freshwater runoff is diverted to another basin?' or 'What are the likely consequences to the fish stocks of dredging a channel through an entrance sill?' or 'How will estuarine dissolved oxygen levels respond to the waste discharge of a proposed industrial installation?' " (CSIRO, 1976). Such questions are not particularly relevant to the management of a watershed in a National Park, which indicates that a basic intention of the Project was the provision of knowledge for predictions not in SWA but in estuaries similar to Port Hacking.

Cuff (1983) presents an evaluation of the Project which includes a history of the thoughts and intentions of its participants as expressed in the minutes and notes of the Project meetings. He supplies an interesting description of the interactions between individuals and their particular expertise in a multi–disciplinary program and draws some useful conclusions for future work. On a more general level, when attempting to evaluate successes and failures of the Port Hacking Estuary Project, it may be useful not to adopt the original aims of the participants as a guide to judgement. Their challenging questions on estuaries in general were probably formulated as bait for the provision of funds. This does not deny them the quality of genuine scientific curiosity, but it implies that the inadequacy of the Project to supply an answer to these questions was known and accepted by the scientists when the Project started. A predictive model for SWA only, on the other hand, is a goal too poor to justify a five–year interdisciplinary study; as mentioned before, SWA is not in danger of being influenced by harmful human activity to a significant extent, and consequently is not in need of prediction. It appears that a reasonable measure for an evaluation of the study lies somewhere in between: the model should guide the program towards improved understanding of some general mechanisms which are at work in Port Hacking and most likely in other estuaries as well.

The hydrographic regime of SWA is of marine type over most of the year, while irregular freshwater run–off following severe rainstorms controls stratification during short intervals of time which add up to 25% of the year. This type of estuary is common in the tropics and sub–tropics but does not seem to have attracted the attention of modellers. Sinclair *et al.* (1983) recall the various modelling efforts for SWA as an example of this type of estuary. The situation which emerges from their description is: data collection, although guided by systems analysis, was insufficient for the purpose of the model; on the other hand, it was excessive for a *qualitative* description which from the point of view of regional management and estuary classification in relation to other estuaries would have been sufficient. In fact, this situation became clear during the end phase of the study, which may have contributed to the disillusionment felt by some participants at that stage. Overall, despite a number of new findings and resulting publications, the final account seems very much in the balance.

When trying to draw lessons from the experiment, the obvious question is: more, or less? Is the proper avenue to follow the road of more observational data for filling the gaps of the model, or the one towards a qualitative description with less data and a model with less demand for quantitative input (for example a word model rather than a mathematical model, still based on systems analysis)? The Port Hacking experiment can provide an answer if its character as basic research is recognized. Two observations emerge. To begin with, the experiment (like every program of basic research) started from an idealized picture of the problem, in this case an estuarine ecosystem. Observation components were designed accordingly and had to be refined in the course of the study. As a result, early parts of observational time series were of less value than later parts, and continuation of the study beyond the anticipated date became likely to improve the data set. A demand for program extension usually results in deterioration of science/administration relations, something the Project could not afford for reasons described below. When the Project terminated in 1978 and participating scientists were assigned new priorities, there was a general feeling that closure of the Project was premature and another year's work could significantly improve results (Sinclair *et al.*, 1983; note however, that Cuff (1983) does not share this opinion). In fact, some of the contributions in this volume (Godfrey, 1983;

Bulleid, 1983) include data collected, on the authors' own initiative, after the Project was formally terminated.

The second observation is that progress in basic research is influenced by decisions outside the area of science and that scientific success or failure is not determined by scientific decisions alone. This is particularly obvious in the Australian context in which the Port Hacking Estuary Project operated. Much of the work done by CSIRO is geared towards support for the needs of the export industry and expected to affect industrial profits positively within short time spans. It is true that industry also has an interest in basic research which does not bring immediate benefit, but Australia's industry is controlled and largely owned by foreign companies which prefer basic research to be persued by organizations of their national governments. This state of affairs is supported by the policy of the Australian Government in general and reflected in its science policy. As a result, basic research in Australia has had, and is having, a difficult time.

In oceanography and marine biology most of the research done is basic research. The Division of Fisheries and Oceanography tried its best to contribute to this field, while at the same time offered practical advice to fishermen and mariculturists. It cannot be denied, however, that at Government and Parliament level the Division was identified with the latter component of its work. This put the Division in a difficult position within CSIRO and resulted in periods of uncertainty with regard to the general direction the Division should take. For some time it seemed that the feedback mechanism: definition of new goals, insufficient time to work up experiments pursued under the earlier goals, experiments, low output in publications, low estime of the Division at CSIRO Executive level, attempt to cure the ills by a redefinition of goals ..., could not be broken. The Port Hacking study fell into that period; in fact, the decision to start the Project had all the marks of a "redefinition of goals" phase (a member of the CSIRO Executive attended the staff meeting at which the Project was first discussed: cf. Cuff, 1983). It set out posing questions related to direct application of results, but then it developed into an undertaking in basic research, justifiable in itself but fitting badly into the CSIRO framework. As a result scientists had to put up not only with the difficulties caused by the interdisciplinary nature of the experiment but also with the possibility of administrative disruption. It is a general truth that observational programs suffer more from disruptions than theoretical models. A model stopped today can be continued next month; an observational time series designed to cover a period of a year risks becoming useless if disrupted for a significant fraction of that time. Given the conditions in Australia in general and in CSIRO at the time in particular, risk that observations would fall short of the model's needs because of administrative disruption was significant from the very beginning. There is evidence that administrative interference became the major factor in 1977 and 1978 when the Project was officially terminated but many papers remained to be written (because of the next redefinition of goals).

Both factors which affect the progress of an estuarine ecosystem model gave the Port Hacking Estuary Project a particularly bad start, but they can be observed everywhere and exert influence on all ecosystem models in various degrees. As a general conclusion it can be said that shortage of input data is likely to occur with all marine ecosystem models. It is more likely for models falling into the category of basic research than for those set up for immediate application. It is also more likely in institutions set up in an environment unfavourable for long–term basic research than in institutions which can rely on stable long–term planning.

3. SOME GENERAL COMMENTS ON THE RELATIONSHIP BETWEEN MODELS AND OBSERVATIONS

We return here to the earlier question (more observations, or less numerical models?). Having been involved in ecosystem studies only very marginally, I will restrict myself to some general comments arising from experience with models in coastal upwelling which was my field of work for a number of years. Numerical models are the amalgamation of elements from two sciences of different character. Mathematics is a formal description of intellectually possible systems; the other science – be it economics, sociology, ecology, physics, medicine, or marine science – is an attempt to understand existing systems. While the existing systems influence the way in which mathematics develops, so the formalisms provided by mathematics influence the way in which we look at existing systems when we study them through numerical models.* In other words, some of the observational necessities of a model may be mathematical necessities. For example, numerical models must be closed, *i.e.* either they must have a complete set of external inputs and outputs, or they have to be confined in closed boundaries. Many experiments guided by models serve only to provide input and output data. Experimental techniques, on the other hand, do not

---

*   Margalef (1979) makes the same point when comparing a model of an ecosystem with a language, the set of parameters representing the vocabulary, and the equations between them representing the grammar. In order to see the analogy, it is worth remembering that language is not simply a tool developed for communication but an expression of our experience with social systems which in turn partly determines our perception of new systems. As an example, European languages have words for relation levels based on procreation and conception and on absolute sex (father, mother, son, daughter, brother, sister). Melanesian languages describe relations based on generation levels and relative sex (the same words are used for father and uncles, for mother and aunts, for sons and nephews etc.; brothers of males and sisters of females are designated by the same word, and the opposite word describes sisters of males and brothers of females). Both systems are incompatible, yet they give a complete description of existing systems and determine the approach of individuals to an analysis of the world.

develop from mathematical necessity, and the resulting discrepancy between experimental ability and model needs is the cause of much frustration. The development of numerical models has led to improved understanding of a variety of processes. The same is true for a large number of experiments performed without reference to numerical models. Both activities are justified in their own right. What has been observed in estuarine and coastal marine ecosystem modelling is a tendency of numerical models to become accepted as the ultimate guide for the evaluation of experimental programs' merits, partly because too much faith has been placed in the predictive capabilities of models.

Since model needs are defined by mathematics, they are basically the same for ecosystem models and others, such as geophysical fluid dynamics models of the ocean: open boundaries with complete input and output information, closed boundaries otherwise. Numerical models of the oceanic circulation expanded greatly during recent years, without establishing a new reference frame for experimental programs. Perhaps it is too obvious that the idea of securing all necessary input data on an oceanic scale by running a particularly designed expedition program is out of range of our possibilities. On the regional scale interaction between observational programs and models has been closer but both activities were still followed rather independently. Coastal upwelling is a good example. Plans for expeditions of the last decade were influenced by results from analytical and numerical models (Tomczak & Hughes, 1980) but drawn up independently, while a series of new models paralleled expedition activities (Niiler, 1975). The fact that the models, for mathematical reasons, had to extend across the ocean did not entail a demand for expeditions reaching from coast to coast. Both observational and modelling activities were basic research, and improved understanding of coastal upwelling resulted from both. Upwelling ecosystem models built on the basis of available information from both sides, without additional observational support, contributed to that understanding (Wroblewski, 1977).

It appears that one reason for the impact of systems analysis on estuarine and coastal marine programs is the restriction in size offered by estuaries which allows a numerical model to be largely closed, on not too large a scale, with only a limited amount of necessary input data. What escapes notice is that this apparent advantage is more than offset by increased internal complexity. Vinogradov & Menshutkin (1977) point out that the simplest object for modelling "and as yet the only one for which the construction of a comprehensive model may be attempted" are pelagic ecosystems. It is puzzling to see that actual model development did not follow this path and that international efforts to improve the situation for the open ocean are a very recent development (SCOR, 1980). Estuarine and coastal marine science is more difficult than open ocean marine science; it is not the best starting point for modelling efforts.

4. SOME CONCLUSIONS

What should be the relationship between models and experiments? Numerical models of the past often tried to improve a qualitative description of observed features into a quantitative picture; they were forced into such close relation by the need or desire to predict. As stated earlier, mathematical and observational techniques are not linked that closely in their development, and more benefit could possibly be derived by bringing the particular advantages of both sides into full play. The purpose of numerical ecosystem models would be to explore concepts and theoretical ideas of ecosystem dynamics. Models should make use of all available information, supplemented by some additional experimental data if this is necessary for basic model assumptions but not for quantitative modifications, and they should be understood as an element of basic research rather than as a tool for prediction. The purpose of experimental programs would be to study particular aspects of the ecosystem and contribute to our understanding of processes, again part of basic research but a different part. Both activities may be built around a common project, enabling them to draw on each other's results, but neither side should completely determine the operation of the other.

Natural systems consist of a hierarchy of identifiable sub–systems, the effects of which on the next higher level can be parameterized if the processes acting in the sub–system are known. Parameterization eliminates the sub–system from a model by converting its effect on the higher system into additional (known) input data or into a modification of the model equations. In geophysical fluid dynamics, a model of the oceanic circulation does not adequately resolve the processes responsible for energy loss in the frictional bottom boundary layer or energy flux by internal waves but both sub–systems are represented by appropriate parameterization of friction. Ecosystem models can be combined from a hierarchy of sub–models as well (Clymer, 1972), and it may be worth looking into the possibilities of parameterization more systematically. This would call for a different type of model–inspired field program, a study of processes which operate on scales the model is unable to resolve and of techniques of their representation in the model, hopefully one which can make better use of available experimental possibilities than the filling in of missing input data.

Of course, the need for prediction in real situations remains. Without prejudice for future possibilities, it seems that the most likely approach to prediction success in the short term is to restrict attention to that part of the ecosystem where prediction is needed and not to try to collate data for a model of the complete ecosystem. There is nothing to guarantee that such a prediction is infallibly correct, but experience with existing complex models shows that their capabilities for prediction do not go much beyond those of much simpler models; in most cases they are limited to a small range of input variation which often does not include the case of interest. It is likely that efforts in basic research towards parameterization of sub–model dynamics will improve the predictive power of restricted models geared at *ad hoc* solutions. In marine engineering, a wide range of parameterizations is used for processes such as wave action or turbulent friction which developed from basic research and result in predictions of acceptable quality although they are still seen as problems in basic research models. Meanwhile, the necessity for detailed quantitative predictions should be carefully judged. Very often a qualitative assessment based on a descriptive model of the situation can tell whether a proposed development makes ecological sense, and the need for quantitative assessment does not arise until the developer wants to prove that effects on the ecosystem are less bad than feared. The idea, of course, is to prove that a *quantitative* difference in adverse environmental effects turns an environmentally senseless proposition into a sensible one, *i.e.* changes its *quality*. Occasions where this is true are fewer than is generally claimed. Only for such occasions are quantitative predictions really necessary.

ACKNOWLEDGEMENTS

I joined the Division of Fisheries and Oceanography after the Port Hacking Estuary Project was finished and was drawn into this presentation of results as an outsider. I am grateful to R.R. Parker for his encouragement to take a fresh look at the Project. I benefitted greatly from discussions with J.S. Godfrey and B.D. Scott. Special thanks go to W.R. Cuff who originally suggested a joint evaluation of the Port Hacking Estuary Project. Some identical sentences in our now separate contributions are remnants of an earlier co–authored draft. Our discussions eventually convinced us to present our conclusions separately, but I still feel that only both papers together give the complete story.

REFERENCES

Bulleid, N.C.: The nutrient cycle of an intermittently stratified estuary. In: W.R. Cuff and M. Tomczak jr, eds *Synthesis and Modelling of Intermittent Estuaries*. Berlin, Heidelberg, New York: Springer (1983)

Clymer, A.B.: Next–generation models in ecology. In: B.C. Patten, ed. *Systems Analysis and Simulation Ecology Vol. 2*. New York: Academic Press (1972)

CSIRO: *Estuarine Project Research Report 1974–1976*. Sydney: CSIRO Division of Fisheries and Oceanography (1976)

Cuff, W.R.: An evaluation of the Port Hacking Estuary Project from the viewpoint of applied science. In: W.R. Cuff and M. Tomczak jr, eds *Synthesis and Modelling of Intermittent Estuaries*. Berlin, Heidelberg, New York: Springer (1983) di Toro, D.M., Thomann, R.V., O'Connor, D.J., Mancini, J.L.: Estuarine phytoplankton biomass models – verification analyses and preliminary applications. In: Goldberg, E.D., McCave, J.N., O'Brien, J.J., Steele, J.H., eds. *The Sea. Vol. 6.* New York: John Wiley & Sons (1977)

Godfrey, J.S.: Tidal flushing and vertical diffusion in South West Arm, Port Hacking. In: W.R. Cuff and M. Tomczak jr, eds *Synthesis and Modelling of Intermittent Estuaries.* Berlin, Heidelberg, New York: Springer (1983)

Ketchum, B.H., ed. *The Water's Edge: Critical Problems of the Coastal Zone.* Cambridge, Mass.: MIT Press (1972)

Kremer, J.N., Nixon, S.W.: *A Coastal Marine Ecosystem, Simulation and Analysis.* Berlin: Springer (1978)

Legovic, T., Kuzmic, M., Jeftic, Lj., Patten, B.C.: Model of the Adriatic regional ecosystem. *Thalassia Jugoslavica* **13**, 125–138 (1977)

Margalef, R.: Some critical remarks on the usual approaches to ecological modelling. *Investigación Pesquera* **37**, 621–640 (1973)

Margalef, R.: Alternative approaches in the modelling of populations. *Investigación Pesquera* **43**, 337–350 (1979)

Niiler, P.P.: A report on the continental shelf circulation and coastal upwelling. *Review of Geophysics and Space Physics* **13**, 609–614 (1975)

Odum, E.P. *Fundamentals of Ecology.* 3rd. edn. Philadelphia: W.B. Saunders (1971)

Parker, R.R., Rochford, D.J., Tranter, D.J.: History and organization of the Port Hacking Estuary Project. In: W.R. Cuff and M. Tomczak jr, eds *Synthesis and Modelling of Intermittent Estuaries.* Berlin, Heidelberg, New York: Springer (1983)

SCOR Working Group 59: Mathematical models in biological oceanography: Report from chairman. *SCOR Proceedings* **16**, 58–59 (1980)

Sinclair, R.E., Cuff, W.R., Parker, R.R.: Ecosystem modelling of South West Arm, Port Hacking. In: W.R. Cuff and M. Tomczak jr, eds *Synthesis and Modelling of Intermittent Estuaries.* Berlin, Heidelberg, New York: Springer (1983)

Swartzman, G.L.: Simulation modelling of material and energy flow through an ecosystem: methods and documentation. *Ecological Modelling* **7**, 55–81 (1979)

Tomczak, M. jr., Hughes, P.: Three–dimensional variability of water masses and currents in the Canary Current upwelling region. *Meteor Forschungsergebnisse* **A21**, 1–24 (1980)

Vinogradov, M.E., Menshutkin, V.V.: The modelling of open–sea ecosystems. In: Goldberg, E.D., McCave, J.N., O'Brien, J.J., Steele, J.H., eds. *The Sea. Vol. 6.* New York: John Wiley & Sons (1977)

Walsh, J.J.: Implications of a systems approach to oceanography. *Science (Washington DC)* **176**, 969–975 (1972)

Wroblewski, J.S.: A model of phytoplankton plume formation during variable Oregon upwelling. *Journal of Marine Research* **35**, 357–394 (1977)

D.A. Ross

Opportunities and Uses of the Ocean

1980. 144 figures, 48 tables. XI, 320 pages
ISBN 3-540-90448-4

This book explores the opportunities that exist for present day and future utilization of the oceans. After a concise introduction to the field of oceanography, the author discusses the legal aspects of the oceans, marine shipping, ocean resources and pollution, military uses and many more topics of crucial concern. A key aim of the author is to focus on those problems that can be solved with available scientific and technical capabilities.
This book will be fascinating and invaluable to a broad range of readers, including all those concerned with our oceans and wise marine policy.

S.A. Gerlach

Marine Pollution

Diagnosis and Therapy

Translated from the German by R. Youngblood, S. Messele-Wieser
1981. 91 figures. VIII, 218 pages
ISBN 3-540-10940-4

Wide-ranging in scope, *Marine Pollution* covers household and industrial wastes, ocean dumping, the effects of oil and radioactivity, the influence of toxic materials on geochemical and biochemical processes, and the dangers of wide-spread pollution by heavy metals and organic substances. This English translation of the highly acclaimed German original has been revised and expanded to keep pace with the rapid process of research in the field.

E. Seibold, W.H. Berger

The Sea Floor

An Introduction to Marine Geology

1982. 206 figures. VII, 288 pages
ISBN 3-540-11256-1

Man's understanding of the composition and evolution of planet Earth has changed radically over the last two decades. This great revolution in geology – now usually subsumed under the concept of plate tectonics and its ramifications – was an outgrowth of the study of the ocean floor, and has had an impact on the earth sciences comparable to the revolution brought about by Darwin more than a century ago.
The Sea Floor is designed to acquaint students and interested laymen with the most important results of this revolution and the scientists who brought them about. Written by two of marine geology's leading exponents, it lays the groundwork for studies in geology, oceanography, and environmental sciences by summarizing modern insights into tectonics and marine morphology, the geologic processes at work on the sea floor, and the Earth's climatic history as recorded in deep sea sediments.
The authors open with an overview of the effects of endogenic forces on the morphology of the sea floor, a topic closely linked to the theory of continental drift and very much in the focus of geologic discussions. They then concentrate on exogenic processes, those which determine the physical, chemical, and biological environment on the sea floor. Closing with a carefully selected bibliography, they provide an important contribution in the search for intelligent uses of the ocean and for the understanding of its role in the evolution of climate and life itself.

Springer-Verlag
Berlin
Heidelberg
New York
Tokyo

Lecture Notes in Coastal and Estuarine Studies

Managing Editors: R. T. Barber,
M. T. Bowman, C. N. Mooers,
B. Zeitzschel

Springer-Verlag
Berlin
Heidelberg
New York
Tokyo

Volume 1

Mathematical Modelling of Estuarine Physics

Proceedings of an International Symposium Held at the German
Hydrographic Institute, Hamburg, August 24–26, 1978
Editors: **J. Sündermann, K.-P. Holz**
1980. 119 figures, 1 table. VIII, 265 pages
ISBN 3-540-09750-3

Contents: Basic formulations and algorithms. – Tides and storm
surges. - Baroclinic motions and transport processes.

Mathematical modelling of hydrodynamic and thermodynamic processes in natural waters has become a valuable tool for coastal engineers and physical oceanographers. Their cooperative efforts are
necessary in the study of estuary dynamics, which has great importance for human life and economy.
The present book contains the contributions of twenty-five experts
from various branches of physical oceanography and coastal engineering to an international symposium held in Hamburg. The topics
they discussed include "Basic Formulations and Algorithms",
"Tides and Storm Surges", and "Baroclinic Motions and Transport
Processes".
This collection of papers gives a good insight into the present state
and future trends of simulation techniques in coastal hydrothermo-
dynamics.

Volume 2
D P. Finn

Managing the Ocean Resources of the United States: The Role of the Federal Marine Sanctuaries Program

1982. IX, 193 pages
ISBN 3-540-11583-8

Contents: Introduction. - Case Studies. - Interagency Coordination
for the Management of Marine Resources. - The Marine Sanctuaries Program. - The Role of Designating Marine Areas for Special
Management. - Recommendations and Conclusions. - Notes. -
Alphabetical List of Major References.

The present study examines the interaction of several major federal
marine programs in the USA to determine whether these programs
do or could provide coordinated management of maritime activities
and marine resources. The emphasis is largely not on the authority
or implementation of particular programs but on whether these programs together can be expected to result in reasonably consistent
and comprehensive results. Because of the fragmentation of federal
authority in the marine field, a number of mechanisms have been
created to help federal agencies coordinate their actions. These
mechanisms, their functioning, and potential changes in their conduct are considered. Finally, one particular federal program - the
marine sanctuaries program established under Title III of the
Marine Protection, Research and Sanctuaries Act 1972 (MPRSA) -
is discussed at length since, according to its proponents, this program provides a means to coordinate the federal effort to manage
certain marine areas that are especially valuable and subject to
threats from human use.
An analysis of the situations in which designation of marine sanctuaries may be warrented is given and recommendations for future
implementation of this program are presented.